IP Telephony

Deploying Voice-over-IP Protocols

IP Telephony: two-book reference set

IP Telephony: *Deploying Voice-over-IP Protocols* is a companion reference to *Beyond VoIP Protocols: Understanding Voice Technology and Networking Techniques for IP Telephony*. More details of this companion text may be found on the last page of this book.

IP Telephony

Deploying Voice-over-IP Protocols

Olivier Hersent
Founder and CTO
NetCentrex

Jean-Pierre Petit
Head of France Telecom R&D Human Interaction
Department (DIS/IPS)
France Telecom

and

David Gurle
WW VP Collaboration Services
Reuters

John Wiley & Sons, Ltd

Other Wiley Editorial Offices

John Wiley & Sons Inc., 111 River Street, Hoboken, NJ 07030, USA

Jossey-Bass, 989 Market Street, San Francisco, CA 94103-1741, USA

Wiley-VCH Verlag GmbH, Boschstr. 12, D-69469 Weinheim, Germany

John Wiley & Sons Australia Ltd, 33 Park Road, Milton, Queensland 4064, Australia

John Wiley & Sons (Asia) Pte Ltd, 2 Clementi Loop #02-01, Jin Xing Distripark, Singapore 129809

John Wiley & Sons Canada Ltd, 22 Worcester Road, Etobicoke, Ontario, Canada M9W 1L1

Wiley also publishes its books in a variety of electronic formats. Some content that appears
in print may not be available in electronic books.

British Library Cataloguing in Publication Data

A catalogue record for this book is available from the British Library

ISBN:10   0-470-02359-7 (HB)
ISBN: 13 978-0-470-02359-4 (HB)

Typeset in 10/12pt Times by Laserwords Private Limited, Chennai, India
Printed and bound in Great Britain by TJ International, Padstow, Cornwall
This book is printed on acid-free paper responsibly manufactured from sustainable forestry
in which at least two trees are planted for each one used for paper production.

Contents

Abbreviations

3GPP	3rd Generation Partnership Project
A/V	Audio-visual
AAD	Average Acknowledgement Delay
AAL2	ATM Adaptation Layer 2
ACD	Automatic Call Distribution
ACELP	Algebraic-Code-Excited Linear-Prediction
ACF	Admission Confirm
ACL	Access Control List
ACM	Address Complete Message
ADEV	Average Delay Deviation
ADPCM	Adaptive Differential Pulse Mode Modulation (MICDA)
ADSL	Asymmetric Digital Subscriber Line
AES	Advanced Encryption Standard
AMR	Adaptive Multi-Rate
AMR-WB	Adaptive Multi-Rate (Wide Band)
ANM	Answer Message
ANSI	American National Standard Institute
AOC	Advice of Charge
APDU	Application Protocol Data Unit
API	Application Programming Interface
ARJ	Admission Reject
ARQ	Admission Request
ASCII	American Standard Code for Information Interchange
ASN-1	Abstract Syntax Notation One
ASP	Application Service Provider or Answer Seizure Ratio
ASR	Automatic Speech Recognition
ATM	Asynchronous Transfer Mode
AUCX	Audit Connection

AUEP	Audit Endpoint
AVC	Advanced Video Coding
AVT	Audio/Video Transport
BASIC	Beginners' All-Purpose Symbolic Instruction Code
BCF	Bandwidth Confirm
BER	Basic Encoding Rule
BNF	Backus-Naur Form
BRJ	Bandwidth Reject
BRQ	Bandwidth Request
CA	Call Agent
CALEA	Communication Assistance for Law Enforcement Act
CallID	Call Identifier
CBC	Cipher Block Chaining
CC	CSRC Count
CCIR	Consultative Committee for International Radio (ITU)
CDMA	Code Division Multiplex Access
CDR	Call Detail Record
CED	CallED
CELP	Code-excited Linear Prediction
CFB	Cipher Feedback
CFU	Call Forwarding Unconditional
CIC	Circuit Identification Code
CID	Conference Identifier
CIF	Common Intermediary Format
CLEC	Competitive Local Exchange Carrier
CLIP	Calling Line Identity Presentation
CLIR	Calling Line Identity Restriction
CMA	Call Management Agent
CMTS	Cable Modem Termination System
CNG	CalliNG; Comfort Noise Generator
CO	Central Office
Codec	COder DECoder
COMEDIA	Connection-Oriented Media Transport in SDP
COPS	Common Open Policy Service
CPE	Customer Premises Equipment
CPG	Call Progress (Message)
CPIM	Common Profile for Instant Messaging
CPL	Call Processing Language
CPU	Central Processing Unit
CR	Carriage Return
CRLF	Carriage Return and Line Feed
CRC	Cyclic Redundancy Check
CRCX	Create Connection
CRV	Call Reference Value

CS-ACELP	Conjugate Structure, Algebraic Code-Excited Linear Prediction
CSRC	Contributing Source
CTI	Computer Telephony Integration
DCF	Disengage Confirm
DCME	Digital Circuit Multiplication Equipment
DCN	Disconnect
DCS	Distributed Call Signaling
DCT	Discrete Cosine Transform
DDNS	Dynamic DNS
DES	Data Encryption Standard
DES/CBC	Data Encryption Standard, Cipher Block Chaining
DHCP	Dynamic Host Configuration Protocol
DiffServ	Differentiated Services
DIS	Digital Identification Signal
DLCX	Delete Connection
DLSR	Delay Since Last Sender Report
DNS	Domain Name System
DNSSEC	Domain Name System Security Protocol
DOCSIS	Data Over Cable Service Interface Specification
DoS	Denial of Service
DRJ	Disengage Reject
DRQ	Disengage Request
DSL	Digital Subscriber Line
DSP	Digital Signal Processor
DSS1	Digital Subscriber Signaling 1
DTMF	Dual-Tone Multi-Frequency
DTX	Discontinuous Transmission
DVMRP	Distance Vector Multi-cast-routing Protocol
ECB	Electronic Code Book
EFR	Enhanced Full Rate
ENUM	"Electronic Numbers" Protocol
EOL	End of Line
EOP	End of Procedure
EPCF	Endpoint Configuration Command
ETSI	European Telecommunications Standardisation Institute
ETSI TIPHON	ETSI Telephony and Internet Protocol Harmonization Over Networks
ETTB	Ethernet to the Building
ETTX	Ethernet to the <anything> (Club, Home, Building)
FCF	Fax Control Field
FCS	Frame Check Sequence
FEC	Forward Error Correction
FIF	Fax Information Field
FIFO	First in First Out

FIPS PUB	Federal Information Processing Standards Publication
FR	Full-rate
FS	FastStart
GCF	Gatekeeper Confirm
GEF	Generic Extensibility Framework
GK	Gatekeeper
GOBs	Group of Blocks
GRJ	Gatekeeper Reject
GRQ	Gatekeeper Request
GSM	Global System for Mobile Communications
GTD	Global Transparency Descriptor
HD	Hang Down (off-hook)
HDLC	High-Level Data Link Control
HR	Half Rate
HTML	Hypertext Markup Language
HTTP	Hypertext Transfer Protocol
HU	Hang Up (on-hook)
IAD	Integrated Access Device
IAM	Initial Address Message
IANA	Internet Assigned Numbers Authority
ICMP	Internet Control Message Protocol
IEC	ISO International Electrotechnical Commission
IETF	Internet Engineering Task Force
IF	Interface
IFP	Internet Fax Protocol
IFT	Internet Fax Transmission protocol
ILS	Internet Locator Service (Microsoft)
IM	Instant Messaging
IMPP	Instant Messaging and Presence Protocol
IMTC	International Multimedia Teleconferencing Consortium
IN	Intelligent Network
INAP	Intelligent Network Application Protocol
IntServ	Integrated Services
IP	Intelligent Peripheral
IP-PBX	Internet Protocol–Private Branch Exchange
IPDC	Internet Protocol Device Control
IPR	Intellectual Property Rights
IPSec	Internet Protocol Security
IRC	Internet Relay Chat
IRQ	Information Request
IRR	Information Request Response
ISDN	Integrated Service Digital Network
ISP	Internet Service Provider
ISUP	SSF ISDN USER PART
ITSP	Internet Telephony Service Provider

ITU	International Telecommunications Union
IVR	Interactive Voice Response
JFIF	JPEG File Interchange Format
JPEG	Joint Photographic Experts Group
LAN	Local Area Network
LCD	Liquid Crystal Display
LCF	Location Confirm
LD-CELP	Low-delay, Code-excited Linear Prediction
LDAP	Lightweight Directory Access Protocol
LF	Line Feed
LNP	Local Number Portability
LRJ	Location Reject
LRQ	Location Request
LSP	Line Spectral Pair
LSR	Last Sender Report
M	Marker Bit (RTP)
mBone	Multicast Backbone of the Internet
MC	Multipoint Controller
MCF	Message Confirmation
MCU	Multipoint Control Unit
MD5	Message Digest 5
MDCX	Modify Connection
MEGACO	Media Gateway Controller
MGCP	Media Gateway Control Protocol
MGCP/L	MGCP Line
MGCP/T	MGCP Trunk
MH	Modified Huffmann
MIME	Multipurpose Internet Mail Extension
MIPS	Millions of Instructions Per Second
MMS	Multimedia Message Service
MMUSIC	Multiparty Multimedia Session Control
MOS	Mean Opinion Score
MP	Multipoint Processor
MP-MLQ	Multipulse Maximum Likelihood Quantization
MPEG	Moving Picture Experts Group
MPLS	Multiprotocol Label Switching
MTP	Message Transfer Part
MTT	Minimum Transmission Time
MTU	Maximum Transmission Unit
MWI	Message Waiting Indication
MX	Mail Exchange
NAPT	Network Address and Port Translation
NAPTR	Naming Authority Pointer Record
NAS	Network Access Server
NAT	Network Address Translation

NCS	Network Based Call Signaling Protocol
NFE	Network Facility Extension
NTFY	Notify
NTP	Network Time Protocol
NTSC	National Television System Committee
OFB	Output Feedback
OGW	Originating Gateway
OID	Object Identifyer
OLC	Open Logical Channel
OO	On–off
OS	Operating System
OSP	Open Settlement Protocol
P-frame	Prediction Frame
PAL	Phase-alternation-line
PBX	Private Branch Exchange
PCM	Pulse Code Modulation
PCMA	Pulse Code Modulation A Law
PCMU	Pulse Code Modulation μ Law
PDU	Protocol Data Unit
PER	Packed Encoding Rules
PGR	Pages Received (Fax)
PGS	Pages Sent (Fax)
PI	Progress Indicator
PIDF	Presence Information Data Format
PIM	Protocol-independent Multicast
POSIX	Portable Open System Interconnect
POTS	Plain Old Telephone Service
PSTN	Public Switched Telephone Network
PT	Payload Type (144×176)
QCIF	Quarter CIF
QoS	Quality of Service
RAI	Resource Availability Indicator
RAS	Registration, Admission, Status Protocol
RC	Reception Report Count
RCF	Registration Confirm
RD	Restart Delay
RED	Random Early Detection
RFC	Request For Comments
RGB	Red–green–blue
RGW	Residential Gateway
RLE	Run Length Encoding
RM	Restart Method
RQNT	Notification Request
RR	Resource Record
RRJ	Registration Reject

RRQ	Registration Request
RRs	Resource Records
RSA	Rivest, Shamir, Adleman (public key algorithm)
RSIP	Restart in Progress
RST	Reset
RSVP	Resource Reservation Protocol
RTC	Return To Command
RTCP	Real-time Control Protocol
RTO	Retransmission Timeout
RTP	Real-time Transport Protocol
RTP/AVT	Real Time Protocol under the Audio/Video Profile
S/MIME Secure	Multipurpose Internet Mail Extension
SAP	Session Announcement Protocol
SCN	Switched Circuit Network
SCP	Service Control Point
SCTP	Stream Control Transport Protocol
SDES	Source Description RTP Packet
SDL	Specification and Description Language
SDP	Session Description Protocol
SECAM	Séquentiel Couleur à Mémoire
SGCP	Simple Gateway Control Protocol
simcap	Simple Capability (SDP Declaration)
SIMPLE	SIP for Instant Messaging and Presence Leveraging Extensions
SIP	Session Initiation Protocol
SIPS	Session Initiation Protocol Secure
SMG	Special Mobile Group (of ETSI)
SMS	Short Message Service
SMTP	Simple Mail Transfer Protocol
SP	Single Space
SQCIF	Sub-QCIF (128 × 96)
SR	Sender Report
SRV	Server DNS Record
SS	Supplementary Service
SS-CD	Supplementary Service: Call Deflection
SS-CFB	Supplementary Service: Call Forwarding on Busy
SS-CFNR	Supplementary Service: Call Forwarding on No Reply
SS-CFU	Supplementary Service: Call Forwarding Unconditional
SS-DIV	All Diversion Supplementary Services
SS7	Signaling System 7
SSF	Service Switching Function
SSL	Secure Sockets Layer
SSW	Softswitch
STP	Signaling Transfer Point

STUN	Simple Traversal of UDP Through Network Address Translators
SUD	Single Use Device
TAPI	Microsoft Telephony API
TCAP	SS-7 Transaction Capabilities
TCF	Training Check Function
TCP	Transport Control Protocol
TCS	Terminal Capability Set
TCS=0	NullCapabilitySet Call Flow in H.323
TDM	Time Division Multiplexing
TFTP	Trivial File Transfer Protocol
TGW	Terminating Gateway
TIA	Telecommunications Industry Association (USA)
TIPHON	Telephony and Internet Protocol Harmonization over Networks (ETSI)
TLS	Transport Layer Security
TLV	Type, Length, Value format
TO	Timeout
TPKT	Transport Packet (RFC 1006)
TTL	Time to Live
TTS	Text to Speech
TURN	Traversal Using Relay NAT
UCF	Unregistration Confirm
UCS	Universal Character Set
UDP	User Datagram Protocol
UDPTL	UDP Transport Layer
UII	User Input Indication
UMTS	Universal Mobile Telecommunication System
UPT	Universal Personal Telephony
URI	Uniform Resource Identifier
URJ	Unregistration Reject
URL	Uniform Resource Locator
URN	Uniform Resource Name
URQ	Unregistration Request
USH	Université de Sherbrooke
UTF	UCS Transformation Format
VAD	Voice Activity Detector
VASA	Value Added Services Alliance
VLAN	Virtual Local Area Network
VoIP	Voice Over Internet Protocol
VPIM	Voice Profile for Internet Messaging
VPN	Virtual Private Network

VSELP	Vector Sum-excited Linear Prediction
WAP	Wireless Application Protocol
WWW	World Wide Web
XML	eXtensible Markup Language
XMPP	eXtensible Messaging and Presence Protocol

Glossary

Abstract Syntax Notation-1 (ASN-1)	Defined in ITU standard X.691.
Access Control List (ACL)	A packet filter on a router.
Admission Confirm (ACF)	A RAS message defined in H.225.0.
Admission Reject (ARJ)	A RAS message defined in H.225.0.
Admission Request (ARQ)	A RAS message defined in H.225.0.
Application Protocol Data Units (APDUs)	See H.450.1.
Associate Session	A related session. Two related sessions must by synchronized (e.g., an audio session can specify a video session as being related). The receiving terminal must perform lip synchronization for those sessions.
Backus–Naur Form (BNF)	See RFC 2234.
Bandwidth Confirm (BCF)	A RAS message defined in H.225.0.
Bandwidth Reject (BRJ)	A RAS message defined in H.225.0.
Bandwidth Request (BRQ)	A RAS message defined in H.225.0.
Basic Encoding Rule (BER)	See ASN.1.
Call Identifier (Call-ID)	A globally unique call identifier.
Call Reference Value (CRV)	A 2-octet locally unique identifier copied in all Q.931 messages concerning a particular call (see also conference identifier).
Conference Identifier (CID)	This is not the same as the Q.931 Call Reference Value (CRV) or the call identifier (CID). The CID refers to a conference which is the actual communication existing between the participants. In the case of a multiparty conference, if a participant joins the conference, leaves, and enters again, the CRV will change, while the CID will remain the same.

The Common Intermediary Format (CIF)	A video format which has been chosen because it can be sampled relatively easily from both the 525-line and 625-line video formats: 352 × 288 pixels
Contributing Source (CSRC)	When an RTP stream is the result of a combination put together by an RTP mixer of several contributing streams, the list of the SSRCs of each contributing stream is added in the RTP header of the resulting stream as CSRCs. The resulting stream has its own SSRC.
Disengage Confirm (DCF)	A RAS message defined in H.225.0.
Disengage Reject (DRJ)	A RAS message defined in H.225.0.
Disengage Request (DRQ)	A RAS message defined in H.225.0.
Dual-Tone Multi-Frequency (DTMF)	Tones composed of two well-defined frequencies that represent digits 0–9, *, #. The combination of frequencies has been selected to be almost impossible to reproduce with the human voice. DTMF tones are used to dial from analog phones and to control IVR servers.
Dynamic Host Configuration Protocol (DHCP)	Used by end points to acquire a temporary IP address and important TCP/IP parameters (router IP address, DNS IP address, etc.) from a server in the network.
End of Line (EOL)	The end of line sequence for group 3 fax (001H).
Energy	For an image on a particular color, the sum of the squared color values of the pixels is called the energy.
ENUM	An E.164 number resolution protocol defined in RFC 2916.
Fast-Connect	A procedure to eliminate media delays after the connection of the call introduced in H.323v2. Another name used for the same procedure is Fast-Start.
Fast-Start	See Fast-Connect.
Gatekeeper Confirm (GCF)	A RAS message defined in H.225.0.
Gatekeeper Request (GRQ)	A RAS message defined in H.225.0.
Gatekeeper Reject (GRJ)	A RAS message defined in H.225.0.
Information Request (IRQ)	A RAS message defined in H.225.0.
Information Request Response (IRR)	A RAS message defined in H.225.0.
Initial Address Message (IAM)	SS7 ISUP message initiating a call set-up.
Inter-mode	Refers to a video-coding mode where compression is achieved by reference to the previous, or sometimes the next, frame.

Interactive Voice Response server (IVR)	A machine accepting DTMF or voice commands, and executing some logic which interacts with the user using pre-recorded prompts or synthetic voice.
Internet Fax Transmission (IFT)	A protocol, see ITU recommendation T.38.
Internet Relay Chat (IRC)	The famous 'chat' service of the Internet, based on a set of servers mirroring text-based conversations in real time.
Intra-mode	Refers to a video-coding mode where compression is achieved locally (i.e., not relatively to the previous frame).
IP-PBX	Private phone switch with a VoIP wide area network interface. Most IP-PBXs have an H.323 WAN interface. See also IPBX.
IPBX	Same as IP-PBX. Some use the term IPBX for private phone switches which use only VoIP (i.e., the phones are also IP phones), whereas an IP-PBX can be a traditional PBX with analog phones and only uses a WAN VoIP interface. See IP-PBX
Jitter	Statistical variance of packet interarrival time. It is the smoothed absolute value of the mean deviation in packet-spacing change between the sender and the receiver. The smoothing is usually done by averaging on a sliding window of 16 instantaneous measures.
jitter	Varying delay.
Location Confirm (LCF)	A RAS message defined in H.225.0.
Location Reject (LRJ)	A RAS message defined in H.225.0.
Location Request (LRQ)	A RAS message defined in H.225.0.
macroblock	For the H.261 algorithm, a group of four 8*8 blocks.
Maximum Transmission Unit (MTU)	The largest datagram that can be sent over the network without segmentation.
Multicast Backbone of the Internet (mBone)	Capable of sending one packet to multiple recipients.
Multipoint Control Unit (MCU)	An H.323 callable end-point which consists of an MC and optional MPs.
Multipoint Controller (MC)	The H.323 which provides the control function for multiparty conferences.
Multipoint Processor (MP)	The H.323 entity which processes the media streams of the conference and does all the necessary switching, mixing, etc.

Naming Authority Pointer Resource of the DNS (NAPTR)	Defined in RFC 2915 and used notably by ENUM, see ENUM.
Network Facility Extension (NFE)	Defined in H.450.1.
Network Time Protocol (NTP)	This defines a standard way to format a timestamp, by writing the number of seconds since 1/1/1900 with 32 bits for the integer part and 32 bits for the decimal part expressed as number of $1/2^{32}$ seconds (e.g., 0x800000000 is 0.5 s). A compact format also exists with only 16 bits for the integer part and 16 bits for the decimal part. The first 16 digits of the integer part can usually be derived from the current day, the fractional part is simply truncated to the most significant 16 digits.
P-frame	Prediction frame obtained by motion estimation or otherwise, and representing only the difference between this image and the previous one.
Packed Encoding Rules (PER)	See ASN.1.
Payload Type (PT)	As defined by RTP.
port	An abstraction that allows the various destinations of the packets to be distinguished on the same machine (e.g., Transport Selectors, or TSELs, in the OSI model, or IP ports). On the Internet, many applications have been assigned 'well-known ports' (e.g., a machine receiving an IP packet on port 80 using TCP will route it to the web server).
Prediction frame (P-frame)	Obtained by motion estimation or otherwise, and representing only the difference between this image and the previous one.
Private Branch Exchange (PBX)	A private phone switch.
Proxy server	An intermediary program that acts as both a server and a client for the purpose of making requests on behalf of other clients. Requests are serviced internally or by passing them on, possibly after translation, to other servers. A proxy interprets, and, if necessary, rewrites a request message before forwarding it.
Q-interface Signaling (QSIG)	Protocol used at the Q-interface between two switches in a private network. ECMA/ISO have defined a set of QSIG standards.

Real-Time Control Protocol (RTCP)	See RFC 1889.
Real-time Transport Protocol (RTP)	As specified by RFC 1889.
Registration Confirm (RCF)	A RAS message defined in H.225.0.
Registration Reject (RRJ)	A RAS message defined in H.225.0.
Registration Request (RRQ)	A RAS message defined in H.225.0.
Registration, Admission, and Status (RAS)	The name of the protocol used between the gatekeeper and a terminal, and between gatekeepers for registration, admission, and status purposes. Defined in H.225.0.
Return To Command (RTC)	Six consecutive EOLs instructing a G3 Fax to return to command mode.
Sender Report (SR)	Used in RTCP and RTP.
Session ID	A unique RTP session identifier assigned by the master. The convention is that the value of the session ID is 1 for a primary audio session, 2 for a primary video session, and 3 for a primary data session. See Associate session.
Single Use Device (SUD)	See H.323 annex F.
SIP dialog	This was defined in RFC 3261 as a peer-to-peer SIP relationship between two UAs which persists for some time. A dialog is established by SIP messages, such as a 2xx response to an INVITE request. A dialog is identified by a call identifier, a local tag, and a remote tag. A dialog was formerly known as a call leg in RFC 2543.
SIP final response	A SIP response that terminates a SIP transaction (e.g., 2xx, 3xx, 4xx, 5xx, 6xx responses). See SIP provisional response.
SIP provisional response	A SIP response that does not terminate a SIP transaction, as opposed to a SIP final response (1xx responses are provisional).
SIP redirect server	A redirect server is a server that accepts a SIP request, maps the address into zero or more new addresses, and returns these addresses to the client. Unlike a proxy server, it does not initiate its own SIP request. Unlike a user agent server, it does not accept calls.
SIP registrar	A registrar is a server that accepts REGISTER requests. A registrar is typically co-located with a proxy or redirect server and may offer location services.

SIP server	A server is an application program that accepts requests in order to service requests and sends back responses to those requests. Servers are either proxy, redirect, or user agent servers or registrars.
SIP transaction	A SIP transaction occurs between a client and a server, and comprises all messages from the first request sent from the client to the server up to a final (non-1xx) response sent from the server to the client. A transaction is identified by the CSeq sequence number within a single-call leg. The ACK request has the same CSeq number as the corresponding INVITE request, but comprises a transaction of its own.
Stream Control Transport Protocol (SCTP)	Defined in RFC 2960.
Supplementary Services (SS-DIV)	Includes all diversion supplementary services, such as SS-CFU, SS-CFB, SS-CFNR, SS-CD.
Switched Circuit Network (SCN)	A generic term for the 'classic' phone network, including PSTN, ISDN, and GSM.
Synchronization Source (SSRC)	Source of an RTP stream, identified by 32 bits in the RTP header. All the RTP packets with a common SSRC have a common time and sequencing reference.
Talkspurt	A period during which a participant usually speaks, as opposed to silence periods.
Time Division Multiplexing (TDM)	The traditional voice transmission and switching technique based on assigning each communication a fixed "time slot" on a communication line between central offices.
TPKT	A TCP connection establishes a reliable data stream between two hosts, but there is no delimitation of individual messages within this stream. RFC 1006 defines a simple TPKT packet format to delimit such messages. It consists of a version octet ('3'), two reserved octets ('00'), and the total length of the message including the previous headers (2 octets).
Transport address	Combination of a network address (e.g., IP address 10.0.1.2) and port (e.g., IP port 1720) which identifies a transport termination point.
Transport Control Protocol (TCP)	The most widely used, reliable transport protocol for IP networks.
Transport Layer Security (TLS)	A secure protocol using TCP, defined in RFC 2246.

Trivial File Transfer Protocol (TFTP)	A very simple file transfer protocol over UDP, frequently used by IP appliances to download their initial configuration parameters.
Uniform Resource Identifier (URI)	Defines a uniform syntax and semantic convention for any resource. The URI is defined in RFC 2396. See also URL, URN.
Uniform Resource Locator (URL)	A specific type of URI identifying a resource by its primary network address. URLs are used by SIP to indicate the originator, current destination, and final recipient of a SIP request, and to specify redirection addresses. See also URI.
Uniform Resource Name (URN)	A specific type of URI required to be universally unique and persistent even if the resource ceases to exist. See also URI.
Unregistration Confirm (UCF)	A RAS message defined in H.225.0.
Unregistration Reject (URJ)	A RAS message defined in H.225.0.
Unregistration Request (URQ)	A RAS message defined in H.225.0.
User Agent Client (UAC)	Also known as a calling user agent. A user agent client is a client application that initiates the SIP request.
User Agent Server (UAS)	Also known as a called user agent. A user agent server is a server application which contacts the user when a SIP request is received and returns a response on behalf of the user. The response accepts, rejects, or redirects the request.
User agent	A SIP end system participating in a SIP transaction. See UAC, UAS.
User Datagram Protocol (UDP)	The most widely used unreliable transport protocol for IP networks. UDP only guarantees data integrity by using a checksum, but an application using UDP has to take care of any data recovery task.
Zone	An H.323 zone is the set of all H.323 end points, MCs, MCUs, and gateways managed by a single gatekeeper.

Preface

VOIP 1998–2004, 6 YEARS FROM R&D LABS TO LARGE SCALE DEPLOYMENTS

Since 1998 Voice over IP, in short VoIP, has been the favorite buzzword of the telecom industry. In 1998, IP was not yet as established and dominant as it is today, and most telecom engineers still believed that only ATM technology would be able to support multimedia applications. Indeed at this time most of us experienced the Internet only through dial-up modems and most ISPs, unable to keep-up with the exploding demand for Internet connections, were providing a level of service that could hardly qualify even for 'best effort'.

But even in this context, the R&D teams that started to work on VoIP were not simply taking a leap of faith. Their bet on VoIP was backed by the last developments of packet networking theory, which proved that properly designed IP networks could provide an appropriate support for applications requiring quality of service. Knowing this, most of these teams felt confident that VoIP could be deployed on a wide scale in the future, and in the mean time tried to evaluate what could be the impact of VoIP, compared to previous technologies.

It took a relatively long time to understand the reasons that would lead a service provider to deploy VoIP instead of traditional switched voice networks. Initially VoIP was presented as a technology that could enable a service provider to transport voice 'for free' over the Internet, because IP transport was 'free', and calls could be routed to local breakout trunks on the far end. The first commercial applications of VoIP focused on prepaid telephony, which was a reasonable target given that potential buyers of prepaid card systems do care about costs, and they are much more tolerant to quality of service issues than any other market segment. VoIP prepaid telephony systems did have a great success—today the majority of international calling card services use VoIP—but not because of cheaper call termination costs (which are regulated independently of the technology in most countries), or cheaper transport costs (traditional voice compression systems are much more efficient than VoIP systems). The reason for the success was mainly because VoIP facilitates the trading of minutes between multiple networks without the constraint of establishing leased lines: on the Internet, virtually all VoIP service providers 'see' each other and can decide to exchange traffic immediately, or to stop as soon as better arbitrage opportunities exist. In addition the central switching system of a VoIP service provider does not process voice streams, but only signaling messages: a call initiated from a gateway in Paris can be routed to a gateway in London by a VoIP call controller located in New-York with very few overhead costs. Only signaling messages make the round trip through the Atlantic, voice packets only cross the Channel.

It is now clear that the key reasons for the success of VoIP are:

- location independence: because of the unique characteristics of VoIP call controllers, or 'Softswitches', many functions that previously required multiple distributed points of presence can now be centralized, reducing administrative overheads and accelerating deployments

- simplification of transport networks: in the example above, service providers no longer need to establish leased lines dedicated to voice prior to exchanging traffic. But the use of standard IP data networks—configured appropriately—is a major breakthrough in many other circumstances: core transport networks no longer need to maintain the dedicated network that was required by SS7 signaling, enterprises moving to new offices can save the significant expenses required by dedicated telephony wiring and use virtual LANs instead.

- Ability to establish and control multimedia communications, e.g. interactive audio and video calls, data sharing sessions, etc.

Because of these unique characteristics, VoIP technology is a very good choice every time a relatively complex call control function would require multiple points of presence close to the end-users in traditional switched technology, and can be centralized with an application softswitch:

- In *residential telephony*, new service providers can deploy centralized VoIP call control servers and use any IP networking technology. For instance FastWeb, in Italy, serves the Italian market from just two PoPs located in Milan and Rome. This is not possible with traditional technology using traditional (TDM) switches (even with V5.2/GR303 ATM gateways used at the edge of the network), because the voice streams need to be physically switched by the backplane of the TDM switch. In addition of course, VoIP technology makes it possible to introduce additional media, like video communications, which differentiate the service and help maintain the ARPU[1].

- *Informal, Distributed contact centers* also become much easier and cheaper to operate with VoIP: the centralized call distribution point no longer needs to switch the voice

---

[1] Average Revenue Per User

streams, and therefore tromboning[2] through the VoIP call distribution server is completely eliminated, which reduces communications costs and minimizes the required bandwidth for the connection of the call distribution platform.

– In general, all applications which previously required a complex *intelligent network* architecture in order to minimize tromboning (call switching occurs at specific nodes in the network, and the applications can be located elsewhere), can be significantly simplified using a centralized call control server which controls voice signaling but optimizes the voice path through the IP network.

Today more and more service providers and enterprises, as they have become confident in the VoIP technology and quality of service of IP networks, deploy VoIP applications in order to enjoy the location independence and greater flexibility of the technology. With more successful deployments, VoIP is gaining in maturity, and the cost of VoIP gateways and IP phones is quickly dropping with the increased volumes. This positive circle should result in massive deployments of VoIP over the next 5 years.

SCOPE OF THIS BOOK

In "IP Telephony", we will also assume, like the pioneers of VoIP, that it is possible to carry multimedia data flows over an IP network with an appropriate quality (i.e. low latency and low packet loss), and we will focus only on the functional aspects of VoIP. Voice coding technology is also presented as a 'black box', with enough information for an engineer who wants to use an existing coder in an application, but without describing the technology in detail. "IP Telephony" will be useful mainly in the lab (development platforms, validation platforms), when designing and troubleshooting new interactive multimedia applications.

The companion book 'Beyond VoIP Protocols' becomes necessary when you deploy these applications in the field, over a real network with limited capacity. 'Beyond VoIP protocols' contains an overview of the techniques that can be used to provide custom levels of quality of service for IP data flows, and guidelines to properly dimension an IP network for voice. It also delves into the details of voice coding technology, and the influence of the selected voice coder and the transmission channel parameters on perceived voice quality.

In theory, it is sufficient to read the VoIP standards in order to become an efficient VoIP engineer. Although reading the standards is always necessary at some point, these documents were never written to be read from A to Z. Not only the mere volume is a problem, hundreds of pages for each standard, but also the structure is inappropriate: all VoIP standards are written as umbrella documents, which point explicitly or implicitly to dozens of other more detailed documents. Sometimes, these documents are also misleading, because some of the recommended methods were discussed in a specific context

---

[2] "Tromboning" refers to a non-optimal media path through the network, compared to the shortest path. It happens when the media streams have to "zigzag" across multiple nodes, reminding of the shape of a bent trombone.

in the standard bodies, but this context was lost or not clearly expressed in the written recommendation (see for instance the issues presented in the advanced topics chapter for call redirection). Last but not least most standards are the result of "diplomatic" agreements between firms, which often results in multiple alternate ways of doing the same thing, very long and cumbersome documents with many 'options' and unclear sentences designed to preserve the agreed compromise, while in practice after a few years, the market forces lead to a "de-facto" standard choice, in general adopted from the practice of the dominant players.

We wrote "IP Telephony" because we believe it is much more efficient to gain first a general overview on VoIP, and only then go into the details of the standard documents, but only when needed and if clarification is required on a specific item. Initially this book was designed as an internal training tool within France Telecom, and over the years it developed by capturing the accumulated experience of the authors and their colleagues, in over 50 voice over IP deployments, among which two of the largest residential VoIP networks in the world: FastWeb for residential telephony (over 450,000 VoIP phone lines, and 1000 new lines every day), and Equant for VoIP Multiservice VPNs (connects over 1,300 sites of 130 multinationals, with a growth rate of 85% per year).

"IP telephony" begins by giving an overview of the techniques that can be used to encode media streams and transmit them over an IP network (chapter 1). If focuses on the functional requirement of transmitting an isochronous data stream over an asynchronous network which introduces delay variations ("jitter"). The media encoding methods themselves are presented very briefly, with just enough details for an engineer who wants to use them and understand the main parameters required for the transmission of the resulting data.

The most popular VoIP standards are presented in chapter 2 (H.323), chapter 3 (SIP) and chapter 4 (MGCP). These chapters do not intend to fully replace the standards, but provide a detailed overview that should be sufficient for most engineers and pointers to relevant normative documents if further reference is required. The value of these chapters comes also from the many discussions on aspects of the standards that are still immature, and descriptions of calls flows or protocol extensions commonly used by vendors but not described in standard documents.

The "advanced topics" chapters (chapter 5 and 6), discusses two issues faced by all service providers when deploying public VoIP services (as opposed to custom services designed for a single enterprise). The first issue comes from the incompatibility of current VoIP protocols with Network Address Translation routers and firewalls, which change the addresses of IP packets on the fly but without properly translating the IP addresses contained in the VoIP messages carried by these packets. The second issue comes from the widespread confusion between private telephony techniques and public telephony techniques for call transfers. In both cases the chapter presents techniques that were deployed successfully, and explains the pros and cons of each possible method.

WHICH PROTOCOLS FOR VOIP ?

As we were writing this book called "IP Telephony: DeployingVoice-over-IP Protocols", it was of course very difficult to ignore the 'protocol wars' which seem unavoidable each time the telecom industry invents a new application or technology. The exercise was made especially difficult by the telecom bubble, during which it seems many manufacturers and many service providers forgot that telecommunications is a science, and more and more strategic or even technical decisions have been made based on misleading marketing campaigns.

In fact even today, almost 100% of what we read in telecom magazines or hear in telecom tradeshows is plain advertising, not only inexact technically, but too often presenting conclusions that are the exact contrary of what any sound technical analysis would lead to. For instance UMTS interactive multimedia applications are always presented as "all IP based", while in fact they are all circuit based[3], and for fundamental technical reasons that will last for years!

This is a very big problem for the telecom industry as a whole, because too many manufacturers or service providers have started to digest and believe their own over-inflated marketing, and this vicious circle leads to inordinate amounts of investment money that will not survive the reality check of deployments. We have seen so many 'concept companies' grow with the bubble, and then fail. Even large companies are still investing massively in programs that sometime seem a bit surrealist and are obviously poised for failure.

As the CEO of one of the largest service providers put it in a recent press conference: "it is high time for us to become a company of engineers again".

There is little a book can do to help sanitize the world of VoIP, but we have tried to discuss openly the pros and cons of each protocol, each time with specific arguments and suggestions for improvements. Our opinion is that all of the protocols described have a future, and each has some unique characteristics that make it unavoidable for the next 5 years:

- H.323 provides the optimal transparency with ISDN endpoints, and is today the only connectivity protocol that can be used with virtually all PABXs of the market. In addition, H.323 provides by far the best support for video communications today and it is the protocol of choice of 3G multimedia applications (under the circuit version 3G-324M) probably for the next 5 years. All pending problems of the protocol have been solved as the protocol matured. In short, it is the dominant protocol today, and it has a brilliant future for PABX interconnect (business trunking), core VoIP networks, and 3G. However it is weak for PC applications.
- SIP undoubtedly benefits from the biggest marketing momentum. However, as we have seen above, it is important to distinguish between marketing and facts. SIP is an excellent choice for PC centric applications: as a UDP based protocol it is relatively easy to get it to work across Network Address Translation devices, which facilitates the deployment of VoIP services independent of the underlying network provider. The use of SIP SIMPLE as a Presence/Instant Messaging protocol, if it prevails over the competitor Jabber, also opens the door for many synergies with PC applications. On

---

[3] Using the 3G-324M protocol adopted by 3GPP UMTS, CDMA 2000, TD-SCDMA. IP based protocols, which introduce an error multiplication effect because a single bit error kills the whole packet, cannot be used on current radio links for interactive media.

the other hand the major weakness of SIP today is its lack of maturity, and overall a very slow convergence process towards a carrier grade version. The most efficient work so far has been done by the 3GPP consortium, which is in fact almost redefining its own version of SIP.

– MGCP has quietly become one of the most successful protocols, both in core networks where it serves to control trunking gateways, and at the edge to control business IP phones (where it has unparalleled features for screen and function key control), and analogue lines (only MGCP allows full transparency with existing PSTN features). MGCP, under the name 'NCS' is also used by the cable industry to deliver VoIP over DOCSIS networks. The protocol had by far the best design quality upfront, and there are very few things to fix in the latest version of MGCP. We believe there is in absolute terms a need for a stimulus protocol like MGCP, at a lower level than SIP or H.323. Therefore we think MGCP is here to stay regardless of the evolution of H.323 or SIP.

We often had the request to describe in details H.248. We decided to not cover it in this edition, as we do not believe H.248 will have a significant market presence in the next 2–3 years, at least at the edge of the network. In the long term, MGCP may be displaced by the ITU version, H.248, we see this happening in the next 2–3 years for the control of core trunking gateway (the protocol used at this level is almost an internal interface between the softswitch and the media gateways, and often what vendors call H.248 is in fact proprietary at this level), but it will take more time in the CPE space where the H.248 offer is virtually inexistent and there is no strong motivations to evolve towards H.248.

Similarly, we did not cover SIGTRAN and SS7 in general. SIGTRAN allows the transport of traditional telephony signaling messages over an IP network. The concept of the protocol is simple: it is a form of tunneling. However a full understanding of SIGTRAN requires a complete introduction to SS7 transport layers (MTP1/MTP2/MTP3) and SS7 application protocols (e.g. SCCP/ISUP), which would be too detailed for most VoIP engineers. In addition the new IP transport protocol introduced by SIGTRAN, SCTP, is in itself a major technology improvement over TCP and UDP, but also much more complex than TCP or UDP. Covering SIGTRAN in details would justify a separate book. In practice, most VoIP engineers will never need to know in details what happens within the "SIGTRAN tunnel".

CONCLUSION

As this book is going to press, the momentum of VoIP seems to be growing every day. VoIP now makes the front page of major economic newspapers that describe it as a technology that will reshape the telecom industry. We hope that this momentum will remain reasonable and will not end up in a new bubble. VoIP does have major advantages and offers a potential for new disruptive business models, but this comes with the challenges of any new, relatively immature, technology. We hope that "IP telephony" and "Beyond VoIP Protocols" will give our readers a comprehensive understanding of VoIP technology and its potential, while keeping the expectations realistic and not forgetting the potential issues.

If we have achieved our goal, our readers will be able to perform a thorough reality check on any elaborate marketing story, and enjoy the benefits of VoIP while avoiding the pitfalls.

1

Voice over Packet

1.1 TRANSPORTING VOICE, FAX, AND VIDEO OVER A PACKET NETWORK

1.1.1 A Darwinian view of voice transport

1.1.1.1 The circuit switched network

The most common telephone system on the planet today is still analog, especially at the edge of the network. Analog telephony (Figure 1.1) uses the modulation of electric signals along a wire to transport voice.

Although it is a very old technology, analog transmission has many advantages: it is simple and keeps the end-to-end delay of voice transmission very low because the signal propagates along the wire almost at the speed of light.

It is also inexpensive when there are relatively few users talking at the same time and when they are not too far apart. But the most basic analogue technology requires one pair of wires per active conversation, which becomes rapidly unpractical, and expensive. The first improvement to the basic "baseband" analog technology involved multiplexing several conversations on the same wire, using a separate transport frequency for each signal. But even with this hack, analog telephony has many drawbacks:

- Unless you use manual switchboards, analog switches require a lot of electromechanical gear, which is expensive to buy and maintain.
- Parasitic noise adds up at all stages of the transmission because there is no way to differentiate the signal from the noise and the signal cannot be cleaned.

IP Telephony   O. Hersent, J.P. Petit, D. Gurle
© 2005 John Wiley & Sons, Ltd   ISBN: 0-470-02359-7

Figure 1.1 Analog telephony, as old as the invention of the telephone, and still in use today at the edge of the network.

For all these reasons, most countries today use digital technology for their core telephone network and sometimes even at the edge (ISDN). In most cases the subscriber line remains analogue, but the analogue signal is converted to a digital data stream in the first local exchange. Usually, this signal has a bitrate of 64 kbit/s or 56 kbit/s (one sample every 125 μs).

With this digital technology, many voice channels can easily be multiplexed along the same transmission line using a technology called **time division multiplexing (TDM)**. In this technology, the digital data stream which represents a single conversation is divided into blocks (usually an octet), and blocks from several conversations are interleaved in a round robin fashion in the time slots of the transmission line, as shown in Figure 1.2.

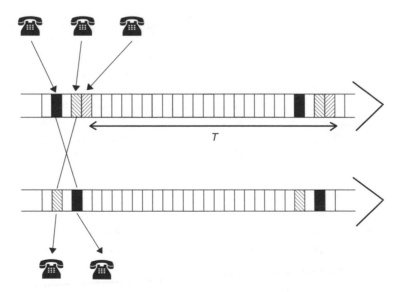

Figure 1.2 TDM switching.

Because of digital technology, the noise that is added in the backbone does not influence the quality of the communication because digital 'bits' can be recognized exactly, even in the presence of significant noise. Moreover, digital TDM makes digital switching possible. The switch just needs to copy the contents of one time slot of the incoming transmission line into another time slot in the outgoing transmission line. Therefore, this switching function can be performed by computers.

However, a small delay is now introduced by each switch, because for each conversation a time slot is only available every T μs, and in some cases in may be necessary to wait up to T μs to copy the contents of one time slot into another. Since T equals 125 μs in all digital telephony networks, this is usually negligible and the main delay factor is simply the propagation time.

1.1.1.2 *Asynchronous transmission and statistical multiplexing*

Unless you really have a point to make, or you're a politician, you will usually speak only half of the time during a conversation. Since we all need to think a little before we reply, each party usually talks only 35% of the time during an average conversation.

If you could press a button each time you talk, then you would send data over the phone line only when you actually say something, not when you are silent. In fact, most of the techniques used to transform your voice into data (known as codecs) now have the ability to detect silence. With this technique, known as voice activity detection (VAD), instead of transmitting a chunk of data, voice, or silence every 125 μs, as done today on TDM networks, you only transmit data when you need to, asynchronously, as illustrated in Figure 1.3.

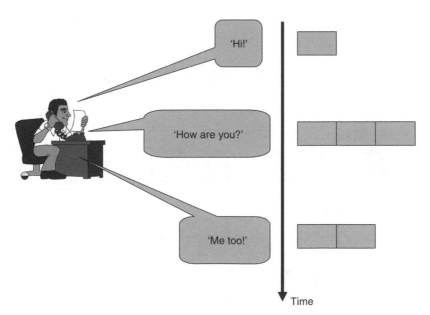

Figure 1.3   Transmitting voice asynchronously.

And when it comes to multiplexing several conversations on a single transmission line, instead of occupying a fraction of bandwidth all the time, 'your' bandwidth can be used by someone else while you are silent. This is known as 'Statistical multiplexing'.

The main advantage of **statistical multiplexing** is that it allows the bandwidth to be used more efficiently, especially when there are many conversations multiplexed on the same line (see companion book, *Beyond VoIP protocols* Chapter 5 for more details). But statistical multiplexing, as the name suggests, introduces uncertainty in the network. As just mentioned, in the case of TDM a delay of up to 125 μs could be introduced at each switch; this delay is constant throughout the conversation. The situation is totally different with statistical multiplexing (Figure 1.4): if the transmission line is empty when you need to send a chunk of data, it will go through immediately. If on the other hand the line is full, you may have to wait until there is some spare capacity for you.

This varying delay is caller **jitter**, and needs to be corrected by the receiving side. Otherwise, if the data chunks are played as soon as they are received, the original speech can become unintelligible (see Figure 1.5).

The next generation telephone networks will use statistical multiplexing, and mix voice and data along the same transmission lines. Several technologies are good candidates (e.g., voice over frame relay, voice over ATM, and, of course, voice over IP).

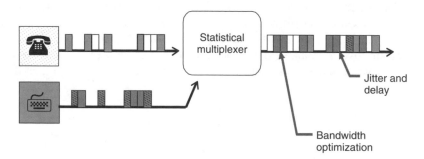

Figure 1.4 Statistical multiplexers optimize the use of bandwidth but introduce jitter.

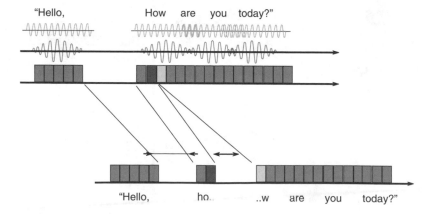

Figure 1.5 Effects of uncompensated jitter.

We believe voice over IP is the most flexible solution, because it does not require setting up virtual channels between the sites that will communicate. VoIP networks scale much better than ATM or frame relay networks, and VoIP also allows communications to be established directly with VoIP endpoints: there is now a variety of **IP-PBXs** (private switches with a VoIP wide-area network interface), or IP phones on the market today that have no ATM or frame relay equivalent.

1.1.2   Voice and video over IP with RTP and RTCP

The Real-time Transport Protocol and Real Time Control Protocol, described in RFC 1889, are the protocols that have been used for the transport of media streams since the first conferencing tools were made available on the Internet. The **visual audio tool (VAT)** used RTP version 0. A description of version 1 is available at ftp://gaia.cs.umass.edu/pub/ hgschulz/rtp/draft-ietf-avt-rtp-04.txt

Since then, RTP has evolved into version 2. RTPv2 is not backward compatible with version 1, and therefore all applications should be built to support RTPv2.

1.1.2.1   Why RTP/RTCP?

When a network using statistical multiplexing is used to transmit real-time data such as voice, jitter has to be taken into account by the receiver. Routers are good examples of such statistical multiplexing devices, and therefore voice and video over IP have to face the issue of jitter.

RTP was designed to allow receivers to compensate for jitter and desequencing introduced by IP networks. RTP can be used for any real-time (or more rigorously isochronous) stream of data (e.g., voice and video). RTP defines a means of formatting the payload of IP packets carrying real-time data. It includes:

- Information on the type of data transported (the 'payload').
- Timestamps.
- Sequence numbers.

Another protocol, RTCP, is very often used with RTP. RTCP carries some feedback on the quality of the transmission (the amount of jitter, the average packet loss, etc.) and some information on the identity of the participants as well.

RTP and RTCP do not have any influence on the behavior of the IP network and do not control quality of service in any way. The network can drop, delay, or desequence an RTP packet like any other IP packet. RTP must not be mixed up with protocols like **RSVP** (Resource Reservation Protocol). RTP and RTCP simply allow receivers to recover from network jitter and other problems by appropriate buffering and sequencing, and to have more information on the network so that appropriate corrective measures can be adopted (redundancy, lower rate codecs, etc.). However, some routers are actually able to parse

IP packets, discover whether these packets have RTP headers, and give these packets a greater priority, resulting in better QoS even without any external QoS mechanism, such as RSVP for instance. Most Cisco® routers support the `IP RTP PRIORITY` command.

RTP and RTCP are designed to be used on top of any transport protocol that provides framing (i.e., defineates the beginning and end of the information transported), over any network. However, RTP and RTCP are mostly used on top of **UDP (User Datagram Protocol)**.[1] In this case RTP is traditionally assigned an even UDP port and RTCP the next odd UDP port.[2]

1.1.2.2 RTP

RTP allows the transport of isochronous data across a packet network, which introduces jitter and can desequence the packets. Isochronous data are data that need to be rendered with exactly the same relative timing as when they were captured. Voice is the perfect example of isochronous data, any difference in the timing of the playback will either create holes or truncate some words. Video is also a good example, although tolerances for video are a lot higher; delays will only result in some parts of the screen being updated a little later, which is visible only if there has been a significant change.

RTP is typically used on top of UDP. UDP is the most widely used 'unreliable' transport protocol for IP networks. UDP can only guarantee data integrity by using a checksum, but an application using UDP has to take care of any data recovery task. UDP also provides the notion of a **'port'**, which is a number between 0 and 65,535 (present in every packet as part of the destination address) which allows up to 65,536 UDP targets to be distinguished at the same destination IP address. A port is also attached to the source address and allows up to 65,536 sources to be distinguished from the same IP address. For instance, an RTP over UDP stream can be sent from 10.10.10.10:2100 to 10.10.10.20:3200:

Source IP address: 10.10.10.10	Source port: 2100	Destination IP address: 10.10.10.20	Destination port: 3200	RTP Data

When RTP is carried over UDP, it can be carried by multicast IP packets, i.e. packets with a multicast destination address (e.g. 224.34.54.23): therefore an RTP stream generated by a single source can reach several destinations, it will be duplicated as necessary by the IP network. (See companion book, *Beyond VoIP Protocols*, Chapter 6. IP multicast routing)

1.1.2.2.1 A few definitions

- RTP session: an RTP session is an association of participants who communicate over RTP. Each participant uses at least two transport addresses (e.g., two UDP ports on the

---

[1] For streaming (e.g., RTSP), since there are is no real-time constraint on transmission delay, it can be used over TCP.

[2] But this is not mandatory, especially when RTP/RTCP ports are conveyed by an out-of-band signaling mechanism.

local machine) for each session: one for the RTP stream, one for the RTCP reports. When a multicast transmission is used all the participants use the same pair of multicast transport addresses. Media streams in the same session should share a common RTCP channel. Note that H.323 or SIP require applications to define explicitly a port for each media channel. So, although most applications comply with the RTP requirements for RTP and RTCP port sharing, as well as the use of adjacent ports for RTP and RTCP, an application should *never* make an assumption about the allocation of RTP/RTCP ports, but rather use the explicit information provided by H.323 or SIP, even if it does not follow the RTP RFC guidelines. This is one of the most common bugs still found today in some H.323 or SIP applications.

- **Synchronization source (SSRC):** identifies the source of an RTP stream, identified by 32 bits in the RTP header. All RTP packets with a common SSRC have a common time and sequencing reference. Each sender needs to have an SSRC, each receiver also needs at least one SSRC as this information is used for **receiver reports** (RRs).

- Contributing source (CSRC): when an RTP stream is the result of a combination put together by an RTP mixer from several contributing streams, the list of the SSRCs of each contributing stream is added in the RTP header of the resulting stream as CSRCs. The resulting stream has its own SSRC. This feature is not used in H.323 or SIP.

- **NTP** format: a standard way to format a timestamp, by writing the number of seconds since 1/1/1900 at 0 h with 32 bits for the integer part and 32 bits for the decimal part (expressed in $\frac{1}{2^{32}}$ s (e.g., $0 \times 800\,000\,00$ is 0.5 s). A compact format also exists with only 16 bits for the integer part and 16 bits for the decimal part. The first 16 digits of the integer part can usually be derived from the current day, the fractional part is simply truncated to the most significant 16 digits.

1.1.2.2.2 The RTP packet

All fields up to the CSRC list are always present in an RTP packet (see Figure 1.6). The CSRC list may only be present behind a mixer (a device that mixes RTP streams, as defined in the RTP RFC). In practice most conferencing bridges that perform the function of a mixer (H.323 calls them 'multipoint processors', or MPs) do not populate the CSRC list.

Here is a short explanation of each RTP field:

- Two bits are reserved for the **RTP version**, which is now version 2 (10). Version 0 was used by VAT and version 1 was an earlier IETF draft.

- A **padding bit P** indicates whether the payload has been padded for alignment purposes. If it has been padded (P = 1), then the last octet of the payload field indicates more precisely how many padding octets have been appended to the original payload.

- An **extension bit X** indicates the presence of extensions after the eventual CSRCs of the fixed header. Extensions use the format shown in Figure 1.7.

- The 4-bit **CSRC count** (CC) states how many CSRC identifiers follow the fixed header. There is usually none.

- **Marker (M):** 1 bit. Its use is defined by the RTP profile. H.225.0 says that for audio codings that support silence suppression, it must be set to 1 in the first packet of each

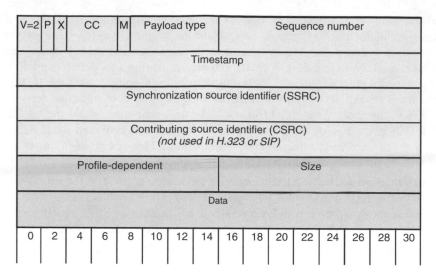

Figure 1.6 RTP packet format.

Figure 1.7 Optional extension header.

talkspurt after a silence period. This may allow some implementations to dynamically reduce the jitter buffer size without running the risk of cutting important words (e.g., by trimming off some silence packets).

- **Payload type** (PT): 7 bits. The payload of each RTP packet is the real-time information contained in the packet. Its format is completely free and must be defined by the application or the profile of RTP in use. It enables applications to distinguish a particular format from another without having to analyse the content of the payload. Some common identifiers are listed in Table 1.1; they are used by H.225 and SIP. These are called **static payload types** and are assigned by **IANA (Internet Assigned Numbers Authority)**; a list can be found a at http://www.isi.edu/in-notes/iana/assignments/rtp-parameters PT 96 to 127 are reserved for **dynamic payload types**. Dynamic payload types are defined in the RTP audio-visual (A/V) profile and are not assigned in the IANA list. The dynamic PT meaning is defined only for the duration of the session. The exact meaning of the dynamic payload type is defined through some out-of-band mechanism (e.g., though Session Description Protocol parameters for protocols like SIP, H.245 OpenLogicalChannel parameters for H.323, or through some convention or other mechanism defined by the application). The codec associated with a dynamic PT is negotiated by the conference control protocol dynamically. Since RTP itself doesn't

Table 1.1 Common static payload types

Payload type	Codec	
0	PCM, μ-law	Audio
8	PCM, A-law	
9	G.722	
4	G.723	
15	G.728	
18	G.729	
34	H.263	Video
31	H.261	

define the format of the payload section, each application must define or refer to a profile. In the case of H.323, this work is done in annex B of H.225.

- A sequence number and timestamp. The 16-bit sequence number and timestamp start on a random value and are incremented at each RTP packet. The 32-bit timestamp uses a clock frequency that is defined for each payload type (e.g., H.261 payload uses a 90-kHz clock for the RTP timestamp). For narrow-band audio codecs (G.711, G.723.1, G.729, etc.) the RTP clock frequency is set to 8,000 Hz. For video, the RTP timestamp is the tick count of the display time of the first frame encoded in the packet payload. For audio, the RTP timestamp is the tick count when the fist audio sample contained in the payload was sampled. Each RTP packet carries a sequence number and a timestamp. RTP timestamps do not have an absolute meaning (the initial timestamps of an RTP stream can be selected at random); even timestamps of related media (e.g., audio and video) in a single session will be unrelated. In order to map RTP packet timestamps to absolute time, one must use the information held in RTCP sender reports, where RTP timestamps are associated with the absolute NTP time. Depending on the application, timestamps can be used in a number of ways. A video application, for instance, will use it to synchronize audio and data. An audio application will use the sequence number and timestamp to manage a reception buffer. For instance, an application can decide to buffer 20 10-ms G.729 audio frames before commencing playback. Each time a new RTP packet arrives, it is placed in the buffer in the appropriate position depending on its sequence number. It is important to note that the protection against jitter allowed by RTP comes with a price: a greater end-to-end delay in the transmission path. If a packet doesn't arrive on time and is still missing at playback time, the application can decide to copy the last sample of the packet that has just been played and repeat it long enough to catch up with the timestamp of the next received packet, or use some interpolation scheme as defined by the particular audio codec in use. The sequence number is used to detect packet loss.

1.1.2.3 RTCP

RTCP is used to transmit control packets to participants regarding a particular RTP session. These control packets include various statistics, information about the participants (their names, email addresses, etc.) and information on the mapping of participants to individual stream sources. The most useful information found in RTCP packets concerns the quality of transmission in the network. All participants in the sessions send RTCP packets: senders send '*sender* reports' and receivers send '*receiver* reports'.

1.1.2.3.1 Bandwidth limitation

All participants must send RTCP packets. This causes a potential dimensioning problem for large multicast conferences: RTCP traffic should grow linearly with the number of participants. This problem does not exist with RTP streams in audio-only conferences using silence suppression, for instance, since people generally don't speak at the same time (Figure 1.8).

Since the number of participants is known to all participants who listen to RTCP reports, each of them can control the rate at which RTCP reports are sent. This is used to limit the bandwidth used by RTCP to a reasonable amount, usually not more than 5% of the overall session bandwidth (which is defined as the sum of all transmissions from all participants, including the IP/UDP overhead).

This budget has to be shared by all participants. Active senders get one-quarter of it because some of the information they send (e.g., CNAME information used for synchronization) is very important to all receivers and RTCP sender reports need to be very responsive. The remaining part is split between the receivers. The average sending rate is derived by the participant from the size of the RTCP packets that he wants to send, and from the number of senders and receivers that appear in the RTCP packets it receives. This is clearly relevant for multicast sessions; in fact, many of the recommendations and features present in the RTP RFC are useless for most VoIP applications, which have a maximum of three participants in most cases. Even for small sessions, the fastest rate at which a participant is allowed to send RTCP reports is one every 5 s. The sending rate is randomized by a factor of 0.5 to 1.5 to avoid unwanted synchronization between reports.

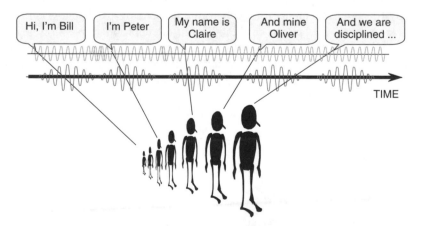

Figure 1.8 Bitrate is self-limiting in audio conferences (at least among polite participants).

Most H.323 and SIP implementations actually use a simplified version of these guidelines, which is not a problem because there is no scaling issue. The RFC recommendations remain applicable for larger conferences, however, such as the conferences using the **H.332** protocol to broadcast information to multiple receivers.

1.1.2.3.1.1   RTCP packet types

There are various types of RTCP messages defined for each type of information:

- SR: sender reports contain transmission and reception information for active senders.
- RR: receiver reports contain reception information for listeners who are not also active senders.
- SDES: source description describes various parameters relating to the source, including the name of the sender (CNAME).
- BYE: sent by a participant when he leaves the conference.
- APP: functions specific to an application.

Several RTCP messages can be packed in a single transport protocol packet. Each RTCP message contains enough length information to be properly decoded if several of those RTCP messages are packed in a single UDP packet. This packing can be useful to save overhead bandwidth used by the transport protocol header.

1.1.2.3.1.2   Sender reports

Each SR contains three mandatory sections, as shown in Figure 1.9.

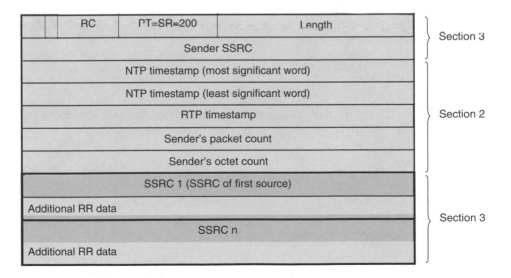

Figure 1.9   RTCP packet format.

The first section contains:

- The 5-bit reception report count (RC), which is the number of report blocks included in this SR.
- The packet type (PT) is 200 for an SR. In order to avoid mixing a regular RTP packet with an SR, RTP packets should avoid payload types 72 and 73 which can be mistaken for SRs and RRs when the marker bit is set. However, normally a UDP port is dedicated to RTCP to eliminate this potential confusion.
- The 16 bit length of this SR including header and padding (the number of 32-bit words minus 1).
- The SSRC of the originator of this SR. This SSRC can also be found in the RTP packets that originate from this host.

The second section contains information on the RTP stream originated by this sender (this SSRC):

- The NTP timestamp of the sending time of this report. A sender can set the high-order bit to 0 if it can't track the absolute NTP time; this NTP measurement only relates to the beginning of this session (which is assumed to last less than 68 years!). If a sender can't track elapsed time at all it may set the timestamp to 0.
- The RTP timestamp, which represents the same time as above, but with the same units and random offset as in the timestamps of RTP packets. Note that this association of an absolute NTP timestamp and the RTP timestamps enables the receiver to compute the absolute timestamp of each received RTP packet and, therefore, to synchronize related media streams (e.g., audio and video) for playback.
- Sender's packet count (32 bits) from the beginning of this session up to this SR. It is reset if the SSRC has to change (this can happen in an H.323 multiparty conference when the active MC assigns terminal numbers).
- Sender's payload octet count (32 bits) since the beginning of this session. This is also reset if the SSRC changes.

The third section contains a set of reception report blocks, one for each source the sender knows about since the last RR or SR. Each has the format shown in Figure 1.10:

- SSRC_n (source identifier)(32 bits): the SSRC of the source about which we are reporting.
- Fraction lost (8 bits): equal to Floor(received packets/expected packets $*$ 256).
- Cumulative number of packets lost (24 bits) since the beginning of reception. Late packets are not counted as lost and duplicate packets count as received packets.
- Extended highest sequence number received (32 bits): the most significant 16 bits contain the number of sequence number cycles, and the last 16 bits contain the highest sequence number received in an RTP data packet from this source (same SSRC).

SSRC of the source	
Lost fraction	Cumulated number of lost packets
Highest received sequence number	
Interrarival jitter	
Last SR (LSR)	
Delay since last SR	

Figure 1.10   Format of a reception report block.

- Interarrival jitter (32 bits): an estimation of the variance in interarrival time between RTP packets, measured in the same units as the RTP timestamp. The calculation is made by comparing the RTP timestamp of arriving packets with the local clock, and averaging the results (as shown in Figure 1.11).
- The last SR timestamp (LSR) (32 bits): the middle 32 bits of the NTP timestamp of the last SR received (this is the compact NTP form).
- The delay since the last SR arrived (DLSR) (32 bits): expressed in compact NTP form (or, more simply, in multiples of 1/65536 s). Together with the last SR timestamp, the sender of this last SR can use it to compute the round trip time.

1.1.2.3.1.3   Receiver reports

A receiver report looks like an SR, except that the PT field is now 201, and the second section (concerning the sender) is absent.

1.1.2.3.1.4   SDES: source description RTCP packet

An SDES packet (Figure 1.12) has a PT of 202 and contains SC (source count) chunks. Each chunk contains an SSRC or a CSRC and a list of information. Each element of this list is coded using the type/length/value format. The following types exist but only CNAME has to be present:

- CNAME (type 1), unique among all participants of the session, is of the form user@host, where host is the IP address or domain name of the host.
- NAME (type 2): common name of the source.
- EMAIL (type 3).

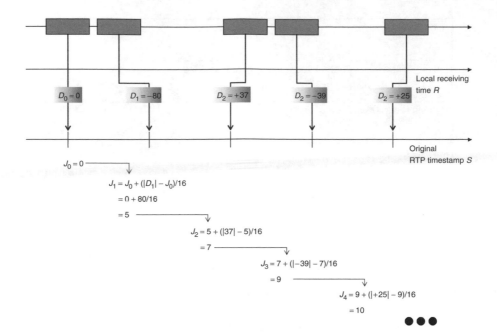

Figure 1.11 Jitter evaluation.

```
0                     1                   2                   3
0 1 2 3 4 5 6 7 8 9 0 1 2 3 4 5 6 7 8 9 0 1 2 3 4 5 6 7 8 9 0 1
|V=2|P|    SC    | PT=SDES=202 |              length           | header
|                      SSRC/CSRC_1                             | chunk
                                                                   1
|                      SDES items                             |
                           ...
|                      SSRC/CSRC_2                             | chunk
                                                                   2
|                      SDES items                             |
                           ...
```

Figure 1.12 SDES message format.

- PHONE (type 4).
- LOC (type 5): location

1.1.2.3.1.5   BYE RTCP packet

The BYE RTCP packet (Figure 1.13) indicates that one or more sources (as indicated by source count SC) are no longer active.

1.1.2.3.1.6   APP: application-defined RTCP packet

This can be used to convey additional proprietary information. The format is shown in Figure 1.14. The PT field is set to 204.

```
 0                   1                   2                   3
 0 1 2 3 4 5 6 7 8 9 0 1 2 3 4 5 6 7 8 9 0 1 2 3 4 5 6 7 8 9 0 1
+-+-+-+-+-+-+-+-+-+-+-+-+-+-+-+-+-+-+-+-+-+-+-+-+-+-+-+-+-+-+-+-+
|V=2|P|   SC   |  PT=BYE=203  |             length              |
+-+-+-+-+-+-+-+-+-+-+-+-+-+-+-+-+-+-+-+-+-+-+-+-+-+-+-+-+-+-+-+-+
|                          SSRC/CSRC                            |
:                             ...                               :
+=+=+=+=+=+=+=+=+=+=+=+=+=+=+=+=+=+=+=+=+=+=+=+=+=+=+=+=+=+=+=+=+
|    length     |              reason for leaving          ... (opt)
+-+-+-+-+-+-+-+-+-+-+-+-+-+-+-+-+-+-+-+-+-+-+-+-+-+-+-+-+-+
```

Figure 1.13 BYE message format.

```
 0                   1                   2                   3
 0 1 2 3 4 5 6 7 8 9 0 1 2 3 4 5 6 7 8 9 0 1 2 3 4 5 6 7 8 9 0 1
+-+-+-+-+-+-+-+-+-+-+-+-+-+-+-+-+-+-+-+-+-+-+-+-+-+-+-+-+-+-+-+-+
|V=2|P| subtype |  PT=APP=204  |             length             |
+-+-+-+-+-+-+-+-+-+-+-+-+-+-+-+-+-+-+-+-+-+-+-+-+-+-+-+-+-+-+-+-+
|                          SSRC/CSRC                            |
|                          name (ASCII)                         |
|                    application-dependent data            ...
+-+-+-+-+-+-+-+-+-+-+-+-+-+-+-+-+-+-+-+-+-+-+-+-+-+-+-+-+-+-+-+-+
```

Figure 1.14 APP message format.

1.1.2.4 Security

Security can be achieved at the transport level (e.g., using IPSec) or at the RTP-level. The RTP RFC presents a way to ensure RTP-level privacy using DES/CBC (data encryption standard, cipher block chaining) encryption. Since **DES**, like many other encryption algorithms, is a block algorithm (for a more detailed description see Section 2.6.2 about H.235), there needs to be some adaptation when the unencrypted payload is not a multiple of 64 bits.

The most straightforward method, padding, is described in RTP (RFC 1889, sec. 5.1). When this method is used the padding bit of the RTP header is set, and the last octet of the RTP payload contains the number of padding bits to remove (Figure 1.15). The last octet can be located because the underlying transport protocol must support framing. There are other encryption methods that do not require padding (e.g., ciphertext stealing); some of these alternative methods are described in Chapter 2 (on H.235).

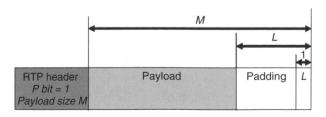

Figure 1.15 RTP payload padding for encryption using block algorithms.

Authentication and negotiation of a common secret is not within the scope or RTP. For instance the negotiation of a common secret can be performed out of band using a **Diffie–Helmann** scheme (see Section 2.6.2.1)

1.2 ENCODING MEDIA STREAMS

1.2.1 Codecs

We have seen already that isochronous (audio, video, etc.) data streams could be carried over RTP. But these analogue signals first need to be transformed into data. This is the purpose of codecs. This section provides a high-level overview of some of the most popular voice and video coding technologies, sufficient in most cases to understand H.323, SIP, or MGCP and to help in the recurring debates about the 'best' codec. The reader wanting more detailed knowledge should read the voice-coding background chapter in the companion book, *Beyond VoIP Protocols*.

1.2.1.1 What is a good codec?

When the International Multimedia Telecommunications Causatium (www.IMTC.org) tried to choose a default low-bitrate codec, a sufficient to promote interoperability, they faced a difficult issue because there wasn't common agreement about what constituted a good codec. The difficulty was so great that other bodies who are also trying to profile VoIP applications are reticent to enter into the debate at all.

Let's look at the criteria that must now be considered when evaluating a voice codec.

1.2.1.1.1 Bandwidth usage

The bitrate of available narrow-band codecs (approximately 300–3400 Hz) today ranges from 1.2 kbit/s to 64 kbit/s. Of course there is a consequence on the quality of restituted voice. This is usually measured by MOS (mean opinion score) marks. MOSs for a particular codec are the average mark given by a panel of auditors listening to several recorded samples (voice samples, music samples, voice with background noise, etc.). These scores range from 1 to 5:

- From 4 to 5 the quality is 'high' (i.e., similar to or better than the experience we have when making an ISDN phone call).
- From 3.5 to 4 is the range of 'toll quality'. This is more or less similar to what is obtained with the G.726 codec (32 kbit/s ADPCM) which is commonly taken as the reference for 'toll quality'. This is what we experience on most phone calls. Mobile phone calls are usually just below the 'toll' quality.
- From 3.0 to 3.5, communication is still good, but voice degradation is easily audible.
- From 2.5 to 3, communication is still possible, but requires much more attention. This is the range of 'military quality' voice. In extreme cases the expression 'synthetic', or 'robotic', voice is used (i.e., when it becomes impossible to recognize the speaker).

There is a trade-off between voice quality and bandwidth used. With current technology toll quality cannot be obtained below 5 kbit/s.

1.2.1.1.2   Silence compression (VAD, DTX, CNG)

During a conversation, we only talk on average 35% of the time. Therefore, silence compression or suppression is an important feature. In a point-to-point call it saves about 50% of the bandwidth, but in decentralized multicast conferences the activity rate of each speaker drops and the savings are even greater. It wouldn't make sense to undertake a multicast conference where there are more than half a dozen participants without silence suppression.

Silence compression includes three major components:

- VAD (voice activity Detector): this is responsible for determining when the user is talking and when he is silent. It should be very responsive (otherwise the first word may get lost and unwanted silence might occur at the end of sentences), without getting triggered by background noise. VAD evaluates the energy and spectrum of incoming samples and activates the media channel if this energy is above a minimum and the spectrum corresponds to voice. Similarly, when the energy falls below a threshold for some time, the media channel is muted. If the VAD module dropped all samples until the mean energy of the incoming samples reaches the threshold, the beginning of the active speech period would be clipped. Therefore, VAD implementations require some lookahead (i.e., they retain in memory a few milliseconds worth of samples to start media channel activation before the active speech period). This usually adds some delay to the overall coding latency, except on some coders where this evaluation is coupled with the coding algorithm itself and does not add to the algorithmic delay. The quality of the implementation is important: good VAD should require minimal lookahead, avoid voice clipping, and have a configurable hangover period (150 ms is usually fine, but some languages, such as Chinese, require different settings).

- DTX, (discontinuous transmission): this is the ability of a codec to stop transmitting frames when the VAD has detected a silence period. If the transmission is stopped completely, then it should set the marker bit of the first RTP packet after the silence period. Some advanced codecs will not stop transmission completely, but instead switch to a silence mode in which they use much less bandwidth and send just the bare minimum parameters (intensity, etc.) in order to allow the receiver to regenerate the background noise.

- CNG, (comfort noise generator): it seems logical to believe that when the caller isn't talking, there is just silence on the line and, when the VAD detects a silence period, it should be enough to switch off the loudspeaker completely. In fact, this approach is completely wrong. Movie producers go to great lengths to recreate the proper background noise for 'silent' sequences. The same applies to phone calls. If the loudspeaker is turned off completely, street traffic and other background noise that could be overheard while the caller was talking would stop abruptly. The called party would get the impression that the line had been dropped and would ask the caller whether he is still on the line. The CNG is here to avoid this and recreate some sort of background noise. With the most primitive codecs that simply stop transmission it will use some random noise with a level deduced from the minimal levels recorded during active speech periods. More advanced codecs such as G.723.1 (annex A) or G.729 (annex B) have options to send enough information to allow the remote decoder to regenerate ambient noise close to the original background noise.

1.2.1.1.3   Intellectual property

End-users don't care about this, but manufacturers have to pay royalties to be allowed to use some codecs in their products. For some hardware products where margins are very low, this can be a major issue. Another common situation is that some manufacturers want to sell some back-end server, while distributing software clients for free. If the client includes a codec, then again intellectual property becomes a major choice factor.

1.2.1.1.4   Lookahead and frame size

Most low-bitrate codecs compress voice in chunks called frames and need to know a little about the samples immediately following the samples they are currently encoding (this is called lookahead).

There has been a lot of discussion (especially at the IMTC when they tried to choose a low-bitrate codec) over the influence of frame size on the quality of the codec. This is because the minimal delay introduced by a coding/decoding sequence is the frame length plus the lookahead size. This is also called the algorithmic delay. Of course, in reality DSPs do not have infinite power and most of the time a fair estimate is to consider the real delay introduced as twice or three times the frame length plus the overhead (some authors improperly call this the algorithmic delay, although this is just an estimate of DSP power).

So, codecs with a small frame length are indeed better than codecs with a longer frame length regarding delay, when if each frame is sent immediately on the network. This is where it becomes tricky, because each RTP packet has an IP header of 20 octets, a UDP header of 8 octets, and an RTP header of 12 octets! For instance, for a codec with a frame length of 30 ms, sending each frame separately on the network would introduce a 10.6-kbit/s overhead. Much more than the actual bitrate of most narrow-band codecs!

Therefore, most implementations choose to send multiple frames per packet, and the real frame length is in fact the sum of all frames stacked in a single IP packet. This is limited by echo and interactivity issues (see companion book, Chapter 3 *Beyond VoIP Protocols*). A maximum of 120 ms of encoded voice should be sent in each IP packet.

So, for most implementations, the smaller the frame size, the more frames in an IP packet. This is all there is to it and there is no influence on delay. Overall, it is better to use codecs that have been designed for the longest frame length (limited by the acceptable delay), since this allows even more efficient coding techniques: the longer you observe a phenomenon, the better you can model it!

We can conclude that in most cases frame size is not so important for IP videoconferences when bandwidth is a concern. The exception is high-quality conferences where interactivity is maximized at the expense of the required bitrate.

1.2.1.1.5   Resilience to loss

Packet loss is a fact of life in IP networks and the short latency required by interactive voice and video applications does not allow us to request retransmissions. Since packets carry codec frames, this in turn causes codec frame loss. However, packet loss and frame loss are not directly correlated; many techniques such as FEC (forward error correction) can be used to lower the frame loss rate associated with a given packet loss rate. These techniques spread redundant information over several packets so that frame information can be recovered even if some packets are lost.

However, the use of redundancy to recover packet loss is a very tricky thing. It can lead to unexpected issues and, can even make the problem worse. To understand this, let's look at what some manufacturers could do (and have done!):

- You prepare a demonstration to compare your product and that of a competitor. You let it be known that you can resist a 50% packet loss without any consequence on voice quality.
- You simulate packet loss by losing one packet out of two.
- You put frames N and $N-1$ into your RTP packet.

Your product can recover all the frames despite one packet out of two being lost. Your competitor is restricted to emitting a few cracks. Bingo! The customer is convinced.

Well, the only problem is that packet loss on the Internet in not so neat. Packet loss occurs in a correlated way and you are much more likely to lose several packets in a row, than exactly one packet out of two. So, this simple RTP redundancy scheme will be close to useless under real conditions and still add a 50% overhead!

The effect of frame erasure on codecs should be considered on a case-by-case basis. If you lose N samples from a G.711 codec (stateless coder) this will just result in a gap of $N * 125$ μs at the receiving end. If you lose just one frame from a very advanced codec it may be audible for much more than the duration of this frame, because the decoder will need some time to resynchronize with the coder. For a frame of 20 ms or so, this may result in a very audible crack of 150 ms. Codecs such as G.723.1 are designed to cope relatively well with an uncorrelated frame erasure of up to 3%, but beyond this quality drops off very rapidly. The effect of correlated loss is not yet fully evaluated. It is possible to reduce the occurrence of consecutive frame loss by interleaving codec frames across multiple RTP packets: unfortunately, this adds a lot of delay to the transmission and therefore can only be used in streaming media transmissions, not in the context of real-time communications.

Apart from the built-in features of the codec itself, it is possible to reduce the frame loss associated with packet loss by using a number of techniques.

FEC-style redundancy (Figure 1.16) can be used to recover from serious packet loss conditions, but it has a significant impact on delay. For instance, if you choose to repeat the same G.723.1 frame in four consecutive IP packets in order to recover from the loss of three consecutive packets, then the decoder needs to maintain a buffer of four IP packets, but this ruins the delay factor. More sophisticated FEC methods use XOR sums instead of simple repetitions, but have the same impact on delay.

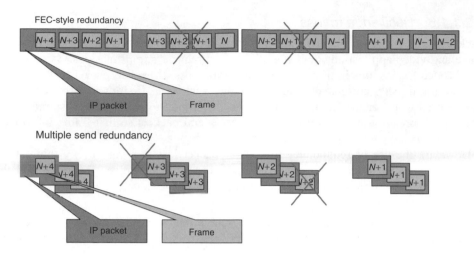

Figure 1.16 Using redundancy to reduce codec frame loss.

It is also possible to send several copies of each frame immediately. But if one packet gets lost, probably all the copies will reach the same congested router at nearly the same time and might get lost as well.

An understanding of the different types of congestion is also important in deciding whether redundancy is useful and which type to use. The network can lose packets because a link is congested or because a router has to route too many small packets per second.

If a link is congested, then any type of redundancy will add to the congestion and increase the overall loss percentage of IP packets. But the frame loss rate of communicating devices that use FEC redundancy will still be reduced.

Some arithmetic proves this. Say we have congestion on a 2-Mbit/s line. It receives 2.2 Mbit/s and the average loss rate is $0.2/2.2 = 9\%$. Part of this is caused by someone using a codec producing a 100 kbit/s stream. The software detects a high loss and decides to use the FEC scheme described above. Now that same application produces a 400 kbit/s stream (the influence of headers is not taken into account for simplicity). The 2-Mbit/s line receives 2.5 Mbit/s and the packet loss rate is increased to 20% for all the users of the link. However, if we assume the congested link never causes the loss of four packets in a row (on average one packet in five is dropped), then the software will recover from *all* loss. However, this would be unacceptable behaviour, because it would be unfair to other users and could destabilize the network. Next-generation IP networks will probably include advanced techniques, such as RED, that will detect the greedy user and drop most of his packets.

If congestion is due to an overrun router (exceeding its packet/s limit), then FEC-style redundancy is not such a bad thing. It increases the size of packets but does not increase the average number of packets that the router has to forward per second. In this case increasing the size of the packets wil not add to the congestion. The other type of redundancy (multiple simultaneous sending) will increase the number of packets through the router and would not work.

1.2.1.1.6 Layered coding

There are several situations in which current codecs are not well suited. For instance, if you want to broadcast the same event to several listeners (H.332 type of conference), some will want high-quality reception (either because they have paid for it or because they have large IP pipes) and others will only be able to receive lower quality. You could send a customized data stream to each listener, but this is not practical for a large audience. The answer is to multicast the data stream to all listeners (for more information on multicast, refer to the multicast chapter of the companion book, *Beyond VoIP Protocols*). Current codecs include complete information in one data stream. If it is multicast, all participants will receive the same amount of data, so you usually have to limit the data rate to the reception capability of the least capable receiver.

Some codecs (most are still at experimental stage) can produce several data streams simultaneously, one with the core information needed for 'military quality' reception and the other data streams with more information as needed to rebuild higher fidelity sound or an image. A crude example for video would be to send black and white information on one channel and colour (chrominance) information on another.

Each part of the data stream can be multicast using different group addresses, so that listeners can choose to receive just the core level or the other layers as well. In a pay-for-quality scheme, you would encrypt the higher layers (this way you have the option of receiving a free low-quality preview and later of paying for the broadcast quality image).

Layered codecs are also very useful when it comes to redundancy: the sender can choose to use a redundancy scheme or a quality of service level for the core layer, so that the transmission remains understandable at all times for everyone, but leave other layers without protection.

H.323v2 was approved with a specific annex on layered video coding (annex B: procedures for layered video codecs).

1.2.1.1.7 Fixed point or floating point

We first need to say a few words about digital signal processors (DSPs). These are processors that have been optimized for operations frequently encountered in signal-processing algorithms. One such operation is $(a * b) + previous\ result$: one multiplication and addition. In a conventional processor, this operation would require multiple processor instructions and would be executed in several clock cycles. A DSP will do it in one instruction and a single clock cycle. Another example is the code book searches frequently used by vocoders. Some conventional processors also have extensions to accelerate signal-processing algorithms (e.g., MMX processors can execute a single instruction simultaneously on several operands as long as they can be contained in a 32-bit register and video algorithms can be accelerated by processing 4 pixels (8 bits each) simultaneously).

There are two types of DSPs: floating point DSPs, which are capable of operating on floating point numbers, and fixed point DSPs. Fixed point DSP operands are represented as a mantissa n and a power p of 2 (e.g., $12345678 * 2^5$), but the DSP can operate on two operands only if the power of 2 is the same for both operands. They are less powerful, but also less expensive, and chosen by many designers for products sold in large quantities. Some codecs have only been specified with fixed point C code. However, many implementations will run on processors or DSPs that are capable of floating point operation, and developers must develop their own version of floating point C code for the algorithm. This often results in interoperability problems between floating point versions.

Therefore, it makes sense for the codec to be specified in floating point C code as well, especially if the code has to run on PCs.

1.2.1.2 Audio codecs

1.2.1.2.1 ITU audio codecs

1.2.1.2.1.1 Choosing a codec at ITU

The choice of a codec at ITU WP3 is typically a very long process. This is not a bureaucracy problem, but rather a problem due to the stringent requirements of ITU experts.

Before a codec is chosen, the ITU evaluates MOS scores and usually requires quality that is equivalent to or better than G.726 ('toll quality'). The ITU also checks that this quality is constant for men and women, and in several languages. The ability to take into account background noise and recreate it correctly is also evaluated. The ITU pays special attention to the degradation of voice quality in tandem operation (several successive coding/decoding processes), since this a situation that is very likely to happen in international phone calls. Last but not least, if the codec has to be used over a non-reliable medium (a radio link, a frame relay virtual circuit, etc.), the ITU checks that the quality remains acceptable if there is some frame loss.

After checking all these parameters, it frequently occurs that no single proposal passes the test! Therefore, many ITU codecs are combinations of the most advanced technologies found in several different proposals. This leads to state-of-the-art choices, but, as we will see, this is a nightmare for anyone who needs to keep track of intellectual property.

1.2.1.2.1.2 Audio codecs commonly used in VoIP

The companion book, *Beyond VoIP Protocols* provides a detailed view on voice coder technology and discrete time signal processing in general. This section's purpose is to provide a quick reference to common VoIP coders found in VoIP systems for engineers uninterested in the details and theory of each coder.

(a) G.711 (approved in 1965)

G.711 is the grandfather of digital audio codecs. It is a very simple way of digitizing analogue data by using a semi-logarithmic scale (this is called 'companded PCM', and serves to increase the resolution of small signals, while treating large signals in the same way as the human hear does). Two different types of scales are in use, the A-law scale (Europe, international links) and the μ-law scale (USA, Japan). They differ only in the choice of some constants of the logarithmic curve. G.711 is used in ISDN and on most digital telephone backbones in operation today.

A G.711-encoded audio stream is a 64-kbit/s bitstream in which each sample is encoded as an octet; therefore, the frame length is only 125 μs. Of course, all VoIP applications will put more than one sample in every IP packet (about 10 ms typically, or 80 samples).

Most sound cards are able to record directly in G.711 format. However, in some cases it is better to record using CD quality, which samples at 44.1 kHz (one 16-bit sample every

23 μs), especially if echo cancelation algorithms are used, since the full performance of some echo cancelation algorithms cannot be achieved with the quantification noise introduced by G.711.

The typical MOS score of G.711 is usually taken as 4.2; it is used as an anchor for other coder tests.

(b)   G.722 (approved in 1988)

Although G.711 achieves very good quality, some of the voice spectrum (above 4 kHz) is still cut. G.722 provides a higher quality digital coding of 7 kHz of audio spectrum at only 48, 56, or 64 kbit/s, using about 10 DSP MIPS. This is an 'embedded' coder, which means that the rate can freely switch between 48, 56, or 64 kbit/s without notifying the decoder.

This coder is very good for all professional conversational voice applications (the algorithmic delay is only 1.5 ms). G.722 is supported by some videoconferencing equipment and some IP phones.

(c)   G.722.1

This more recent wideband coder operates at 24 kbit/s or 32 kbit/s. It has been designed by Picturetel, which also sells a 16-kbit/s version (Siren™). The coder encodes frames of 20 ms, with a lookahead of 20 ms. The 16-kbit/s version is supported by Windows® Messenger.

(d)   G.723.1 (approved in 1995)

In the early days of VoIP, the VoIP Forum chose the G.723.1 codec as the baseline codec for narrow-band H.323 communications. It is also used by the video cellphones of UMTS 99 (H.324M standard).

Technology

G.723.1 uses a frame length of 30 ms and needs a lookahead of 7.5 ms. It has two modes of operation, one at 6.4 kbit/s and the other at 5.3 kbit/s. The mode of operation can change dynamically at each frame. Both modes of operation are mandatory in any implementation, although many VoIP systems have an incorrect implementation that works on only one of the two modes.

G.723 is not designed for music and does not transmit DTMF tones reliably (they must be transmitted out-of-band). Modem and fax signals cannot be carried by G.723.1.

G.723.1 achieves an MOS score of 3.7 in 5.3-kbit/s mode and 3.9 in 6.4-kbit/s mode. Table 1.2 compares the performance of G.723.1 (6.4 kbps) and the ADPCM released by Bell Labs in March 94.

Table 1.2   Impact of frame-erasure and tandeming quality

	G.723.1, 6.4 kbit/s	32 kbit/s ADPCM
Clear channel, no errors or frame erasure	3.901	3.781
3% frame erasure	3.432	—
Tandeming of two codecs	3.409	3.491

The main effect of frame erasures is to desynchronize the coder and the decoder (they may need many more frames to resynchronize). In practice, networks should always have a frame error rate below 3% (and below 1% ideally).

G.723.1 is specified in both fixed point (where it runs at 16 MIPS on a fixed point DSP) and floating point C code (running on a Pentium 100, it takes 35% to 40% of the power of the processor). The fixed point implementation runs on VoIP gateways, the floating point version runs on all Windows® PCs.

Beyond VoIP systems, G.723.1 is used in the H.324 recommendation (ITU recommendation for narrow-band videoconferencing on PSTN lines) and will also be used in the new 3G-324M (3GPP, 3GPP2 organizations) standard for 3G wireless multimedia devices.

Silence compression

G.723.1 supports voice activity detection (VAD), discontinuous transmission (DTX), and comfort noise generation (CNG) (defined in annex A of the recommendation).

Silence is coded in very small, 4-octet frames at a rate of 1.1 kbit/s. If silence information doesn't need to be updated, transmission stops completely.

Intellectual property

G.723.1 is one of the codecs that resulted from many contributions and, therefore, uses technology patented from several sources. About 18 patents currently apply to G.732.1 (the precise number is hard to keep track of), from eight different companies.

The main licensing consortium, which is made up of AudioCodes, DSP Group, FT/CNET, Université de Sherbrooke, and NTT, oversees the patents. The rights are managed by the DSP Group and SiproLabs (www.sipro.com) for all members of this consortium. Other patents are held by AT&T (1), Lucent (3), British Technology Group (1, formerly held by VoiceCraft), Nokia Mobile Phone (1, formerly held by VoiceCraft). Patent applications have also been made by Siemens, Robert Bosch, and CSELT. The source code is copyrighted by four companies.

There are typically several licensing agreements for this codec (the details hinge, of course, on the company involved), depending on whether the application is for a single user or multiple users, whether it is going to be a paying or free application, and on the volume licensed.

Of course, exact prices have to be negotiated with both patent owners and implementers, but some data can be gathered from conferences and newsgroups, *although they must be taken cautiously*. For instance, here are some price indications for acquisition of the intellectual rights of G.723.1:

- A license for a single-user client is said to be worth an initial payment of around $50,000 plus $0.8 per unit.

- A license for a server is said to be worth an initial payment of about $20,000 plus $5 per port.

- A license for unlimited distribution of a single-user application is said to be worth about $120,000.

Then, unless you do your own implementation (which is not recommended if you are not an expert!), be prepared to approximately double the previous fees to license a well-optimized implementation.

Here is a quote from a company trying to license these codecs, picked from a mailing list:

> *We have been trying to negotiate licensing arrangements with the patent holders for more than six months. As of today, we have received terms and conditions from six of the holders, and little to no response from the rest. The costs proposed by the first six strongly imply a substantial initial investment, and a per port cost in excess of $20.00.*
>
> *Our concern, however, extends far beyond the cost. The Internet's success is due to its readily available standards and lack of non-essential rules and constraints. The time requirements and logistics of establishing contact with 12 parties and negotiating licensing are significant barriers to growth in the industry. The legal risks associated with not doing so are an impediment to the rapid evolution of the industry.*

The reality is not quite so bad, as many IPR rights are now managed by Sipro Labs (www.sipro.com). The investments needed to produce the technology of standardized coders such as G.723.1 indeed justify a fee. But the question is how much is reasonable? When patented technology becomes a standard, the temptation is high to use this monopoly situation to maintain high licence prices. This underlies the so-called 'codec wars' that periodically break out in VoIP standard bodies.

(e)   G.726 (approved in 1990)

G.726 uses an ADPCM technique to encode a G.711 bitstream in words of 2, 3, or 4 bits, resulting in available bitrates of 16, 24, 32, or 40 kbit/s.

G.726 at 32 kbit/s achieves a MOS score of 4.3 and is often taken as the benchmark for 'toll quality'. It requires about 10 DSP MIPs of processing power (full duplex) or 30% of the processing power of a Pentium 100. This is a low-delay coder: 'frames' are 125 μs long and there is no lookahead. There is also an embedded version known as G.727.

(f)   G.728 (approved in 1992-94)

G.728 uses an LD-CELP (low-delay, code-excited linear prediction) coding technique and achieves MOS scores similar to that obtained by G.726 at 32 kbit/s, but with a bitrate of only 16 kbit/s. Compared with PCM or ADPCM techniques, which are waveform coders (i.e., they ignore the nature of the signal), CELP is a coder optimized for voice (vocoder). These coders specifically model voice sounds and work by comparing the waveform to encode with a set of waveform models (linear predictive code book) and find the best match. Then, only the index of this best match and parameters like voice pitch are transmitted. As a result music does not transmit well on CELP coders, and it is only at 2.4 kbit/s that fax or modem transmission can succeed with G.728 compression. G.728 is used for H.320 videoconferencing and some H.323 videoconferencing systems.

G.728 needs almost all the power of a Pentium 100 and 2 Kb of RAM to implement. It is a low-delay coder (between 625 μs and 2.5 ms).

(g)   G.729 (approved in 1995–1996)

(i)   Technology

G.729 is very popular for voice over frame relay applications and V.70 voice and data modems. Together with G.723, it has become the most popular voice coder for VoIP, but is still not supported natively on the Windows® platform. It uses a CS-ACELP (conjugate structure, algebraic code-excited linear prediction) coding technique. G.729 is not designed for music and does not transmit DTMF tones reliably (they must be transmitted out-of-band). Modem and fax signals cannot be carried by G.729.

G.729 produces 80-bit frames encoding 10 ms of speech at a rate of 8 kbit/s. It needs a lookahead of 5 ms. It achieves MOS scores around 4.0. There are two versions:

- G.729 (approved in December 1996) requires about 20 MIPS for coding and 3 MIPS for decoding.
- G.729A (approved in November 1995): annex A is a reduced complexity version of the original G.729. It requires about 10.5 MIPS for coding and 2 MIPS for decoding (about 30% less than G.723.1).

(ii)   Silence compression

Annexes A and B of G.729 define VAD, CNG, and DTX schemes for G.729. The frames sent to update background noise description are 15 bits long and are only sent if the description of the background noise changes.

(iii)   Licences

Both G.729 and G.729A are the result of about 20 patents belonging to six companies: AT&T, France Telecom, Lucent, Université de Sherbrooke (USH, Canada), NTT, and VoiceCraft. NTT, France Telecom, and USH have formed a licensing consortium managed by SiproLabs, but not all patents (notably AT&T) are covered by this consortium. The source code is copyrighted by five companies.

As with G.723, there are several ways of getting a several licence for this codec, but the prices of the G.729 IPR pool managed by SiproLabs (www.sipro.com) are in the public domain.

1.2.1.2.2   ETSI SMG audio codecs

The ETSI SMG11 (European Telecommunications Standardization Institute Special Mobile Group) standardized the speech codecs given in Table 2.3. In addition the new AMR coder has been standardized for use in UMTS, but is not used yet in VoIP systems.

1.2.1.2.2.1   GSM full rate (1987)

GSM full rate, also called GSM 06.10, is perhaps the most famous codec in use today and runs daily in millions of GSM cellular phones. It provides good quality and operates well in the presence of background noise. It uses an RPE-LTP technique to encode voice

Table 1.3 Performance of GSM coders in various environmental conditions

Codec	MOS in clean conditions	Vehicle noise	Street noise
GSM FR	3.71	3.83	3.92
GSM HR	3.85	3.45	3.56
GSM EFR	4.43	4.25	4.18
Reference with no coding	4.61	4.42	4.35

Source: TR 06.85 v2.0.0 (1998). Reproduced by Permission of the European Telecommunications Standards Institute - ETSI.

in frames of 20 ms at a rate of 13 kbit/s. It needs no lookahead. GSM-FR achieves MOS scores slightly below toll quality.

GSM-FR is not extremely complex and requires only about 4.5 MIPS and less than 1 Kb of RAM.

The GSM full-rate patent is held by Philips and the license is free for mobile phone applications.

1.2.1.2.2.2 GSM half-rate (1994)

Also called GSM 06.20, this coder aims at using less bandwidth while preserving the same or slightly lower speech quality as GSM-FR. This codec uses VSELP and encodes speech at a rate of 5.6 kbit/s. The frames are 20 ms long and there is a lookahead of 4.4 ms. The GSM-HR algorithm requires approximately 30 MIPS and 4 Kb of RAM. This coder has not been very successful, due to its high sensitivity to background noise. The patent is also held by Philips; ATT patents on CELP and NTT patents on LSP may also apply.

1.2.1.2.2.3 GSM enhanced full rate (1995)

This high-quality coder exceeds the G.726 'wireline reference' in clear channel conditions and in background noise. It is also called GSM 06.60. It was selected as the base coder for the PCS 1900 cellular phone service in the US and was standardized by TIA in 1996. This codec uses a CD-ACELP technique and encodes 20-ms frames at a rate of 12.2 kbit/s. Optional VAD/DTX functions with comfort noise generation have been defined and there is also an example implementation for error concealment.

AT&T patents for CELP and NTT patents for LSP may apply.

1.2.1.2.3 Other proprietary codecs

1.2.1.2.3.1 Lucent/Elemedia SX7003P

The SX7003P is another popular codec. Although used in Lucent hardware it is licensed to other manufacturers as well. This codec has a frame size of 15 ms, which contains 4 control octets and 14 data octets. Silence frames have 2 octets of data.

In many VoIP implementations, two frames are packed in each IP packet (overhead of 40 bytes), leading to an IP bitrate of 20.3 kbit/s during voice activity periods and only 13.6 kbit/s during silence periods.

1.2.1.2.3.2   RT24 (Voxware)

The RT24 is one of the ultra-low-bitrate coders. Unfortunately, it is spoiled by the IP overhead. It has a bitrate of 2,400 bit/s and achieves an MOS of 3.2. It has a frame size of 22.5 ms (54 bits) which results in a measured IP-level bitrate of 16.6/9.5/7.1/6 kbit/s with 1/2/3/4 frames per IP packet.

1.2.1.2.4   Future coders

Both the AMR and AMR-WB (G.722.2) coders (described in detail in the companion book, *Beyond VoIP Protocols*) will probably be implemented in VoIP systems. Their ability to dynamically reduce the bitrate to adapt to the conditions of the transmission channel is not as useful in VoIP as it is over radio links. Over radio links, there are bit errors, but AMR makes it possible to add redundancy information dynamically to the media stream without requiring more bandwidth when network conditions degrade. Because of this, the AMR coder can offer better voice quality than any of the current narrow-band coders, over a much wider range of transmission network quality.

Over IP transmission links, there are only frame erasure errors, because each IP packet (containing one or more coder frame) is protected by a CRC code[3]. Redundancy can only be added by repeating each frame in multiple packets (forward error correction and inter-leaving). This has a significant, often unacceptable impact on end-to-end delay. Therefore, AMR will mainly be used to avoid any transcoding (Tandeming) when communicating with UMTS and CDMA2000 systems, thereby improving end-to-end voice quality.

Widespread use of both coders is not expected before 2005, because they are closely associated with the deployment of UMTS and CDMA2000 3G systems. Both coders will not only require more DSP processing power, they will require more powerful DSPs to achieve the densities of current G.723.1/G.729 systems.

1.2.1.3   ITU video codecs

1.2.1.3.1   Representation of colours

The representation of colours is derived from the fact that any colour can be generated from three primaries. From an artist's point of view, the three primaries are red, yellow, and blue. These colours are called subtractive primaries, because any colour can be generated from a white beam passed through a sequence of red, yellow, and blue filters. When an artist puts a layer of yellow paint of a sheet of paper, this layer acts as a filter that allows most of the yellow component of the white light to be reflected, but filters out most other colours.

But, video monitors use additive primaries: red, green, and blue. By mixing three beams of red, blue, and green light with various intensities, it is possible to generate any colour. Therefore, any colour can be represented by its barycentric co-ordinates (representing the intensity of each primary colour, not necessarily positive as illustrated in Figure 1.17!) in

---

[3] Some VoIP networks have disabled the UDP checksum mechanisms in order to be more tolerant to bit ends. This could open the way to a more efficient use of the capabilities of AMR

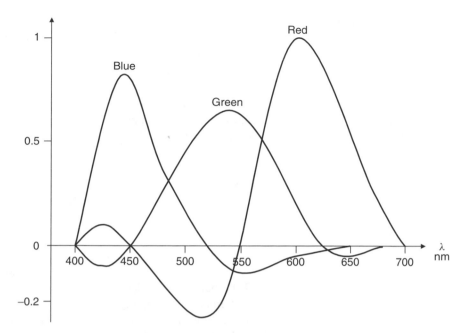

Figure 1.17 Red–green–blue components of visible colours in the 400–700-nm wavelength range.

$$\begin{pmatrix} Y \\ U \\ V \end{pmatrix} = \begin{pmatrix} 0.299 & 0.587 & 0.114 \\ -0.1687 & -0.3313 & 0.5 \\ 0.5 & -0.4187 & -0.0813 \end{pmatrix} \begin{pmatrix} R \\ G \\ B \end{pmatrix}$$

Figure 1.18 RGB to YUV conversion (JFIF).

a triangle with a primary colour at each edge: this is the RGB (red–green–blue) format. The weight of each colour usually ranges from 0 to 255 in the RGB format: each pixel is described using 8 bits for each colour weight, which leads to 24 bits per pixel.

Another common representation is to use luminance (brightness represented by Y) and chrominance (hue represented by U and V, or Cr and Cb). Several conventions exist for this conversion (JFIF for JPEG, CCIR 601 for H.261, and MPEG). Figure 1.18 shows how JFIF converts from an RGB format to a YUV format.

Y, U, and V cover the range from 0 to 255 (U and V are often shifted to take values between -128 and $+127$). For CCIR this range is from 16 to 235.

Experiments have shown that the human eye is more sensitive to the luminance information. Because of this U and V values can be sampled at reduced frequency without inducing significant loss in the quality of the image. Typically, U and V are only sampled for a group of 4 pixels. Coding an image in this way leads to a 2:1 compression (i.e., instead of 24 bits per pixel, we now have 8 bits for Y and $(8 + 8)/4$ pixels for U and V).

1.2.1.3.2 Image formats

Several image formats are commonly used by video codecs. CIF (common intermediary format) defines a 352*288 image. This size has been chosen because it can be sampled relatively easily from both the 525- and 625-line video formats and approaches the popular 4/3 length/width ratio.

In addition to the resolution of CIF being below that for TV quality, it is still relatively difficult to transmit over low-bandwidth lines, even with efficient coding schemes such as H.261 and H.263. For this reason two other formats with lower resolutions have been defined. At half the resolution in both dimensions, quarter CIF (QCIF) is for 176*144 images, and SQCIF is only 128*96.

For professional video application, CIF is clearly insufficient, and images need to be coded using 4CIF (704*576) or 16CIF (1,408*1,152) resolution (see Table 1.4).

1.2.1.3.3 H.261

H.261 is a video codec used in H.320 videoconferencing to encode the image over several 64-kbit/s ISDN connections, but in video over IP applications the bitstream is encoded in a single RTP logical channel. The H.261 codec is intended for compressed bitrates between 40 kbit/s and 2 Mbit/s. The source image is normally 30 (29.97) frames per second, but the bitrate can be reduced by transmitting only 1 frame out of 2, 3, or 4. The image formats shown in Table 1.5 can be encoded by H.261. The 4CIF and 16CIF formats are not supported by H.261.

The H.261 coding process involves several steps. After initial *YUV* coding of the original image using CCIR parameters, as described above, the image is divided in 8*8

Table 1.4 Uncompressed bitrate for various video formats

| | | | Uncompressed bitrate (Mbit/s) | | | |
| | Luminance | | 10 frames/s | | 30 frames/s | |
Picture format	Pixels	Lines	Grey	Colour	Grey	Colour
SQCIF	128	96	1.0	1.5	3.0	4.4
QCIF	176	144	2.0	3.0	6.1	9.1
CIF	352	288	8.1	12.2	24.3	36.5
4CIF	704	576	32.4	48.7	97.3	146.0
16CIF	1,408	1,152	129.8	194.6	389.3	583.9

Grey images are obtained by transmitting only the *Y* luminance component. Colour images are obtained by also transmitting the *U*, *V* chrominance components sampled at half the resolution.

Table 1.5 Video image sizes supported by H.261

SQCIF	128*96	Optional
QCIF	176*144	Required
CIF	352*288	Optional

luminance pixels blocks for the luminance plane. The same surface is coded with only 4∗4 chrominance pixels for each chrominance plane. Four luminance blocks are grouped with two chrominance blocks (one for U, one for V) in a structure called a **macroblock**.

Each macroblock can be coded using the 'intra' method or the 'inter' method (Figure 1.19). The intra method codes by means of a local compression method (just using information relative to macroblocks that have already been encoded in the same image), while the inter method codes adjacent frames relatively in time. The coding method can be defined for a macroblock, for a 'group of blocks' (GOB's), or for a full frame. In general, video coders use the same method within a frame; hence the name intraframe (I-frame) or interframe (P-frame) frequently used when discussing video applications. The inter method is much more efficient, but leads to error accumulation; therefore, it is necessary to send intraframes intermittently.

Intraframes (I-frames) use a coding similar to the one used by JPEG, which involves DCT (discrete cosine transform), quantization, run length encoding and entropy encoding (Figure 1.20).

For interframes (P-frames), the algorithm follows these steps:

- Motion detection: comparison of the image to be coded with the last coded image trying to find those parts of the image that have moved. This results in a representation of the difference between the motion-compensated image and the real one.
- Coding of the difference image using DCT transform and run length encoding.
- Entropy encoding to further reduce the image size.

1.2.1.3.3.1 *Motion detection*

The second stage of the H.261 P-frame coding process uses the fact that most images in a video sequence are strongly related. If the camera angle changes, many pixels will simply shift from one image to another. If an object moves in the scene, most of the

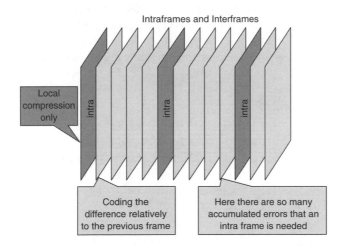

Figure 1.19 Intra and inter coding methods.

Figure 1.20 JPEG-style encoding used for intra frames.

pixels representing the object in a frame can be copied from the preceding frame with a shift. For each macroblock of the image to be encoded, the algorithm tries to discover whether it is a translated macroblock of the previous image. The search is done in the vicinity of ± 15 pixels and only considers luminance. The difference between the original macroblock of the $n + 1$ frame and each translated block of the n frame in the search area is the absolute value of pixel-to-pixel luminance difference throughout the block. The translation vector of the best match is considered the motion compensation vector for that macroblock (Figure 1.21). The difference between the translated macroblock and the original block is called the motion compensation macroblock.

If the image has changed completely (e.g., a new sequence in a movie), interframe coding is not optimal. Further reason, the H.261 coding process must decide at each frame which coding is better for the macroblock: intra or interframe. The decision function is based on the energy and variance of the original macroblock and the motion-compensated macroblock.

1.2.1.3.3.2 DCT transform

The pixel values of the image difference that we obtained at the previous step vary slowly within a macroblock. Let's take such a macroblock and repeat it in two dimensions so that we obtain a periodic function (Figure 1.22). Such a function can be reproduced efficiently using just a few coefficients from its Fourier transform.

This transformation is called a bidimensional DCT. The formula used by H.261 to calculate the DCT of an 8*8 block is:

$$F(u, v) = \frac{1}{4} C(u) C(v) \sum_{i=0}^{7} \sum_{j=0}^{7} f(i, j) \cos \left((2i + 1) u \frac{\pi}{16} \right) \cos \left((2j + 1) v \frac{\pi}{16} \right) \quad (1.1)$$

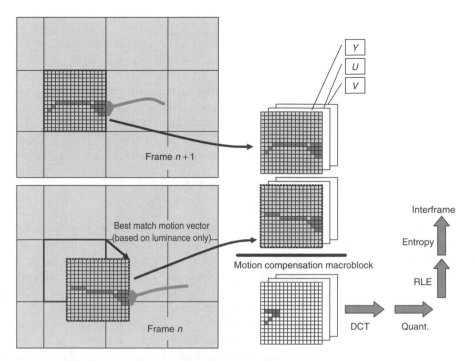

Figure 1.21 Motion prediction and compensation of residual error.

Figure 1.22 Construction of the periodic function used for the DCT transform from the reference macroblock.

where $C(0) = 1/\sqrt{2}$ and $C(x \neq 0) = 1$. The DCT is a 'frequency' representation of the original image. The coefficient in the upper left corner is the mean pixel value of the image. Values in higher row positions represent higher vertical frequencies and values in higher column positions represent higher horizontal frequencies.

The DCT is very interesting because most high-frequency coefficients are usually near 0. At the decoder and, the inverse of the DCT is obtained with:

$$f(i, j) = \frac{1}{4} \sum_{u=0}^{7} \sum_{v=0}^{7} C(u)C(v)F(u, v) \cos\left((2i + 1)u\frac{\pi}{16}\right) \cos\left((2j + 1)v\frac{\pi}{16}\right) \quad (1.2)$$

1.2.1.3.3.3 Quantization

So far the representation of the image that we have is still exact. We could obtain the original image by reversing the DCT and repeatedly adding the resulting block to the shifted block of the previous frame.

Quantization is the lossy stage in H.261; it consists in expressing each frequency domain $F(u, v)$ value in coarser units, so that the absolute value to be coded decreases and the number of zeros increases. This is done using standard quantization functions: one is used for the constant component (DC) coefficient and another is selected for a macroblock. Depending on the amount of loss that can be tolerated, the coder can choose fine or very coarse functions.

1.2.1.3.3.4 Zigzag scanning and entropy coding

Once the DCT coefficients are quantized, they are rearranged in a chain with the DC coefficient first and then they follow the sequence shown in Figure 1.23. This concentrates most nonzero values at the beginning of the chain. Because there are long series of consecutive zeros, the chain is then run length-encoded. This uses an escape code for the most frequently occurring sequences of zeros followed by a nonzero coefficient and variable escape codes for other less frequently occurring combinations.

This chain can be further compressed using entropy coding (similar to Huffmann coding), which creates smaller code words for frequently occurring symbols.

Huffmann coding first sorts the values to be encoded according to frequency of appearance, then constructs a tree by aggregating the two least frequent values in a branch, then repeating the process with the two values/branches that have the smallest occurrence values (counting the occurrence of a branch as the sum of the occurrences of its leaf nodes). Once the tree is complete, a '1' is assigned to each left side of any two branches and a '0' to each right side. Any value can be identified by its position in the tree as described by the sequence of digits encountered when progressing from the root of the tree to the value.

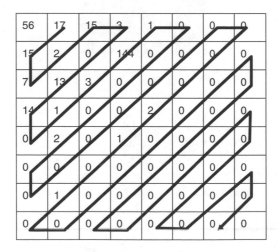

Figure 1.23 Zigzag scanning.

The output of the H.261 encoder consists of entropy-encoded DCT values. This bit-stream can be easily decoded once the decoder has received the Huffmann tree. In the case of H.261, the tree calculation is not done in real time; the recommendation itself provides codes for the most frequently occurring combinations.

1.2.1.3.3.5 Output format

The H.261 bitstream is organized in GOBs (a group of blocks) of 33 macroblocks (each encoding 16∗16 luminance pixels and 8∗8 *U* and *V* pixels). A PAL CIF image has 12 GOBS, and a PAL QCIF image has 3 GOBS. A CIF picture cannot be larger than 256 kbits, and a QCIF picture cannot be larger than 64 kbits.

The output bitstream will consist of alternating inter-coded macroblocks and intra-coded macroblocks. The receiver can force the use of intra coding to recover from cumulative or transmission errors. Otherwise, a macroblock should be updated in intra mode at least once every 132 transmissions to compensate for error accumulation.

1.2.1.3.3.6 Conclusion on H.261 video streams

The description of H.261 found in the previous sections is not complete, but it is enough to allow a network expert to understand the nature of video traffic. The most important conclusion is that video traffic using H.261-style coding (this is also valid for H.263 and MPEG) is extremely bursty. A typical network load profile is represented in Figure 1.24. For instance, Microsoft Netmeeting sends an intraframe every 15 seconds. A videocon-ferencing MCU will send an intraframe for all macroblocks each time the speaker, and therefore the image, changes ('videoFastUpdate'). In other circumstances some implemen-tations will not send all intra macroblocks simultaneously, in order to avoid the occurrence of large traffic peaks in the network.

It is also important to remember that H.261 only specifies a decoder. In fact, a very bad implementation could choose to use only intraframes if it was not capable of doing motion vector searches for interframes and still be H.261-compliant. This explains why not all video boards and not all video/conferencing software are equal, despite claiming

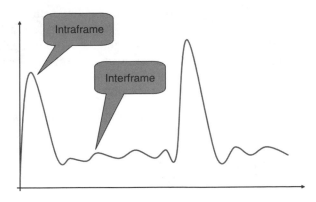

Figure 1.24 Video traffic can be very bursty due to intraframes.

Table 1.6 Image sizes supported by
H.263

SQCIF	128*96	Required
QCIF	176*144	Required
CIF	352*288	Optional
4CIF	704*576	Optional
16CIF	1,408*1,152	Optional

they are using H.261 or H.263. A network engineer should always try to measure the actual bandwidth used by these devices.

1.2.1.3.4 H.263

Table 1.6 lists the image formats that can be encoded with H.263. H.263 was designed for low-bitrate communication, as low as 20 kbits/s. The coding algorithm of H.263 is similar to that used by H.261, but involves some changes to improve performance and error recovery. H.263 is more recent, more flexible, and about 50% more bitrate-effective than H.261 for the same level of quality. It will replace H.261 in most applications. The main differences between H.261 and H.263 are:

- Half-pixel precision is used by H.263 for motion compensation, whereas H.261 used full-pixel precision and a loop filter. This accounts for much of the improved efficiency.
- Some parts of the hierarchical structure of the data stream are now optional, so the codec can be configured for a lower data rate or better error recovery.
- There are now four negotiable options included to improve performance: unrestricted motion vectors, syntax-based arithmetic coding, advance prediction, and forward and backward frame prediction (similar to MPEG) called P-B frames. Backward frames are added to allow motion vectors to refer not only to past frames, but also to future frames (e.g., when a partly hidden object becomes visible in a future frame).
- H.263 supports five resolutions. In addition to QCIF and CIF, which were supported by H.261, there is SQCIF, 4CIF, and 16CIF. SQCIF is approximately half the resolution of QCIF. 4CIF and 16CIF are 4 and 16 times the resolution of CIF, respectively. Support of 4CIF and 16CIF means the codec can now compete with other higher bitrate video-coding standards, such as the MPEG standards.

With these improvements, H.263 is a good challenger to MPEG-1 and MPEG-2 for low to medium resolutions and bitrates. They have comparable features (such as B frames in MPEG and P-B frames in H.263) which are just as good for moderate movements. H.263 even has some options not found in MPEG, like motion vectors outside the picture and syntax-based arithmetic coding. MPEG has more flexibility, but flexibility means overhead. For videoconferencing applications, with little movement and a strong bandwidth constraint, H.263 is a very good choice.

1.2.1.3.5 H.264

H.264 is the latest ITU-standardized video coder and the latest video compression profile for MPEG-4 (part 10). Its production required more than 7 years of work. H.264, or **advanced video coding (AVC)**, requires only one-half to one-third of the video bandwidth necessary for an equivalent MPEG-2 channel when using all the possible optimizations of H.264 (Figure 1.25). Broadcast quality video becomes possible at a rate of 1.5 Mbit/s. This is likely to trigger an accelerated development of on-demand video over IP, in the same way that the "MP3" format made musical applications popular on the Internet. With the traditional rate of 3.75 Mbit/s for MPEG2 movies, delivering video over ADSL is restricted only to the shortest copper lines and densely populated areas. Below 2 Mbps it is possible to add video streaming to many more ADSL lines. It also makes it easier to provide video content over wireless links (H.264/AVC is one of the standard video coders of 3GPPv6).

With H.263, it is possible to have a business quality videoconference at about 386 kbit/s. With H.264, an equivalent conference can be achieved at about 192 kbit/s. The downside of H.264 is that it requires much more CPU power for compression than H.263, and therefore will probably not be usable for interactive video before the end of 2005, leveraging Moore's law and the ever-increasing power of PC processing power. In addition, some of the optimizations introduced by H.264 (e.g., the ability to encode interframes referring to future frames) can only be used in non-real-time mode.

In line with the other MPEG standards, II.264 only describes the format of the encoded bitstream and gives no indication of the algorithms that should be used to generate the encoded data. Prediction, DCT, quantization, and entropy encoding are not fundamentally different from the previous standard, but they have been enhanced.

Figure 1.25   H.264/AVC encoding extends the reach of video over ADSL. Reproduced with permission from Envivio, Inc.

Each frame is processed in 16∗16-pixel macroblocks, each one being encoded in intra or inter mode. In intra mode, the encoded macroblock contains interpolation data using previously encoded macroblocks of the same frame. In inter mode, the encoded macroblock contains motion compensation information based on previous or future frames (up to two previous or subsequent frames).[4] One of the improvements of H.264 over its predecessors is that it allows intra or inter mode to be selected, not at each image, but in groups within the image called 'slices'.

In inter mode, the ability to refer to frames not immediately adjacent to the frame currently encoded is one of the major optimizations of H.264 compared with the previous generation of video coders. The difference between the predicted macroblock and the macroblock to encode is then computed, block-transformed, quantized, and the reordered coefficients are then entropy-encoded. The bitstream is formed from entropy-encoded coefficients, the quantizer step size, and the information required to recreate the predicted macroblock (motion-compensated vector, etc.).

In intra mode, prediction data describing a block can be built for either 4∗4 or 16∗16 luminance macroblocks, and for the corresponding 8∗8 chrominance macroblock. Prediction block data are built from already-encoded pixels (light gray bands in Figure 1.26), using one of eight extrapolation modes, each based on a characteristic extrapolation direction angle (Figure 1.26 shows mode 4, diagonal to the right of the P-frame, for a 4∗4 luminance macroblock). The mode resulting in the smallest sum of absolute errors compared with the original macroblock is selected.

Both in intra and inter modes, H.264 uses a new 'deblocking' filter that considerably reduces the differences between macroblocks in the reconstructed image, which were clearly visible with coders of the previous generation. This filter operates on the reconstructed image just before the differences between the reconstructed macroblocks and the

Figure 1.26   Prediction in H.264's intra mode (mode 4).

---

[4] In the baseline profile, which is more suitable for interactive videoconferencing, only P-frames and I-frames are supported (no backward prediction).

original image are encoded: it smoothes the reconstructed image by reducing the differences across adjacent macroblocks, thereby eliminating in the next phase the brutal changes in difference compensation values between the original image and these adjacent macroblocks.

1.2.2 DTMF

Strictly speaking, DTMF tones that are generated by a touchtone telephone when you press a key are part of the media stream. They are just another sound transmitted by the telephone. In the circuit-switched network this sound is digitized by the G.711 codec as part of the media stream and played back at the receiving end of the line. This does not cause any problem because G.711 does not assume that the signal is voice.

But, some narrow-band codecs that achieve much higher compression rates do use the fact that the signal is voice. Others do not assume the signal is voice, but distort it in such a way that the pure frequencies composing the DTMF tone cannot be correctly recognized when the signal is regenerated. DTMF will not get through these codecs.

Whenever a communication involves an IVR system, it is very important to be able to reliably transmit DTMF tones. In most cases the IVR system just asks a question and waits for a DTMF response. It just cares about which key has been pressed, the exact duration and timing of the tone is not so important. In other cases the IVR system will need more accuracy in the timing (e.g., when the system reads a list and asks you to press the star key when you hear something of interest).

In order to interwork properly with these IVR systems, it was necessary to develop special procedures to handle DTMF:

- H.323 generally uses the signaling channel (H.245 UserInputIndication) to convey DTMF tones (in fully decoded form). This method is sufficient in most cases and works with application servers that need to implement switching functions (e.g., contact centers), without accessing the media stream. Alternatively, since H.323v4 it is also possible to use RFC 2833, which transmits the DTMF tone in fully decoded form, but over the RTP channel. RFC 2833 mandates implementations to be able to handle this telephony event channel as a separate channel (i.e., it should not necessarily be sent to the destination address of other media streams). Unfortunately, most current implementations cannot do this, thereby preventing the service provider from being able to implement application servers in the network. The use of RFC 2833 should be discouraged unless the implementation can correctly send the DTMF information to the application server.

- SIP mainly uses two methods: a signaling method, based on the INFO message or the NOTIFY message (see Chapter 3 for details), which is still not well standardized, or RFC 2833. The same comments apply to RFC 2833. Most implementations do not allow the sending of DTMF information to the application server. De facto it is very difficult in current SIP networks to implement a standards-based application server that is not accessing the media stream.

- MGCP uses a sophisticated out-of-band mechanism that allows the transmission of most telephony events to the call agent in the signaling stream, and implements filtering and accumulation capabilities as well. The mechanism uses the request notification (RQNT) and notify (NTFY) messages. See Chapter 4 for more detail.

1.2.3   Fax

1.2.3.1   A short primer on Group 3 fax technology

The purpose of facsimile transmission is to transmit one or several pages of a document across the telephone network. The first fax systems used Group 1 or Group 2 technologies, which scanned the document line by line and converted each line in black or white pixels. The data were then transmitted without compression over the phone line at the rate of 3 lines per second for Group 1 and 6 lines per second for Group 3.

Because this took over 3 min for an A4 document (1,145 lines of 1,728 bits) even in the best case, Group 3 technology was introduced. Group 3 faxes use a more efficient image-coding mechanism known as modified Huffmann coding (MH). MH coding uses the fact that each line is composed of large sequences of white pixels and large sequences of black pixels. Instead of sending data for each pixel. MH coding just sends a short code for the sequence. Now the transmission time depends on the document, but is usually much shorter than 3 min: no wonder Group 3 faxes today rule the fax market.

With the advent of ISDN, Group 4 faxes have been introduced. The main difference from Group 3 is that ISDN can transmit raw data, so Group 4 technology need not care about the many hacks that are needed to carry data over an analog line. However, Group 4 has not succeeded in gaining a significant market, and the probability of having a Group 4 fax talking to another Group 4 fax is so low that this case has not so far been considered in the ITU's SG16 which is in charge of H.323.

1.2.3.1.1   Transmitting a line (Group 3)

Most faxes are physically linked to a printer. Because of compression, it is now possible to transmit a single line very quickly if the line is simple, so quickly that the receiving fax may not have enough time to print it. Of course the fax could buffer it in memory, but most faxes are very simple devices with very little memory. Therefore Group 3 supports a minimum transmission time, as represented in Figure 1.27. If a line does not contain enough compressed data to take more than the MTT to be transmitted, a filling sequence of zeros will be added before the end of line sequence.

1.2.3.1.2   Transmitting a page

As Figure 1.28 shows, the transmission of a page is quite simple. Each line is transmitted in sequence, separated by an EOL, and the whole page is terminated by six consecutive EOLs, which means the fax has to return to command mode (RTC).

Figure 1.27 Fax line transmission.

Figure 1.28 Fax page transmission.

1.2.3.1.3 Complete fax transmission

The calling fax dials the destination number, then sends a special sequence called CNG (CalliNG tone), which consists of a repetition of 1,100-Hz tones sent for 0.5 seconds separated by 3 seconds of silence (Figure 1.29). Faxes manufactured before 1993 may not send this tone.

Figure 1.29   Overview of a fax transmission.

When an incoming connection arrives at the receiving fax, it first sends a special 2,100-Hz tone called CED for 3 seconds. After a short pause, the receiving fax (1) begins to send commands using V.21 modulation (quite slow at 300 bit/s), (2) to transmit synchronizing flags for 1 second (called a preamble), (3) may transmit some non-standardized data (NSF) and its local identity (CSI), and (4) must transmit its capabilities (DIS, or digital identification signal). Each of these data elements is an HDLC frame that consists of:

- A starting flag (7Eh).
- An address field (always set to FFh).

- A command field which is set to C8h for a final frame and C0h otherwise.
- A fax control field (FCF): 02h for CSI, 01h for DIS, etc.
- A variable length fax information field (FIF).
- A checksum (FCS, or frame check sequence).

Transmitting NSF, CSI, and DIS may take up to 2.5 seconds.

The sending fax selects a mode of transmission (DCS) and replies by sending its own capabilities and its identity (TSI). As soon as the receiving fax is ready the sending fax begins the actual transmission phase and will use a faster modulation scheme, such as V.27 (4,800 bits/s) or V.29 (9,600 bits/s). This requires a training phase which is used by the receiving side to compensate for phase distortions and other issues. At the end of the training phase the sending fax sends zeros for 1.5 seconds (called a training check, or TCF). If the called fax receives this sequence correctly it considers the training phase successful and sends a CFR (ConFirmation to Receive) command to let the transmitting fax know it has succeeded. After another training sequence, the sending fax transmits the actual page data as formatted above. This takes approximately 30 seconds in V.29 mode and 1 minute in V.27 mode.

When this is finished, the modem can send an MPS (multi-page signaling) message to send another page or an EOP (end of procedure message) when it has transmitted the last page. The receiving fax acknowledges it with an MCF (Message ConFirmation) which means that the image data have been correctly received, and the sending fax sends a disconnection message DCN (DisCoNnect).

1.2.3.1.4 Detection of fax for VoIP gateways

It is important to reliably detect faxes on VoIP gateways, since fax modulation is not reliably transmitted across low-bitrate voice codecs. On the originating gateway, the T.30 calling tone can be detected, but it is an optional signal. Therefore, detecting CNG is not a reliable way to detect a fax signal. This can be resolved at the terminating gateway by detecting the V.21 preamble flag sequence which follows the called station identification tone (CED), when the CED is present. The CED itself cannot be used because it is also used by modems (V.25 ANS modem tone).

As soon as the signal is detected as a fax, the gateway should stop using regular audio encoding and switch to T.38 encoding.

1.2.3.1.5 Error conditions

If the training is not successful, the receiving fax can send an FTT command to ask for another try at a lower speed.

If an error is present in a line, the receiving fax will find it by counting how many pixels are present in the decoded line. If there are not exactly 1,728 (A4 format), the line is ignored or copied from the previous line, depending on manufacturer preference.

A fax can request the retransmission of a command at any time by sending a CRP command.

1.2.3.2   Fax transmission over IP (T.38 and T.37)

1.2.3.2.1   Store-and-forward fax and the challenge of real-time fax

Sending faxes over the Internet is not something new. Many companies have been offering this service, called store-and-forward fax, for some time. The idea behind store-and-forward fax is quite simple. When computer A receives a fax, the fax data are represented as a set of bitmaps. This set of bitmaps is a file that can be transmitted to another computer (B) closer to the destination. Once this computer has received the file, it just needs to dial the receiving fax machine and emulate a fax machine to send the bitmap.

This technique is also used for bulk faxing, in which the original document is faxed once to a computer, and the computer is then provided with a list of fax numbers and sends a copy of this fax to each of them. Store-and-forward fax transmission is now standardized at ITU in recommendation T.37. However, since this book focuses of real-time applications we choose to put the emphasis on the real-time standard, T.38.

The problem with store-and-forward fax technology is that many faxes report back on the transmission of the document. Usually, they keep the result code in memory and then print it on demand. Many people tend to rely on these transmission reports: for fax-to-fax transmission this is confirmation that the fax has been correctly received, with a timestamp and the identity of the receiving fax machine.

When using store-and-forward, this report is only a confirmation that the fax has been sent, because the receiving machine is in fact computer A.

When it receives the file containing the document, computer B will dial the number indicated. But, the fax could be busy or, even worse, it could be a wrong number. So any company providing a store-and-forward service needs to report back to the sender, via email or fax. When receiving a negative acknowledgement the sender needs to know whether it is a problem with the receiving fax or the provider. This leads to potential conflicts and increases the cost of providing the service.

It is much easier for a service provider to be completely transparent in the transmission. In other words, the success report that is received by the originating fax machine should appear as a success report from the distant fax machine: such a service is real-time fax.

Real-time fax is much more complex than store-and-forward fax. There are many timers in the T.30 protocol. Once computer A has picked up the line, computer B has only a limited time budget to dial the other fax machine and get an answer. During the call, when A's fax machine has sent a command, it expects a reply within 3 seconds. So, during this limited time budget A must send the command to B over the Internet, B must send it to the receiving fax machine, receive the reply, and forward it to A.

Fortunately, the ITU had a human operator in mind when setting the value of these timers; so, all are expressed in seconds. Moreover, as we saw in the preceding section 1.2.3.1.5 there are many ways of recovering from error conditions, which can be used to spoof the sending fax and get it to wait a little more if needed. These techniques are quite difficult to implement reliably with all brands of faxes. However, some manufacturers have built up a lot of experience and have announced they could transmit real-time faxes over IP networks with a round trip latency of up to 2 seconds!

Lately, many carriers have been tempted to do IP trunking without telling their customers. VoIP gateways make this quite simple. But, without support for real-time fax, whenever a subscriber tries to send a fax that is routed through this IP trunk, it will fail miserably. Of course, it is possible to tell subscribers not to use their faxes or even to dial a special prefix for faxes; but, this significantly complicates their lives. Real-time fax is the only appropriate answer to these issues: all VoIP gateways should be able to dynamically recognize a fax call and switch to T.38 transport mode.

1.2.3.2.2 T.38

1.2.3.2.2.1 IFP

T.38 is the approach of the ITU's SG16 to the problem of real-time fax. Its title is "Procedures for real-time Group 3 facsimile communication between terminals using IP networks". This recommendation is limited to Group 3 only and describes real-time fax transmission using VoIP gateways over an IP network, between faxes and computers connected on the Internet, or even between computers (the latter may not seem useful, but in some cases the receiving computer will be identified by an H.323 alias or even a phone number, and you may not know this is a computer). Usage of the T.38 protocol within the framework of H.323 is defined in H.323 annex D and was included in H.323v3.

T.38 uses a special transport protocol called IFP. IFP packets can be carried over TCP or UDP. Most gateways support UDP, but TCP transport has also been made mandatory in H.323v4. UDP transport includes a forward error correction mechanism.

IFP packets contain a type field and a data field (Figure 1.30) both encoded using ASN.1 syntax. The type field can have three values:

Figure 1.30   IFP packet formats for the first, middle, and last HDLC frames.

- T30_INDICATOR: the value of this indicator gives information on received CED and CNG tones, V.21 preambles, and V.27, V.29, and V.17 modulation training.
- T30_DATA: the value of this indicator tells us over which transport (V.21, V.17, or V.29) the data part of the message has been received.
- DISCONNECT: used to disconnect the session normally or after a failure, the value describes the error code.

The DATA part of the IFP message contains T.30 control messages as well as the image data. This DATA element is organized in fields that contain a field type and field data. Examples of these fields are:

- Type HDLC data: the data part of the field contains one, or part of an HDLC data frame, not including the checksum (FCS). This is coded as an ASN-1 octet.
- Type FCS OK: indicates that an HDLC frame is finished and the FCS has been checked. There are sill other HDLC frames after an FCS OK.
- Type FCS OK-sig-end: same as FCS OK, except that this is the last HDLC frame.
- T4-non-ECM: the data part contains the actual image data including filling and RTC.

1.2.3.2.2.2   IFP over TCP or UDP

IFP messages can be carried as TCP payload or can be encapsulated in UDP, as shown in Figure 1.31.

An additional redundancy mechanism has been defined on top of UDP in order to make the delivery of IFP packets more reliable. As shown in Figure 1.31, the payload part contains one or more IFP messages, and the sequence number that appears in the header is the sequence number of the first IFP message in the payload, which is also called the primary message. The first message sent by a gateway should have a sequence number

Figure 1.31   IFP transport methods and error correction modes.

of 0. After this primary message other messages are inserted for error/loss recovery purposes; two modes can be used for this: redundancy mode and FEC (forward error correction) mode.

The control header indicates whether the secondary messages are redundancy messages (bit 3 set to 0) or FEC messages (bit 3 set to 1).

(a)   Redundancy mode

In redundancy mode, copies of previous IFP messages are simply inserted after the primary message (Figure 1.32). The number of copies is the number of frames minus one. By adding *n* copies to the message, the transmission is protected against loss of up to *n* consecutive packets.

A gateway is not required to transmit redundancy packets, and receiving gateways that do not support them may simply ignore the presence of redundancy packets.

(b)   FEC mode

FEC mode is more complex. Each FEC message is the result of a bit-per-bit exclusive-OR performed on *n* primary IFP messages. Before performing the OR, shorter messages are

Figure 1.32   Redundancy mode.

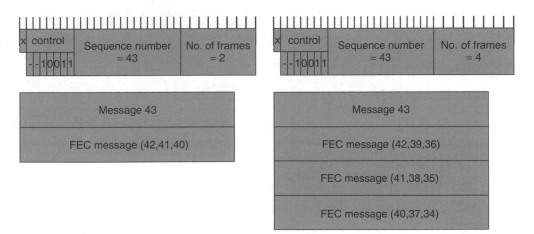

Figure 1.33   FEC mode.

right-padded with zeros, so the resulting FEC message is as long as the longest n primary message. The value of n is indicated in the four last digits of the control field, '3' in Figure 1.33.

When several FEC messages are added, as in the left part of our example, the primary messages used for each FEC message are interleaved. When n FEC messages are added, transmission is protected against the loss of n consecutive UDP packets.

1.2.3.2.2.3 T.38, H.323, and SIP

The use of T.38 by SIP is explained in T.38 annex D 'SIP/SDP call establishment procedures'. The use of T.38 with the H.323 protocol is described in II.323 annex D and T.38 annex B. H.323 annex D mandates the use of IFP transport over TCP, but transport over UDP is still allowed as an optional mode. In reality, all vendors seem to use IFP over UDP. These capabilities (T38-TCP and T38-UDP) have been added in the DataApplicationCapability of DataProtocolCapability of H.245. IFP is transmitted over two logical channels (sender to receiver and vice versa).

2

H.323: Packet-based Multimedia Communications Systems

2.1 INTRODUCTION

H.323 is now the dominant protocol for voice and multimedia communications over IP, and carries well over 95% of VoIP minutes worldwide. Most VoIP equipment, from gateways to IP phones (or IP-PBXs), now support H.323, and the interoperability between vendors is excellent, with many multi-vendor networks in production all over the world. It has taken, over 6 years to get there.

H.323v1 had little ambition. Noting the growing success of IP, IPX, and AppleTalk-based local area networks in all kinds of companies, **Study Group 16** of **ITU-T** decided to create **H.323**, 'Visual telephone systems and equipment for local area networks which provide a non-guaranteed quality of service', a LAN-only standard for audiovisual conferences. SG16 leveraged the know-how of SG15, which had already acquired a lot of experience during the development of **H.320**, 'Multimedia conferencing for ISDN-based networks'. This background led to many benefits for H.323, such as seamless inter-working with H.320 and H.324 systems (videoconferencing over POTS lines) and, in general, comprehensive support for interactive video, but it also led to some drawbacks in certain areas.

H.323 did not attract major interest from the market until VocalTec and Cisco founded the Voice over IP Forum to set the standards for VoIP products. At that time the focus in the VoIP Forum was given to the specification of endpoints using non-H.323, UDP-based signaling protocols. When major software and hardware firms realized the potential of Internet telephony they pushed the VoIP Forum to become part of the IMTC (International

IP Telephony   O. Hersent, J.P. Petit, D. Gurle
© 2005 John Wiley & Sons, Ltd   ISBN: 0-470-02359-7

Multimedia Teleconferencing Consortium) and simultaneously changed the focus of the VoIP activity group to profiling H.323 for use over the Internet, as opposed to creating a new protocol. Indeed, with a few minor adaptations, H.323 appeared to be just as usable over the Internet as on LANs.

Soon the ITU's SG16 acknowledged that the success of H.323v1 called for a much broader scope, and the title of H.323v2 was changed to 'Packet-based multimedia communications systems'.

2.1.1   Understanding H.323

H.323 is an umbrella specification that refers to many other ITU documents. It describes the complete architecture and operation of a videoconferencing system over a packet network. H.323 is not specific to IP. In fact, there are sections on the use of H.323 over IPX/SPX or ATM. The framework of H.323 is complete and includes the specification of:

- Videoconferencing terminals.
- Gateways between a H.323 network and other voice and video networks (H.320 video-conferencing, POTS, etc.).
- Gatekeepers, the control servers of the H.323 network, performing registration of terminals, call admission, and much, much more.
- **MCUs (multipoint control units)**, which are used for multiparty conferencing and include a control unit called an **MC (multipoint controller)**, and one or more media-mixing units called **MPs (multipoint processors)**.

2.1.1.1   Core specifications

In addition to the H.323 ITU recommendation itself, the H.323 standard references several other ITU recommendations and IETF RFCs. The most important normative documents are:

- **IETF RTP/RTCP**   (Real Time Transport Protocol, Real Time Control Protocol) is described in RFC 1889. RFC 1889 describes a general framework enabling the transport of real-time (or, more precisely, isochronous) data over IP. RTP allows a level of tolerance for packet jitter and detection of packet loss by using sequence numbers and timestamps. Some profiling work is needed on top of RFC 1889 in order to build a specific application, as RFC 1889 does not describe the transport of specific media types within the RTP stream.
- **ITU recommendation H.225.0**   does this profiling work in the context of H.323 video-conferencing applications (in fact, the entire specification of RTP/RTCP is annexed to it, as the ITU wanted to guarantee some stability in its references to the IETF standard). In particular, H.225.0 defines which identifiers are to be used for each type of codec recognized by the ITU and discusses some conflicts and redundancies between RTCP

and the H.245 media-control protocol that is used in H.323 to negotiate, open, and close media channels. H.225.0 also describes the **RAS (registration, admission, status)** protocol, which is used between a terminal and a gatekeeper (see Section 2.2.2). The RAS protocol is used mainly for the management of endpoint IP addresses and their mapping to aliases (e.g., telephone numbers—'registration'), but can also be involved in the call process, mainly for authorization and admission purposes. It can also be used to query the endpoint about some local statistics. Last but not least, H.225.0 defines the **call-signaling channel** protocol used in H.323. The call-signaling channel is used during the establishment and break-off phases of the call (see Section 2.2.1) and will look familiar to anyone used to **ISDN** networks. Like ISDN, it also uses ITU **Q.931** call control messages, but these messages are extended in order to support multimedia communications. The extended information is packed into the Q.931 user-to-user information element. It also describes how this information is to be transported over **TCP**, and more recently UDP and event SCTP (H.225.0v5).

- **ITU recommendation H.245** is mainly a library of **ASN-1** messages (ASN.1 has a formal syntax defined by the ITU for data structures and their serialization in messages) used by the H.245 control channel, which is opened at the beginning of the call to negotiate a common set of codecs and remains in use throughout the call to perform all media-related control functions. H.245 also defines the protocol-state machines that are used in H.323 and many other ITU standards for the management of media streams (in particular, video). H.245 is also used by the H.320 ISDN videoconferencing standard, the H.324 POTS videoconferencing standard, and the new videoconferencing standard for 3G mobile phones H.324M.

2.1.1.2 Abstract Syntax Notation 1

Use of the ASN.1 syntax by H.323 is the reason for its reputation of being a 'complex' protocol. ASN stands for 'abstract syntax notation'. ASN.1 is defined in ITU X.680 ('Abstract Syntax Notation-1'), and its serialization—the actual bit-level representation of the structured data for transport over a network used to code H.323 protocol data units (PDUs)—is defined in ITU X.691 ('ASN-1 encoding rules, specifications of packet encoding rules'). A small summary can be found at the end of H.245 specification.

ASN.1 has a very similar syntax to **XML** and can be used to the describe almost any data structure. Although it is less popular than XML these days, ASN.1 is much more powerful than XML: it can describe a greater variety of types, has better support for constraints, and is much less ambiguous when used to specify data types. It is also a bit harder to learn than XML. Everything comes at a price!

ASN.1 also defines two ways of serializing data for transport over a network: **BER (basic encoding rules)** and **PER (packed encoding rules)**. BER is simple but not optimized, PER is complex but very efficient (typically data can be stored using ten times less space than data encoded in XML). The ASN.1 data description can be used by compilers that produce highly optimized BER/PER encoders automatically from the ASN.1 definition of the message set. ASN.1 is used extensively in telecommunication applications, as the use of ASN.1 automatic encoders greatly improves the robustness of applications, minimizing the sensitivity of telecom applications to malformed packets compared with manually designed parsers. ASN.1 data structure specifications are also free from any ambiguity, which facilitates interoperability across applications.

2.1.2   Development of the standard

2.1.2.1   H.323v1

Work on H.323v1 began in May 1995, and this version was approved in June 96. This first version of H.323 still had a lot of issues; notably, the connection of audio streams was very slow and for the first few seconds of each H.323v1 phone call it was almost impossible to hear anything. Moreover, H.323v1, with its focus on LAN environments, lacked any mechanism for security. Despite these issues, H.323v1 still enjoyed great success due to the early introduction of NetMeeting® by Microsoft, an H.323v1-capable communication software. Unfortunately, too much flexibility was allowed for terminals implementing H.323v1 leading to interoperability issues, notably when endpoints also implemented the T.120 data-sharing protocol.

All these issues are still remembered today, and in many trade shows you can still hear complaints about H.323 delay or interoperability issues. This is very misleading, as all these issues have been solved by the more recent versions of H.323.

2.1.2.2   H.323v2

H.323v2 was approved in February 1998 and fixed the major issues of H.323v1. Post-connect audio delay was completely eliminated using a new procedure known as 'fast connect'. H.323v2 was extended to enable the use of enhanced security procedures. These procedures were defined in the new **H.235** standard, addressing the need for **authentication** (making sure people in a conference really are who they pretend to be), **integrity** (making sure the modification of the content of H.323 messages by a third party cannot go unnoticed), **non-repudiation** (the ability to prove that someone participating in a conference was there), and **privacy** (making sure that information exchanged between individuals remains unaccessible to third parties). The early deployment of H.323v1 had revealed other weaknesses that appeared in very specific call flows:

- In Germany and some other countries the size of phone numbers is not known in advance and there is a need to send the dialing digits one by one to the network until the network decides the number is complete. This requires a capability called **overlapped sending**.
- When calling some announcement servers in a network you may have to stop playing the ring-back tone to play specific announcements, without connecting the call first. The announcement server uses the **progress message** to inform the network of coming in-band prompts. The same interaction with interactive voice response servers required a better means of transporting **DTMF** tones.

- Even more fundamentally, people realized that the true benefit of VoIP was not so much 'voice compression', but the ability to control media channels from a server without requiring the server to be on the path of media channels. For instance, in a prepaid application, after the initial prompts to authenticate the user and ask for a destination, once the calling party *A* half-call has been connected to the called party *B*, the prepaid application server can cause media streams to be exchanged directly between endpoints *A* and *B*, while remaining in the call control path so that it can cut off the communication once credit has expired. In traditional telephony, the same server needs to relay these media streams for the entire duration of the communication. This key feature required a new procedure called the 'empty-capability-set' (or third-party rerouting, see Section (2.7.1.2.2)) and makes VoIP much more scalable and easier to deploy than traditional TDM voice for services such as prepaid or hosted contact centers.

All these enhancements, and many more, were introduced in H.323v2, making it suitable for widespread deployment. At that time the new **H.450** standard series, providing supplementary services for H.323, was also introduced. H.450 is based on the QSIG extensions of ISDN for use by PBXs. **H.450.1** defined the general framework for exchanging supplementary service commands and responses for use by supplementary services, **H.450.2** defined the **call transfer** procedure (**blind call transfer** and **call transfer with consultation**), and **H.450.3** defined the **call diversion** procedures (*call forwarding unconditional, call forwarding on no answer, call forwarding on busy*, and **call deflection**).

Even though new versions of H.323 have been approved, H.323v2 is still the protocol powering the vast majority of voice over IP networks worldwide and serves this function very well.

2.1.2.3 H.323v3

H.323v3 was approved in September 1999. Of all the enhancements introduced in H.323v3, only one was really needed: the ability to support the CLIR (calling line identity restriction) in the same way as the traditional PSTN network. This was a legal requirement in most countries. Most of the other enhancements introduced in this version, such as the ability to reuse signaling connections, the ability to use UDP for the transport layer (annex E), an interdomain-routing protocol (annex G), have not really made it for commercial products, following the well-known pragmatic approach of 'if it works, don't fix it'.

The H.450 standard series was also expanded to include call hold, call park, call pickup, call waiting, and message-waiting indication (MWI). Out of these only MWI, H.450.7, is widely supported today by IP phones and residential gateways.

2.1.2.4 H.323v4

H.323v4 was approved in November 2000. It includes some useful modifications, such as the ability to start H.245 procedures in parallel was fast connect. Prior to this the fast-connect procedure (described in Section 2.3.3) was used to accelerate the establishment of media streams, but DTMF tones could not be transmitted before the call fully connected, which created interoperability issues in some call flows involving the PSTN intelligent network. Another very useful addition is the description of how protocols that cannot be fully mapped to H.323 can be transported in H.323. Annex M1 describes the encapsulation of QSIG (a supplementary services protocol for ISDN PBXs), and annex M2 describes the encapsulation of ISUP (the standard protocol for establishing calls in core telephone networks). H.323v4 also introduces an official format for an H.323 URL, which took almost 2 years to stabilize and be approved by all parties![1]

Throughout the life of the H.323 standard, each addition of a new feature required an edit of the ASN.1 description of control messages, as the new feature was explicitly described in ASN.1. This was powerful, as interoperability was guaranteed at least for the parsing of new features, but the growing ASN.1 syntax posed problems to developers wanting to support only a few of the features available. Since version 4, a generic extensibility framework allows the indication of features that are supported, desired, or required, without the need for further edits of ASN.1 syntax.

There were also a few more additions to the H.450 series (**H.450.8**: 'Name identification service'; **H.450.9**: 'Call completion'; **H.450.10**: 'Call offer'; **H.450.11**: 'Call intrusion'). As with the other members of the H.450 series, these additions have not received much support from the industry. They are only being employed in some private networks of IP-PBXs that use H.323 as a PBX-to-PBX protocol, where the H.450 series plays the role of QSIG (which is used in private networks of PBXs connected through ISDN).

In fact, instead of attempting to precisely define a standard for each feature of a business phone, the industry has now taken another approach: phones offer a stimulus-based control protocol for all of their user interface components (screen, buttons, lamps) and media streams. The network optimises use of these resources to provide value-added services. H.323 proposes two approaches for this:

- Using an HTTP control channel, which provides an arbitrary user interface, and making the network responsible for the execution of services.
- Annex L (stimulus control).

The industry has not adopted these methods and seems unlikely to do so in the medium term. Instead MGCP has become the de facto standard for stimulus control of IP phones. This will be covered in detail when we discuss the MGCP protocol in Chapter 4.

2.1.2.5   H.323v5

H.323v5 was approved in July 2003. This version does not introduce any major new feature but does correct some remaining problems with the former H.323 specifications, such as the missing 'hop count' parameter that prevents call loops, the much awaited H.460.6 (extended fast connect), which makes it possible to redirect and renegotiate media streams while in fast-connect mode. Another nice addition is the concept of digit maps (H.460.7), borrowed from MGCP, which makes it possible to reduce the post-dial

---

[1] For instance, see h323:someone@domaine.com

delay of a call without using the overlapped sending procedure, when the digit pattern of the expected number is known in advance (e.g., in virtual private networks).

Most of the new additions use the generic extensibility framework that was introduced in H.323v4; this helps stabilize the ASN.1 files.

Although conversion of existing systems is likely to take some time, H.323v5 adds the possibility of using SCTP as a transport protocol, in addition of TCP and UDP, providing a very robust option with combined latency control and reliability.

2.1.3 Relation between H.323 and H.245 versions, H.323 annexes, and related specifications

Where are we today? Essentially, the core H.323 protocol is complete and only minor additions are required from time to time to cover specific details required in the context of a specific service. Since H.323v3, there has been a growing divide between the status of the standard and the reality of its deployment. No one is in any hurry to implement every detail of the newer standards. Rather, these documents are viewed by vendors as a pool of standard approaches to certain issues and implemented as and when customers request them. This book will cover in detail many of the H.323v2 and v3 features, and select only the features of H.323v4 that really have something to offer for real-life deployment.

Each version of H.323 corresponds to a version of the H.225.0 call control protocol and must be used with specific versions of the H.245 media control protocol (see Table 2.1). The protocol version is indicated in the protocolIdentifier information element of the messages (e.g., {itu-t (0) recommendation (0) h (8) 2250 version (0) 2}). The H.245 version can change dynamically during a call if third-party rerouting is used.

2.1.3.1 H.323 annexes

Multiple annexes to H.323 have been defined, each specifying additional details for specific needs (Table 2.2).

2.1.3.2 H.323-related specifications

Beyond the core set of specifications—H.225.0, H.245, and the annexes—many other specifications exist which relate to specific applications or aspects of H.323:

Table 2.1 Relationships between versions H.323, H.225, and H.245

H.323	H.225	H.245
v1	v1	v2
v2	v2	v3 or higher
v3	v3	v5 or higher
v4	v4	v7 or higher
v5	v5	v9 or higher

Table 2.2 List of annexes to H.323

Annex A*	H.245 messages used by H.323 endpoints
Annex B*	Procedures for layered video codecs
Annex C*	H.323 on ATM
Annex D*	Real-time facsimile over H.323 systems
Annex E*	Framework and wire protocol for multiplexed call-signaling support
Annex F*	Simple endpoint types
Annex G*	Text conversation and text set
Annex J*	Security for H.323 annex F
Annex K*	HTTP-based service control transport channel
Annex L*	Stimulus control protocol
Annex M1*	Tunelling of QSIG in H.323
Annex M2*	Tunelling of ISUP in H.323
Annex M3	Tunelling of DSS1 through H.323
Annex N	Quality of service
Annex O	Use of DNS
Annex P	Transfer of modem signals over H.323
Annex Q	Far-end camera control
Annex R	Robustness methods for H.323 entities

*These annexes are now in the main H.323 document.

- **H.235** specifies a secure mode of operation for H.323 terminals and refers to the SSL (secure sockets layer) specification.

- **H.246** describes in more detail the operation of H.323 gateways and specifies how to map SS7 ISUP call-control messages onto H.323 messages to maximize transparency of call flows initiated and terminated on a traditional SS7 network but traversing an H.323 network.

- **H.332** (loosely coupled conferencing) profiles H.323 and extends it for use in the context of a large conference with few speakers but a large audience. H.332 is a bridge between the world of conferencing and the world of broadcasting (see the companion book, *Beyond VoIP Protocols*, chapter 6 on multicast technology).

- **H.450** is a series of standards defining messages and call flows for supplementary services, such as call transfers or how to set up the message-waiting indication on an IP phone (Table 2.3). These supplementary services mimic the services of QSIG and most are targeted for private telephony networks.

H.460 is a series of more recent recommendations, all of which use the generic extensibility framework (GEF) introduced in H.323v4:

- H.460.1: 'Overview of the generic extensibility framework and "author's guide"'.
- H.460.2: 'Number portability (GEF)'.
- H.460.3: 'Circuit status map (GEF)'.
- H.460.4: 'Call priority designation (GEF)'.
- H.460.5: 'Transport of duplicate Q.931 IEs (GEF)'.

Table 2.3   List of H.450 specifications and services

H.450.1	02/1998	Generic functional protocol for the support of supplementary services
H.450.2	02/1998	Call transfer supplementary service
H.450.3	02/1998	Call diversion supplementary service
H.450.4	05/1999	Call hold supplementary service
H.450.5	05/1999	Call park and call pickup supplementary services
H.450.6	05/1999	Call-waiting supplementary service
H.450.7	05/1999	Message-waiting indication supplementary service
H.450.8	02/2000	Name identification service supplementary service
H.450.9	03/2001	Call completion supplementary service
H.450.10	03/2001	Call offer supplementary service
H.450.11	07/2001	Call intrusion supplementary service

- H.460.6: 'Extended fast connect (GEF)'.
- H.460.7: 'Digit maps (GEF)'.
- H.460.8: 'Querying for alternate routes (GEF)'.
- H.460.9: 'QoS monitoring and reporting (GEF)'.

2.1.4   Where to find the documentation

All **ITU** documents can be purchased on the ITU website (www.itu.int). However, H.323 is a living standard and the latest specifications only become available some time after they have been approved. For those needing detailed and up-to-date technical information, the best option is to read the working documents of SG16, together with the interesting discussions of standard details and implementation guides at http://www.packetizer.com. This excellent site is well maintained and really presents all the useful information a developer needs to start an H.323 project. Another related site has a similar focus on H.323: http://www.h323forum.org. It offers interesting discussions and forums on the ongoing development of the H.323 standard.

It is also interesting to monitor the discussions of the **TIPHON** (Telephony and Internet Protocol Harmonization over Networks) project of the **European Telecommunications Standards Institute (ETSI)**. Originally, ETSI was a members-only, Europe-centric organization, but TIPHON triggered a revolution. The focus of TIPHON is now truly international, and its working documents and specifications are available on the Web (http://www.etsi.org/tiphon).

Many more standard bodies are involved in VoIP; for a more comprehensive view of the most active organizations on VoIP and voice quality of service over packet networks, see Figure 2.1.

H.323 is a complex standard. Although it is well understood and very well defined, there is still room for new interpretations. The live discussions held at SG16 and TIPHON are invaluable for the expert who is trying to keep track of updates. Much of the material presented in this chapter was gathered from these discussions and real-deployment experience. It reflects the state of the art at the time of publishing, but the reader who

Figure 2.1 Some of the many organizations involved in VoIP standardization.

wants a broader picture is encouraged to refer directly to ITU recommendations, TIPHON specifications, and the various working documents available on the Web.

2.2 H.323 STEP BY STEP

H.323 did not invent videoconferencing over IP. Researchers and students did this for years on the mBone network using RTP/RTCP (refer to the companion book, *Beyond VoIP Protocols*) multicast chapter. However, RTP/RTCP has very basic signaling capabilities, as we saw in **Chapter 1**, and cannot be used for common telephony.

H.323 mainly defines the signaling needed to set up calls and conferences, choose common codecs, etc. RTP/RTCP is still used to transport isochronous streams and get feedback on the quality of the network, but fancy RTCP features like email alias distribution are not normally used by H.323.

As H.323 tackles a very complex problem, it is consequently complex itself, as we have already stated. The set of documents that an H.323 engineer needs as a reference (Q.931, H.323, H.225, H.245, H.235, H.332, ETSI TIPHON and other profiles, etc.) is extensive and takes a while to read. Therefore, we have chosen here not to paraphrase H.323, but rather to illustrate the behaviour of H.323 entities in various configurations.

We did our best to track inaccuracies, and various engineers have checked these lines. This new edition also benefits from the many emails I have received from readers in the last 3 years. We appreciate any feedback that can help improve future editions (mailto: book@netcentrex.net).

2.2.1   The 'hello world case': simple voice call from terminal A to terminal B

For our first example we assume that two users would like to establish a voice call, both using IP endpoints with fixed and well-known IP addresses. This is an important assumption, because most of the time IP addresses are dynamic and cannot be used directly to reach a user. Calls can also be established with regular phones not directly connected on IP: this more general situation will be studied in Section 2.2.2. We will be using the basic H.323v1 connection sequence, without security and without any of the optimizations of H.323v2, v3, or v4.

Establishment of a point-to-point H.323 call requires two TCP connections between the two IP terminals: one for call set-up and the other for media control and capability exchange:

- Call set-up messages are sent during the initial TCP connection established between the caller and a well-known port (defined by the standard, usually port 1720) at the callee endpoint. This connection carries the call set-up messages defined in H.225.0 and is commonly called the Q.931 channel, or call-signaling channel.

Media-control messages are carried during a second TCP connection. On receipt of the incoming call, the callee starts listening, during the second TCP connection to a dynamic port, and waits for the media control connection to be established; the callee communicates this port in the new-call acceptance message. The caller then establishes the second TCP connection to that port. The second connection carries the control messages defined in H.245 and is used by the terminals to exchange audio and video capabilities and to perform a 'master–slave' determination; this is useful in very specific call flows (i.e., the simultaneous opening of a bidirectional data-sharing channel) which require a notion of priority of one endpoint over the other to resolve the race condition. It is then used to signal the opening of 'logical channels' for audio and video streams (each corresponding to an RTP session), fax data (the media is then exchanged using the IFP protocol described by T.38), or even a data-sharing T.120 channel. The H.245 channel remains open for the duration of the conference.

Once the H.245 channel is established, the first connection is no longer necessary and may in theory be closed by either endpoint, and re-opened only for sending additional call control messages (e.g., to bring the call to an end). In practice, though, since TCP connections take significant resources and time to get established, we do not know of any endpoint in the market that closes call control connections.

2.2.1.1   First phase: initializing the call

H.323 uses a subset of the Integrated Service Digital Network (ISDN) Q.931 user-to-network interface that signals messages for call control. The following messages belong to the core H.323 and must be supported by all terminals:

- SETUP.
- ALERTING.

- CONNECT.
- RELEASE COMPLETE.
- STATUS FACILITY.

Other messages, such as CALL PROCEEDING, STATUS, STATUS ENQUIRY, are optional. Support for the Q.931 PROGRESS message has been added in H.323v2 to support the interworking of call flows with the PSTN, notably when the PSTN signals the presence or absence of in-band media before making the connection. Regarding supplementary services, only the FACILITY message is supported; all others, such as HOLD, RETRIEVE, SUSPEND, are forbidden (they have been replaced by H.450 equivalents). Moreover, the ISDN RELEASE and DISCONNECT messages are not supported in H.323.

As we will section 2.2.1.6, each time an ISDN message has been removed to make H.323 simpler, it was subsequently found to be a mistake and the message was either added later on (PROGRESS) or other messages were extended to support an equivalent feature (e.g., DISCONNECT is in some cases replaced by a PROGRESS message).

In our example John, logged on terminal *A*, wants to make a call to Mark, knowing Mark's IP address (10.2.3.4). Terminal *A* sends to terminal *B* a SETUP message on the well-known CallSignalingChannel port (port 1720 as defined by H.225.0 appendix D), using a TCP connection (see Figure 2.2). This message is defined in H.225.0 and contains the following fields, which have been borrowed from Q.931:

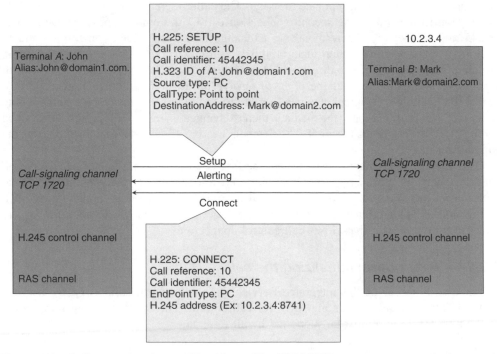

Figure 2.2 Call set-up to a known IP address. The CONNECT message returns the transport address for H.245 signaling.

- A protocol discriminator field set to 08h (Q.931 defines this as a user network call-control message).

- A 2-octet, locally unique call reference value (CRV) chosen by the originating side which will be copied in each further message concerning this call. Here John's terminal has picked CRV = 10.

- A message type (05h for SETUP as specified in Q.931 table 4.2).

- A bearer capability, a complex field that can indicate, among other things, whether the call is going to be audio-only or audio and video. ISDN gateways can place in this field some elements copied from the ISDN SETUP message.

- A called party number and sub-address, which must be used when the address is a telephone number. This field contains a numbering plan identification. When it is set to 1001 (private numbering plan) it means that the called address will be found in the user-to-user information element of the SETUP message (see below). If John knows Mark by his transport address only (10.2.3.4:1720), the numbering plan will be set to 1001.

- A calling party number and sub-address, which will be present if the caller has a telephone number.

- A user-to-user H.323 PDU (H323-UU-PDU) which encapsulates most of the extended information needed by H.323. In this case it is a SETUP information element that contains:

  - A protocol identifier (which indicates the version of H.225.0 in use).

  - An optional H.245 address if the sender agrees to receiving H.245 messages before connection. In the normal procedure, as used in the example, the callee allocates a TCP port for H.245 and waits for a H.245 connection from the caller.

  - A source address field listing the sender's aliases (e.g., John@myhouse.uk) (as indicated above, in case the sender only has an E.164 phone number then it should be in the Q.931 calling party information element).

  - A source information field can be used by the callee to determine the nature of the calling equipment (MCU, gateway, ...).

  - A destination address which is the called alias address(es). Several types are defined in H.323v2: E.164 which is a regular phone number using only characters in the set ≪0123456789#∗,"; H323-ID which is a unicode string; url-ID (a URL like those you can type on your browser, but this type in unused in practice); transport-ID (e.g., 10.2.3.4:1720), and Email-ID (e.g., Mark@domain.org). H.323v4 renamed type 'e164' into 'dialedDigits', as E.164 refers to a precise number format (country code, plus national number) which in general will not be used by end-users, who use their national numbering conventions or private numbers. H.323v4 also added a specific format for an H.323 URL, which must begin with "h323:" followed by a username and hostname (e.g., h323:mark@mydomain.org).

  - A unique Conference identifier (CID). This is not the same as the Q.931 CRV described above or the call identifier described below. The CID refers to a conference which is the actual communication existing between the participants. In the case of a multiparty conference, all participants use the same CID, and if a participant joins the conference, leaves and enters again, the CRV and CallID will change, while the CID will remain the same. Refer to Section 2.4 for more details.

- A conferenceGoal which indicates if the purpose of this SETUP message is to create a conference, invite someone in an existing conference, or join an existing conference. In this simple scenario, we simply want to create a conference.

- A call identifier (CallID) which is set by A, and should be the globally unique identifier of the call, not only locally unique like the Q.931 CRV. It is also used to associate the call-signaling messages with the RAS messages (RAS is used in the next call scenario, see Section 2.2.2). In the gatekeeper scenario (also in the next example), the call leg to the gatekeeper and from the gatekeeper to the called endpoint should have the same CallID.

Note that TCP is a stream-oriented protocol and does not provide framing (delimitation of individual messages). For this reason the Q931 messages are not transported directly over TCP, but are first framed using a 'length data' type of structure known as **TPKT** and defined in RFC1006 (ISO transport service on top of the TCP). This structure can be see in the network capture of Figure 2.3, and in Figure 2.4.

Figure 2.3   Capture of a SETUP message (using Microsoft Network Monitor).

Vrsn (8 bit) = 3	Reserved (8 bit)	Packet length (16 bit)	Data

Figure 2.4 RFC 1006 framing using TPKT structure.

Either CALL PROCEEDING, ALERTING, CONNECT, OR RELEASE COMPLETE must be sent by Mark's terminal immediately on receipt of a SETUP message. One of these must be received by John terminal's before its set-up timer expires (in general, 4 s). After Alerting is sent, indicating that 'the remote phone is ringing', the user has up to 3 min to accept or refuse the call.

Finally, as Mark picks up the call, his terminal sends a CONNECT message with:

- The Q.931 protocol discriminator, the same call reference (10), and message type 07h.
- In the H323-UU-PDU there is now a CONNECT user-to-user information element with:
  - The protocol identifier.
  - The IP address and port that B wishes A to use to open the H.245 TCP connection.
  - Destination information, which allows A to know if it is connected to a gateway or not.
  - A conference ID copied from the SETUP message.
  - The call identifier copied from the SETUP message.

Note that, *the procedure we just described is called the 'en bloc' procedure. The destination address information is sent at once. This method is always used when the destination address is not a phone number (email alias, IP address, etc.). When the destination address is a phone number the 'en bloc method' is also used by cellular phones that have a 'send' button. For a normal phone without a 'send' button, however, it is not obvious to know when the number is complete and so it should be sent in the SETUP message. Most IP phones use a timer, which fires a few seconds after the last digit key is pressed. If this waiting time is inconvenient, or when the calling device is an existing PBX, a more sophisticated procedure exists in ISDN and H.323: 'overlapped sending'. With overlapped sending, the calling endpoint sends partial numbering information in the SETUP message (with a canOverlapSend flag), and if the number is incomplete the gatekeeper (see the next example for more information on routing the signaling messages through the gatekeeper) will respond with a SETUP ACKNOWLEDGE message instead of a CALL PROCEEDING or ALERTING message. The calling device then continues to send digits in 'INFO' messages, until it receives a CALL PROCEEDING message, meaning that enough digits have been accumulated.*

Since H.323v5 (H.460.7), the 'DigitMap' function enables the gatekeeper to configure the endpoint with a set of patterns that can trigger an 'en bloc' call immediately the pattern is recognized, resolving the timer problem.

2.2.1.2 Second phase: establishing the control channel

2.2.1.2.1 Capability negotiation

Media control and capability exchange messages are sent on the second TCP connection, which the caller establishes to a dynamic port on the callee's terminal. The messages are defined in H.245.

The caller opens this H.245 control channel immediately after receiving the Alerting, CALL PROCEEDING, or Connect message, whichever specifies the H.245 transport address to use first. It uses a TCP connection which must be maintained throughout the call. Alternatively, the callee could have set up this channel if the caller had indicated an H.245 transport address in the SETUP message. The H.245 control channel is unique for each call between two terminals, even if several media streams are involved for audio, video, or data. This channel is also known as logical channel 0.

The first message sent over the control channel is the TerminalCapabilitySet (Figure 2.5), which carries the following information elements:

- A sequence number.
- A capability table, which is an ordered list of codecs the terminal can support for the reception of media streams, each codec being identified by an integer, the CapabilityTableEntryNumber. Up to 256 codecs can be described. Not all combinations

Figure 2.5 Capability negotiation over the H.245 channel using TerminalCapabilitySet messages.

of codecs can be supported, and the CapabilityDescriptors structure describes which combinations of codecs can be supported.

- CapabilityDescriptor. This is a rather complex structure (Figure 2.6) which describes precisely the combinations of codecs a terminal can support. The **CapabilityDescriptor** structure is a list of supported codec configurations. Each supported codec configuration is of the form (Codec 1 *or* Codec 2 *or* Codec 3) *and* (Codec 4 *or* Codec 5) *and* ... where *or* is exclusive. The *and* structure is called a **SimultaneousCapabilities** block, and the *or* substructures are called **AlternativeCapabilitySets**. Each codec is represented by its number in the capability table.

For instance, a terminal could declare the following for its capability descriptors:

1 (G.723 *or* g729) *and* T.120.

2 G.711 *and* T.120 *and* (H.261 *or* H.263).

This would mean that the endpoint has a limited CPU and cannot support video compression (H.261 or H.263) simultaneously with audio compression (G.723 and G.729). If video is used, then only simple voice coders (G.711) can be used. In all cases, T.120 data sharing can be used.

This structure is also very useful for simultaneous presence video applications, where the capabilities structure can be used to indicate how many instances of the video decoder can be used simultaneously: the video codec is repeated in a SimultaneousCapabilities structure, (e.g., 'H263 and H263 and H263').

The terminals send this **terminalCapabilitySet** message to each other simultaneously (a common bug in early H.323 endpoint implementations was to wait for the other endpoint to send its capabilities before sending its own) and must acknowledge the reception of the other endpoint capabilities with a *terminalCapabilitySetAck* message.

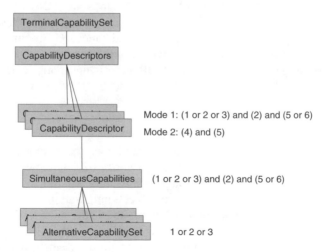

Figure 2.6 TerminalCapabilitySet structure.

When troubleshooting audio problems on an H.323 network, the terminalCapabilitySet is one of the most useful messages to look at, in conjunction with the subsequent openLogicalChannel messages and the RTP streams. The problem is most likely a mismatch between the codec parameters (codec type, frame size) advertised by the terminalCapabilitySet, the parameters chosen by the OpenLogicalChannel, and the actual parameters streamed in the RTP flow, caused by a wrong parsing or use of the H.245 messages.

2.2.1.2.2 Master/slave determination

The notion of master and slave is useful when the same function or action can be performed by two terminals during a conversation and it is necessary to choose only one (e.g., when choosing the active MC on the opening of bidirectional channels). In H.235, the master is responsible for distributing the encryption keys for media channels to other terminals.

The determination of who will be the master is done by exchanging **masterSlaveDetermination** messages which contain a random number and a **terminalType** value reflecting the terminal category: multipoint control units, the H.323 name for a multimedia conferencing bridge (**MCU**); gatekeeper; gateway; simple endpoint. The terminalType values specified in H.323 prioritize MCUs over gatekeepers over gateways over terminals, and multipoint control (MC, multipoint conference-signaling control features)+multipoint processor (MP, media-mixing feature) capable units over MC-only units over units with no MC or MP.

2.2.1.3 Third phase: opening media channels

Now terminal A and terminal B need to open media channels for voice, and possibly video and data. The digitized media data for these media channels will be carried in several 'logical channels' which are unidirectional except in the case of T.120 data channels.

In order to open a voice-logical channel to B, A sends an H.245 *OpenLogicalChannel* message which contains the number that will identify that logical channel, and other parameters like the type of data that will be carried (audio G.711 in our example of Figure 2.7). In the case of sound or video, which will be carried over RTP, the *OpenLogicalChannel* message also mentions the UDP address and port where B should send RTCP receiver reports, the type of RTP payload, and the capacity to stop sending data during silences.

The codec type and configuration (number of frames per packet), *must* be selected from one of the supported configurations advertised by the other endpoint in its terminalCapabilitySet message. If prior channels have been opened, then the endpoint should check the SimultaneousCapabilities of the other endpoint to verify that the new coder is supported in conjunction with the other coders. Although this is not a requirement in the standard, it appears that most implementations attempt to select configurations in the order in which they appear in the CapabilitiesDescriptor structure, and if the other endpoint has already opened channels to this endpoint it also attempts to use symmetrical coders. This is in no way mandatory, and assymetrical communications were the A to B and B to A streams use different coders are valid.

Figure 2.7   Opening media channels using H.245 OpenMediaChannel messages.

B sends an *OpenLogicalChannelAck* for this logical channel as soon as it is ready to receive data from A. This message contains the IP address and UDP port number where A should send the RTP data and the UDP port where A should send RTCP sender reports. Meanwhile, B also opens a logical channel to A following the same procedure.

2.2.1.4   Handling of DTMF tones

In H.323, there are several ways to transport DTMF tones:

- The special H.245 User Input Indication (UII) message, which must be supported by all H.323 systems. It has the advantage of using a reliable TCP connection, and therefore the message cannot be lost. But because TCP will try to retransmit the packet if it has been lost in the network, information might get delayed and get to the receiver too late. Two modes can be used: 'alphanumeric' and 'signal'. The most widely used mode is alphanumeric, this can be taken as the default in most gateways and H.323 phones. The UII message in this mode can carry all numeric characters, 'A', 'B', 'C', 'D', '*' and '#'. In H.323v2, the UserInputIndication message was updated to also include other information, such as the length and signal level of a tone, and synchronization information with the RTP stream: this is the signal mode. Here is an extract of the H.245 User Input Indication ASN.1 definition showing the added parameters:

```
UserInputIndication ::=CHOICE
{
        nonStandard NonStandardParameter,
        alphanumeric GeneralString,
        ...,
```

```
        userInputSupportIndication CHOICE
        {
                nonStandard     NonStandardParameter,
                basicString     NULL,
                iA5String       NULL,
                generalString   NULL,
        ...
        },
        signal                  SEQUENCE
        {
                signalType      IA5String (SIZE (1) ^ FROM
("0123456789#*ABCD!")),
                duration        INTEGER (1..65535) OPTIONAL,   --
milliseconds
                rtp                             SEQUENCE
                {
                timestamp       INTEGER (0..4294967295)
OPTIONAL,
                expirationTime  INTEGER (0..4294967295)
OPTIONAL,
                logicalChannelNumber    LogicalChannelNumber,
                ...
                } OPTIONAL,
                ...
        },
        signalUpdate    SEQUENCE
{
                duration        INTEGER (1..65535), --
milliseconds
                rtp                             SEQUENCE
                {
                        logicalChannelNumber
LogicalChannelNumber,
                        ...
                } OPTIONAL,
                ...
        }
}
```

- The ISDN 'Keypad Facility' Information Element, which can be included in the SETUP or INFORMATION message. This is inherited from ISDN and used only in conjunction with the overlapped sending call flow (see note in Section 2.2.1.1).

- More recently (i.e., since H.323v4), H.323 can also use RFC 2833 for DTMF signaling (see Chapter 3 for details on RFC 2833); this requires H.245v7 and is an optional call flow. RFC 2833 can be used in conjunction with UserInputIndication (in this case the UserInputIndication message should have an **rtpPayloadIndication** flag). RFC 2833 also decodes the DTMF tone and includes it in a packet, but this time the packet is an RTP packet, not a signaling link packet. RFC 2833 can encode many telephony events (e.g., 'flash-hooks'), in addition to just DTMF tones. Note that RFC 2833 requires implementations to be able to send the telephony events to a destination that may not be the destination of the rest of the media streams. The inability to do so is a serious bug, as it prevents any DTMF-driven application from being built without accessing the media channel (e.g., a contact center). Unfortunately, this is a very frequent bug.

- A special RTP logical channel can be opened to carry the RTP DTMF payload. This payload is formatted as indicated in Figure 2.8. The unit used for the duration is the same as the unit used for the timestamp. If a separate logical channel is opened the sampling rate will be considered to be 8,000 Hz. For more details on RFC 2833, see Chapter 3. This method had been proposed for H.323v2 by the VoIP Forum. This was before H.245v7 was published. The original opening procedure was described as follows: It is possible to insert a DTMF RTP packet in the same logical channel as voice. In this case the payload type should be formed as follows to avoid confusion with dynamic or fixed RTP PT (these should be less than 128): 'chosen voice PT (e.g., 8)' 'DTMF PT' + 128. This PT should be used in the OpenLogicalChannel. If the remote terminal doesn't understand this meta-type, it means it doesn't support this method. This method should be scorned and replaced by the H.323v4 and H.245v7 procedure.

Overall, it seems that the User Input Indication method is preferred, since packet loss is typically very small for signaling links on well-engineered networks, and very few IVRs are sensitive to DTMF timing. In our various deployment experiences, this method always worked correctly. However, for international calls with large round trip times and time-sensitive IVR systems, it might prove necessary to use the second method. RFC 2833

Figure 2.8 RFC 2833 RTP packet format for DTMF transport.

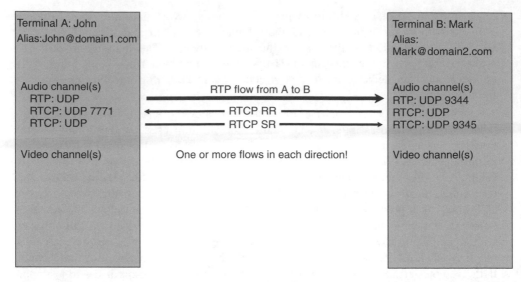

Figure 2.9 RTP flows and associated RTCP receiver report and sender report messages.

implementations which are not capable of sending telephony events to an application server should be avoided. In addition, gateways should be extremely careful to mute in-band DTMF and convert it to an H.245 UserInputIndication or a special RTP payload type, since simultaneous transmission of in-Band DTMF and the special H.245 or RTP messages might cause the egress gateway to first render the RTP DTMF packet or H.245 UserInputIndication and then transmit the DTMF tone contained in the audio stream, duplicating the original tone.

2.2.1.5 Fourth phase: dialogue

Now John and Mark can talk, and see each other if they have also opened video-logical channels. The media data are sent in RTP packets as shown in Figure 2.9.

RTCP receiver reports (RRs) enable each endpoint to measure the quality of service of the network: RTCP messages contain the fraction of packets that have been lost since the last RR, the cumulative packet loss, the inter-arrival jitter and the highest sequence number received. In theory, H.323 terminals should respond to increasing packet loss by reducing the sending rate, possibly by changing the audio coder dynamically ... but, in practice, RTCP information is not used by most endpoints.

Note that H.323 mandates the use of only one RTP/RTCP port pair for each session. There can be three main sessions between H.323 terminals: the audio session (session id 1), the video session (session id 2) and the data session (session id 3), but nothing in the standard prevents a terminal from opening more sessions.

For each session there should be only one RTCP port used (i.e., if there are simulta-neously RTP flows from A to B and from B to A, then the RTCP sender reports and receiver reports for both flows will use the same UDP port).

2.2.1.6 Clearing the call

How do we go about ending an H.323 call? Well, it is not that simple. If John hangs up, terminal A must send an H.245 **CloseLogicalChannel** message for each logical channel that A opened. B acknowledges those messages with a CloseLogicalChannelAck.

After all logical channels have been closed A sends an H.245 **endSessionCommand**, waits until it has received the same message from B (B will also close all its media channels before sending the endSessionCommand) and closes the H.245 control channel.

Finally, A and B must send an H.225 **ReleaseComplete** message over the call-signaling channel if it is still opened, then close this H.225 channel. The call is now cleared.

Needless to say, many software endpoints are not so polite, and terminate rather than close calls.

Note also that the call release sequence is different in ISDN. An ISDN endpoint releasing a call would first send a DISCONNECT message, which would be acknowledged at the other end by a RELEASE message, and the call would be over after the releasing endpoint sends a RELEASE COMPLETE. H.323 takes a short cut approach and sends only a RELEASE COMPLETE message.

In most cases this is fine, but this causes interoperability problems with the PSTN when the PSTN, instead of just releasing the call, wants to send an audio message to the caller first (this situation occurs frequently when calling mobile phones). An example of such an announcement is 'The party you are calling is currently not reachable on the network'. In this case the ISDN **DISCONNECT** message may contain a 'progress indicator' information element, with value 1 or 8 meaning that audio information is being provided to the caller. After a timer value of about 30 s or if the calling party hangs up before this timer, the calling party will send the **RELEASE** message to the network. In H.323, in order to respect the standard, a possible solution is to convert the DISCONNECT (progress indicator = 1 or 8) message into an H.323 PROGRESS (same **progress indicator** value) message in the PSTN gateway, with a special indication that this is really a release indication and the call should not be maintained longer than 30 s. This solution has been implemented in Cisco® gateways, for instance (see http://www.cisco.com/warp/public/788/voip/busytone.html for details). It is important to signal in the PROGRESS message that this is really a DISCONNECT, in order to provide the proper information to automated equipment that cannot interpret the in-band tone.

This progress indicator can also occur in other phases of the call, within the SETUP, ALERTING, CALL PROCEEDING, PROGRESS, CONNECT messages, but for these messages it does not cause problems because H.323 can transport it (these messages all exist in H.323). The only caveat is to make sure, when testing H.323 gateways, that the PI is properly supported. The PI can have the following values:

- Progress indicator = 1: the call is not end–end ISDN. Further call progress information may be available in-band.
- Progress indicator = 2: destination address is non-ISDN. This may be found in PROGRESS and CONNECT messages.
- Progress indicator = 3: origination address is non-ISDN. This is used in a SETUP ISDN message to signal that the calling party device is expecting in-band messages.

This is the case for most devices using analogue or CAS connections. A VoIP gateway receiving an ISDN SETUP with PI = 3 should provide in-band ring-back, as the calling device is unable to generate the ring-back locally.

- Progress indicator = 8: in-band information or an appropriate pattern is now available.
- No progress indicator in a message assumes that the originating device will provide the appropriate tone signaling to the calling party.

The commonly used progress indicators are '1' and '8'. A good test is to send a PRO-GRESS message in state 'alerting' with the progress indicator = 8. The originating gate way should play the in-band audio and stop ringing. If a new PROGRESS message is sent with no PI, since the state is 'alerting', local ring-back should be provided.

2.2.2 A more complex case: calling a public phone from the Internet, using a gatekeeper

In the simple case described above, Mark called John directly on his current IP address 10.2.3.4. This situation is very convenient to show the basics of H.323, but very unlikely to happen in reality. If nothing else, a plain IP address is very hard to remember. In many cases it will even change—most ISPs allocate a dynamic IP address to their subscribers.

Our next example is more realistic: Mark now wants to call his grandmother, who only owns a regular phone and doesn't have the slightest idea of what an IP address is. This example will show the need for a new H.323 entity, called the **gatekeeper**.

The gatekeeper is the most complex component of the H.323 framework. It was first introduced in H.323v1, but at that time most people didn't really understand how useful it would be. At best, the gatekeeper was considered to be a sort of directory mapping friendly names to IP addresses. Some companies found alternative ways of doing this: some now-obsolete 'H.323-compliant' software and hardware used proprietary mechanisms ranging from IRC servers to LDAP servers to find the transport address of another VoIP phone or gateway.

H.323v2 has clarified the role of the gatekeeper, and now it is widely acknowledged that the gatekeeper is responsible for most network-based services (i.e., services which need to be performed independently of the terminal or when the terminal is turned off). These services include registration (the ability to know that someone has logged on and can be reached at a particular terminal, sometimes called 'presence'), admission (checking the right to access resources), and status (monitoring the availability of telephone-related network resources, such as gateways and terminals). Finding the transport address to use to reach a particular alias is naturally also part of the gatekeeper's role, since this transport address might depend on the status of the called party (e.g., if the person is not logged on, the call should be redirected to an answering machine or a regular phone through a gateway), the identity of the caller (not everybody might be allowed to call Mark, such as in the case of a do-not-disturb service), or the status of a particular resource (if all ports on a gateway are busy, then it might be necessary to use another gateway). Therefore, the gatekeeper is also in charge of routing all VoIP calls in the H.323 network, and the implementation of services like call forward on no answer.

The set of all H.323 endpoints, conference servers (MCUs) or gateways managed by a single gatekeeper is called a *zone*. In our example John's terminal and the gateway belong to the same zone.

In this section 2.2.2 we consider that the caller has access to a gatekeeper, and show some of the gatekeeper features in action. The terminal and the gatekeeper use a specific protocol for registration, admission, and status purposes, which has logically been named RAS. This protocol is also defined in H.225.0.

2.2.2.1 *Locating the gatekeeper*

In simple configurations, the gatekeeper's IP address might simply be configured manually or automatically in the VoIP terminal. This is the most frequent case in real-life H.323 networks. This IP address is usually acquired when the VoIP endpoint boots: it first acquires an IP address and basic configuration parameters through the Dynamic Host Configuration Protocol (DHCP), one of the configuration parameters is the name of a TFTP server and a configuration file. The endpoint then downloads a configuration file using the TFTP protocol, which specifies, among other parameters, the address of a gatekeeper (and most of the time, a back-up gatekeeper). If such a configuration mechanism cannot be used or is not suitable, H.323 has developed a mechanism to dynamically find a gatekeeper on the network. This has a number of advantages (e.g., when someone has got a laptop and roams between several office locations). This mechanism also provides a way to introduce redundancy and load balancing between several gatekeepers in the network.

In order to find a gatekeeper, a H.323 terminal should send a multicast **Gatekeeper Request (GRQ)** to the group address 224.0.1.41 on UDP port 1718 (for more information on multicast, see companion book, *Beyond VoIP Protocols*). Within the GRQ message, it can specify whether it is willing to contact a particular gatekeeper. The terminal also mentions its aliases, allowing a gatekeeper to reply only to specific groups of terminals. Eventually, a GRQ can also be sent in unicast to port 1718, or preferably 1719, this is the default for unicast RAS messages, but obviously in this case the endpoint should know the possible gatekeeper IP addresses in advance.

The GRQ message should be sent with a very low TTL (time to live) initially in order to reach the gatekeepers on the local network first, and then use expanding ring search (Figure 2.10). This GRQ message tells the GK on what address and port the terminal expects to receive the answer, which type of terminal it is and what the terminal alias(es) is (are).

Each gatekeeper should be a member of group 224.0.1.41 and listen on port 1718. Therefore, one or more of these gatekeepers will reply on the address specified by the terminal with a **Gatekeeper Confirm (GCF)** message which indicates the name of the gatekeeper, and the unicast address and port that this gatekeeper uses for RAS messages. It can also include the names and transport information of other back-up gatekeepers.

The use of multicast for gatekeeper discovery has raised much controversy. In fact, not many IP networks support multicast today. Multicast routing is not activated by default on routers, and many network administrators feel comfortable with static routes and are not

Local GK

GRQ sent using multicast

GRQ will cross this router
only if its TTL > 1

External GK

Figure 2.10 Locating the gatekeeper using multicast GRQ messages.

really willing to experiment with a dynamic multicast routing protocol, such as DVMRP or PIM. Moreover, many of the Ethernet hubs and switches installed today still do not support multicast—sometimes some of these devices turn multicast into broadcast traffic. Ironically, if you are still using an old Ethernet coax network, multicast will work on your LAN segment and you don't have to do anything. Obviously, the most recent switches support multicast too (see the multicast section of the companion book, *Beyond VoIP Protocols* to see what it means exactly), so if you upgraded your network recently it is very likely that multicast will work on your LAN too! Multicast capability is a key requirement if you plan on buying a new switch for your organization LAN. In the near future, all switches will support multicast, and if you plan to support videoconferencing and video broadcasting efficiently across your IP network beyond the LAN, you will also have to turn on some multicast protocol in your routers.

2.2.2.2 Registration

If it received more than one answer from a gatekeeper in the discovery process, the terminal chooses one and registers this with the selected gatekeeper by sending a unicast **Registration Request (RRQ)** message (usually on UDP port 1719). This message carries an important additional piece of information compared with GRQ: the transport address which is to be used for call signaling. The registration can be 'soft state' if the terminal so desires, in which case it also specifies a **time to live** for the registration and will refresh its registration periodically by sending more RRQs, also called **lightweight RRQs** or

keep-alive RRQs. These RRQs have a special parameter 'keepAlive' set and do not include the full registration information.

The gatekeeper replies with a **Registration Confirm (RCF)** message in which the gatekeeper assigns a unique identifier to this terminal which must be copied in all subsequent RAS messages. The GK can also assign an alias to the requesting endpoint in this RCF. Whether or not the terminal chose to use the 'keepAlive' registration, the gatekeeper can also request keepAlive RRQs by specifying a maximum time to live in its response.

Since the advent of H.323v4 it is also possible to use *additive registrations*, in order to register many aliases which would not fit in a single RRQ message (this can be used by IP-PBXs when registering many extensions to a core network). Such RRQs have a specific 'additiveRegistration' flag. They are also acknowledged by an RCF.

2.2.2.3 *Requesting permission to make a new call*

Now that John's terminal has found a gatekeeper and is registered, John still needs to request a permission from the gatekeeper for each call he wants to make. In this case he wants to reach his grandma at +33 123456789.

His terminal will first send an **Admission Request (ARQ)** message to its gatekeeper. The ARQ message includes:

- A sequential number.
- The GK-assigned terminal identifier.
- The type of call (point to point).
- The call model that the terminal is willing to use (direct or gatekeeper-routed—see Section 2.2.2.2).
- The destination information (in this case the E.164 address +33 123456789 of grandma, but it could also have been Mark's email alias). Note that we used '+' to denote the country code, but this character is not transported in the ARQ message. In reality John would probably use the local dialing convention, and not a full number in international format.
- A Call Reference Value (CRV), which should be copied in the SETUP message.
- A globally unique CallID.
- An estimation of the bidirectional bandwidth that will be used for this call for media streams. This includes audio and video that will be sent from the called party and is measured excluding network overhead. This is a *very* rough estimation in most cases, since the codecs will be negotiated later. For instance, an audio-only terminal might indicate 128 kbit/s as a worst case if the two terminals negotiate a G.711 codec (64 kbit/s) for the incoming and outgoing audio logical channels. The endpoint may use **Bandwidth Request (BRQ)** messages later to ask for additional bandwidth (e.g., if it needs to open video channels).

The two possible call models refer to the way the call-signaling channel (carrying Q.931 messages) and the H.245 channel are set up between the endpoints. The calling endpoint can establish these channels directly with the called endpoint (the **direct mode**), or it can establish these channels with the gatekeeper which will relay the call-signaling and call control information to the called endpoint (there might be several gatekeepers routing the Q.931 and H.245 channels between the two endpoints). The later mode is the gatekeeper **routed mode**.

In this example we will use the direct model; we will discuss the GK-routed mode later. As we will see, the GK-routed mode is much more powerful and is the only model that works in carrier-class deployments.

If it decides to accept the call, the gatekeeper replies with an **AdmissionConfirm (ACF)** message which specifies:

- The call model to use (regardless of what was previously indicated by the calling endpoint)
- The transport address and port to use for Q.931 call signaling. This address can be the IP address of the called terminal directly (or the IP address of a gateway when calling a regular phone number) in the direct model, or it might be the gatekeeper itself if it decides to route the call. In our example the gatekeeper replies with the IP address of a gateway.
- The allowed bandwidth for the call.
- The GK can also request the terminal to send IRR (Information Request) messages from time to time to check whether the endpoint is still alive.

Note that this admission phase is really redundant if the gatekeeper wishes to use the routed mode, because the gatekeeper keeps full control of the call in routed mode. In H.323v2, the admission phase using ARQ/ACF messages can be skipped if the gatekeeper grants the **preGrantedARQ** right to the endpoint during the registration phase (see Section 2.3.6).

2.2.2.4   Call signaling

The Admission Confirm message has provided John's terminal with the information it needed to complete the call (Figure 2.11). Now the terminal can establish a call-signaling connection to the call-signaling address and port specified by the gatekeeper, in our case a gateway to the phone network, and send a Q.931 SETUP message. Before proceeding, the gateway may itself be required to ask the gatekeeper if it is authorized to place the call using and ARQ/ACF sequence. The ARQ will mention both the calling endpoint alias/call-signaling address and the called endpoint alias/call-signaling address, and a field indicating that this is an ARQ related to the termination of a call. In this receive side ARQ, the CRV (Call Reference Value) will be locally generated and, therefore, will differ from the CRV of the calling side ARQ. But the CallID should be copied from the SETUP message.

The gateway knows from the called party number information element of the H.323 SETUP message which phone number it must call. If it is connected to an ISDN phone line, it will simply send an ISDN Q.931 SETUP message on the D channel to initiate the connection on the ISDN. If it is connected to an analog line, it will go off-hook and

Figure 2.11   Direct mode call authorized by the gatekeeper.

dial the number using DTMF. If it is sending the call to an SS7 ISUP network, it will convert the SETUP message to an ISUP **Initial Address Message (IAM)**. Note that the format of the phone number may need to be changed (e.g., the country code may need to be removed). In the direct mode, this needs to be done by the gateway, whereas in the routed mode this would typically be done by the gatekeeper, centralizing the numbering plan and routing management.

The gateway will send an H.225 ALERTING message to the caller as soon as it has received an indication from the phone network that Grandma's phone is ringing, and send the CONNECT message as soon as she has picked up the handset. If the gateway was connected through an ISDN line, these events will be signaled by the phone network using similar Q.931 ALERTING and CONNECT messages. If it is an analog line, the gateway needs to detect the appropriate ring/busy/connect conditions.

The ALERTING or CONNECT message contains a transport address to allow John's terminal to establish an H.245 control channel on which it can negotiate codecs and open media channels. This procedure is identical to the procedure used above when John was calling Mark. The media channels are then opened between the gateway and John's terminal as in the previous example.

2.2.2.5   Termination phase

Whoever hangs up (e.g., the gateway in Figure 2.12) first needs to close its logical channels using the H.245 CloseLogicalChannel message. The gateway then sends an H.245

Figure 2.12   Call released at H.245 and H.225 level (Q.931 and RAS).

endSessionCommand message to John's terminal and waits to receive the same message from John's terminal. The gateway then closes the H.245 channel. If the Q.931 channel is still open, each terminal must send a Q.931 ReleaseComplete message before closing it, then the terminal and the gateway must send a **Disengage Request (DRQ)** message to the gatekeeper, enabling the gatekeeper to know that the call and associated network resources have been released. The gatekeeper replies to each with a **Disengage Confirm (DCF)**. At this stage, if the terminal or the gateway had been sending IRR messages to the gatekeeper, they must stop.

If there is more than one gatekeeper and the gateway and the terminal are registered to different gatekeepers, each one sends a DRQ to its own gatekeeper.

The terminal or the gateway have no reason to unregister (done by sending an **UnregistrationRequest** or **URQ** to the gatekeeper), unless, for instance, John decides to close his IP telephony software. A terminal should remain registered as long as it can make or receive calls.

If during the communication the gatekeeper wants to clear the call it can also send a DRQ to one or both endpoints. On receiving the DRQ the endpoint must send an H.245 endSessionCommand to the other endpoint, wait to receive an endSession command, close the Q.931 channel with a release complete, and send a DCF to the gatekeeper. Of course, this is dependent on the endpoint implementation and cannot be used reliably for applications like prepaid calls if at least one side of the call is not a trusted device. For such applications, if the endpoints are not fully trusted, the routed call model must be used.

In order to prevent a terminal from pretending it is closing a connection with a gateway without sending an endSessionCommand/release complete to the gateway, when a gatekeeper receives a DRQ from the terminal, it will wait until it has received a DRQ from the gateway before replying with a DCF. If the gatekeeper receives a DRQ from the gateway (as in our example), it will wait until it has received a DRQ from the terminal before sending a DCF to the gateway. In case the gatekeeper doesn't receive a DRQ from the terminal within a few seconds, it will ask the terminal to disconnect by sending a DRQ. The terminal is supposed to disconnect and send back a DCF, but if it doesn't within a few seconds (the PC might have crashed), the gatekeeper will send a DCF to the gateway anyhow. This procedure minimizes fraud and unwanted operation due to unstable or non-conformant terminals, but the routed mode is still more reliable.

In the direct mode, Call Detail Records must be generated by the gateways, the RAS information available at the gatekeeper level is not accurate enough for most purposes. For instance, it does not have access to the call release causes (a Q850 release code is provided by the network for each released call, specifying the reason for dropping the call: normal clearing, network congestion, user busy, switch failure, etc.). Also, the timing information is loosely coupled with the timing of the actual call release, and the start and stop information are available at different machines in the case of a network with multiple gatekeepers. This makes gatekeeper-level accounting records approximate at best. In real-life deployments with multi-gatekeeper networks, where unfortunately gateways or routers sometimes crash or show unexpected behaviour, the direct mode also makes it quite difficult to detect so-called 'zombie calls', calls that for some reason remain active in the network, out of control, for days or months.

All these issues are resolved by using the more complex gatekeeper-routed mode.

2.2.3   The gatekeeper-routed model

Initially, virtually all gatekeeper implementations were using the direct call model. This model, where the gatekeeper is used really only as a sort of directory, seems very attractive at first glance:

- Very simple implementation, very few messages must be supported.
- The implementation can be made almost stateless if the accounting functions are external.
- The established calls are not affected if the gatekeeper fails.
- And, more importantly for marketing purposes, since the gatekeeper really does not do much, the manufacturer can claim the great performance figure of several hundred calls per second!

The direct gatekeeper model is still very important today, not only because most vendors still publish their performance in direct mode (without mentioning it most of the time), but also because the majority of H.323 networks worldwide still use direct mode routing.

The situation is changing rapidly, however, because in fact the direct model has many shortcomings that do not allow VoIP networks to get to the same level of quality of service as traditional TDM networks. In any case, direct mode remains acceptable for enterprise networks.

2.2.3.1 Major issues of the direct mode

2.2.3.1.1 Poor termination rates

In direct mode, the calling endpoint and the called endpoint communicate directly with one another, once the IP address of the called endpoint has been discovered. This is fine as long as the call succeeds. But if the first attempt to terminate the call fails (Figure 2.13), then the call is released.

A call attempt can fail for many reasons:

- Instability of gateways, resulting in their unavailability when the call arrives.
- Congestion of gateway resources.
- Congestion somewhere in the terminating PSTN network (as shown in Figure 2.13).

In the same situation, if a traditional TDM network had been used, then one of the class 4 central offices of the service provider in the path of the call would have detected the failure by analysing the **Q.850 release cause** included in the ISDN or SS7 release message. It would not have released the call on the calling side, but would have rerouted the terminating leg to other trunks. It is only in the unlikely situation where no trunk in the network can terminate the call that the call would have been released; and, even then, instead of just dropping the call, the call would have been routed to an announcement

Figure 2.13   Direct mode gatekeeper cannot improve call termination rates.

server explaining to the calling party that a temporary failure is occurring. Such a situation would also generate alarms at the service provider supervision center, and someone would verify the network dimensioning.

By comparison, the direct model in VoIP is not only very poor, it is in fact completely unacceptable as soon as some real traffic is carried. Many VoIP networks started by just providing low-quality prepaid termination, a market segment not particularly noted for its quality of service. But, as soon as the traffic started to diversify, many service providers were faced with complaints from users that the termination rate was poor. In fact, this poor termination rate quickly become a show-stopper because it provided traffic termination for professional users, one of the most profitable segments of the market.

2.2.3.1.2 Attempts to improve the direct model: Resource Availability Indicators (RAIs)

Since the routed model is significantly more complex than the direct model, the initial response of the H.323 developer community to the poor performance of the direct mode was to attempt to avoid some of the causes for failed calls. For this purpose, the new RAI (Resource Availability Indicator) was introduced. The goal of this message was to let the gatekeeper know when a gateway was becoming congested. Above a certain threshold, the gateway will indicate to the gatekeeper that it is 'almost out of resources', and the direct mode gatekeeper is expected to divert traffic to other termination gateways.

This seems a good fix at first glance, but does it really solve the problem? Unfortunately, it doesn't:

- As we have just seen, most of the congestion situations occur in the PSTN, not locally at the gateway. For some destinations where the telephone network is not well developed the congestion rate can be as high as 50%! Also, some niche service providers specialized in low-cost termination have a poor quality of service. In order to save an termination fees, it is nice to be able to route traffic to them, but only if failures can be recovered by routing calls to alternative service providers in the event of a failure. Obviously, the RAI message only monitors resources at the gateway level and does not help for PSTN congestion.

- The RAI doesn't really help either for gateway congestion. Let's take two extreme situations: if the gateway average usage level is very low, say 50%, the RAI threshold level can be put very low (60%), despite obviously not needing the RAI to avoid gateway congestion. On the other hand, if the gateway usage rate is very high (a desirable situation given the cost of gateways), say 95%, then RAI on–off thresholds will be very high (e.g., 95% RAI 'OK' and 98% for RAI 'out of resources'). Unfortunately, a race situation occurs between RAI messages and the incoming calls from the PSTN. As each gateway has few T1/E1 ports, there will be an average of about two new call events and two call release events per second, when the difference between the two RAI thresholds represents only about four calls. This means that the RAI will continually change status, and the RAI status may be obsolete as soon as it is sent to the gatekeeper, if new calls arrive. Therefore, the RAI improves the situation only in networks where gateway usage is not above 80%, which is not very good from a capital utilization perspective. If you have a low-cost service provider where you make a margin of a fraction of a cent a minute, and an alternative service provider where your margin may be negative, you really want the gateway to the low-cost service provider to be used at 100% capacity at all times!

- The RAI is an RAS message that is not routed if there are multiple gatekeepers; therefore, it is only useful at the last hop (last gatekeeper). But in many situations you would like rerouting to occur before the last hop.

- As a consequence of the previous limitation, the RAI doesn't work across administrative boundaries. If you are exchanging traffic with another VoIP service provider, it is almost certain that the other service provider will have its own gatekeeper, and you will not receive any RAI indication.

Less importantly, RAI is an H.323-only message with no SIP equivalent. If you deploy a mixed H.323/SIP network you will end up with a management of resources that is different between H.323 and SIP devices, which can quickly lead to some serious headaches; and, if you plan to migrate from H.323 to SIP, you will have to completely redesign network routing and congestion management.

If you have no other choice, you can use RAI when you can, but you should not expect major improvements of your network quality. RAI only works in marginal cases. As we will see in the coming paragraphs, the real solution to the issue is nothing new; it is the same solution as used on current TDM networks: full routing of the signaling messages by the switches (not the media streams in the case of VoIP), analysis of the Q.850 release codes which are also present in H.323 (and have SIP equivalents), and dynamic rerouting of calls.

2.2.3.1.3 Centralized routing

Although most gateways have some internal call-routing logic, using these capabilities quickly becomes very hard to manage as the number of gateways increases. A network of five gateways will need at least five routes to be configured on five gateways, a network of 100 gateways will need 100 routes on 100 gateways. Entering these 10,000 routes is a daunting task for a network manager.

Using a direct mode gatekeeper to control the routing of calls significantly simplifies the management process, but is still not ideal:

- Most gateway internal routing engines can fall back from one destination gateway to another in the case of congestion on other cause of call failure. This feature disappears when using the direct mode gatekeeper, possibly resulting in a reduced perceived quality of service by network users.

- Centralized routing really covers two tasks: selecting the proper destination, and changing the format of call aliases. A call initiated in San Jose, California to +1 212 xxx xxxx must be rewritten as a call to xxx xxxx if the destination gateway is in New York. Similar changes must be made to the calling party number. The direct mode gatekeeper can manipulate the destination alias with the **CanMapAlias** feature of H.323, but very few gateway vendors support it. In addition, the source alias cannot be changed. As a consequence, it is fair to say that as soon as the service becomes complex, with multiple vendors, or requires manipulation of the calling party number (if the number presentation service is required), with the direct model the alias format management must remain distributed at gateway level (all gateways must convert local alias formats to/from an agreed network-wide 'pivot' format).

2.2.3.1.4   Centralized accounting

Another frequent issue faced by service providers is the management of accounting information. In first-generation VoIP networks, the accounting information was generated by the edge gateways. It was either collected by batch processes by a central accounting function, or sent in real time by gateways using protocols, such as Radius.

While this works well for closed VoIP networks built from a single vendor, it becomes problematic if:

• The network is open to partners (clearing houses, termination partners, etc.) who do not provide access to their gateways.

• The network is open to customers (IP-PBXs, ASPs, etc.), who obviously cannot be trusted for billing information.

• The network uses multiple vendors, each having its own format for CDRs (Radius is only the transport protocol, the actual accounting information is always proprietary to each vendor).

A direct mode gatekeeper has only limited access to call information: it knows approximately the timing of the call start by using the ARQ messages and the timing of the call stop through the DRQ message. It does not know the call release causes (Q.850). Obviously, if the network involves multiple direct mode gatekeepers, this model also becomes complex because part of the RAS information is provided to different gatekeepers. It also does not work if the edge devices cannot be trusted (they could potentially send DRQ messages while continuing a conversation). These limitations do not allow the direct mode gatekeeper to be a reliable device to generate accounting records centrally in a network.

2.2.3.1.5   Security issues

The last issue of the direct mode in an open network relates to **security**. Since the direct mode gatekeeper lets endpoints exchange signaling directly, any endpoint on the network can learn the IP addresses of other devices (this in itself is not a security problem), but more importantly can send signaling at any moment to any endpoint. This makes denial-of-service attacks trivial. Because of this, VoIP networks using direct mode gatekeepers cannot be opened up to third parties. They cannot be used to connect IP-PBXs and cannot send traffic directly to other VoIP networks. In fact, today, as routed mode gatekeepers are still relatively new, the major VoIP clearing houses still interconnect their various partners through traditional TDM central offices!

2.2.3.2 The gatekeeper-routed model

A gatekeeper using the routed model handles all call-signaling information and does not let endpoints establish calls directly. Some gatekeepers can be configured to use the routed model or the direct model on a per-route basis.

The routed model is exactly identical to the way traditional TDM switches handle phone calls, with one exception: when using the routed model, the media streams are still exchanged directly by endpoints. The routed model provides all the advantages of full class 4 routing (ability to analyse release causes, reroute calls, better security), while still not requiring dedicated telecom hardware since no TDM switching matrix is required. Because of this the density- and hardware-related cost of softswitches is far better than their TDM counterparts.

All the issues described above for the direct mode are solved:

- Congestion, whether at the gateway level or anywhere in the PSTN network, is detected by analysing the Q.850 release cause. The call can be dynamically rerouted to other termination routes (Figure 2.14). This works regardless of the number of softswitches and across administrative boundaries (clearing houses or terminating VoIP partners can be used). Since the calls are rerouted dynamically in the event of congestion, the least costly routes can be used at 100% capacity without affecting the perceived quality of service of the network. With a routed mode gatekeeper, the failure rate perceived by call sources is equal to the product of the failure rates of all termination routes for a given destination. If the network has two partners each experiencing a 50% failure rate to a country, the perceived failure rate seen by service provider customers is only 25% (compared with 50% in the direct model). This drops to a 13% perceived call loss with three partners each losing half of the calls. If the routed mode gatekeeper has

Now the call is properly completed.
True class IV resolves network congestion cases, both in the VoIP network and in the PSTN.
This allows to peer with less reliable PSTN partners, but still offer the best call completion rates

Figure 2.14 The gatekeeper can interpret Q.850 release causes and redirect the call as appropriate on the fly.

the least costly routing features, a low-cost partner route losing 20% of the calls can be used at 100% capacity, while a high-cost partner losing only one call in a thousand can be used only in the event a call is dropped by the low-cost partner. This optimizes costs, while still providing a perceived call failure rate of less than one in a thousand to service provider customers. Note that with this model gateways do not need to support the RAI feature. In fact, the RAI message becomes completely useless with a routed mode gatekeeper.

- If calls cannot be completed due to congestion or any other reasons, they can be routed to a network announcement server (simply defined as the last-resort route for all destinations), terminating calls gracefully rather than just dropping them.

- Centralized routing now handles properly not only the selection of the proper destination, but also the conversion of alias formats. The gateways only need to support the basic H.323 call flow, with no local logic for routing or the manipulation of call aliases. Everything is provisioned centrally in the routed mode gatekeeper-routing engine. Since both the source and the destination alias can be manipulated, the calling line ID features can be provided. The routed mode gatekeeper has complete access to the alias information, which also contains the caller ID blocking status (**Q.931 octet 3A**): it can provide caller ID blocking for certain routes (e.g., international routes to ensure privacy), and caller ID forced delivery for emergency calls. The routed mode also enables more sophisticated features (e.g., virtual private networks) if the gatekeeper can translate between private and public numbering plans. This does not require any capability at the endpoints besides support for an H.323 basic call and can be provided to any endpoint, including IP phones or IP-PBXs.

- Centralized accounting information can be provided by the routed mode gatekeeper. The gatekeeper now has access to all signaling information including call release causes. Gateway-level accounting features can be disabled. The endpoints do not need to be trusted, as the gatekeeper can provide reliable accounting for IP-PBXs or simple IP phones. This enables service providers to provide VoIP business trunking services, replacing traditional E1/T1 lines connected to PBXs with VoIP-enabled broadband connections. With such a service, IP-PBXs do not need a local PSTN gateway in the customer premises: the service provider routed mode gatekeeper is defined as the default route and appears as a regular gateway to the IP-PBX. The only requirement is that the IP-PBX should support H.323 connections toward the public network, but this is the case of most IP-PBXs on the market today.

- Connectivity with third-party networks and customers is secured because the signaling is relayed by the routed mode gatekeeper. It may be useful to use a dedicated gatekeeper for connections with third parties. If it is attacked, the worst that can happen is that connectivity with those partners may be lost, but the rest of the network is not compromised. Note that media streams (RTP) can still flow directly between partners. With proper access lists on edge routers (RTP filters, UDP ports above 1024 only, anti-spoofing filters), this is secure. Some firewall vendors recommend relaying media streams on dedicated devices in core networks; this is very costly, degrades quality of service (added delays), and affects IP network design (tromboning is introduced). These techniques should be reserved for very specific situations (e.g., clearing houses wanting to hide the identity of their partners, or when there are incompatible IP-addressing plans that need to be converted).

Besides resolving all the issues that cannot be addressed with direct mode gatekeepers, routed mode gatekeepers offer many more possibilities. For instance, they can act as multiprotocol softswitches acting both as an H.323 routed mode gatekeeper and as a SIP proxy with access to enough information to convert between signaling protocols (e.g., H.323 and SIP). Note that this requires SIP to support true out-of-band DTMF signaling though **INFO** or **NOTIFY** messages (major SIP gateway vendors already support these messages, but support in IP phones is still scarce).

2.2.4   H.323 calls across multiple zones or administrative domains

As the initial title of H.323 implied, the first version of H.323 did not consider issues that would occur in a wide area environment. It was more or less assumed that the gatekeeper would get a complete view of the network and would be controlling all endpoints and gateways. In this context there was not much effort spent on defining the call flows to be used if the network was controlled by multiple gatekeepers or if multiple VoIP networks were connected. The VoIP industry had to solve this issue very quickly because real VoIP networks required multiple gatekeepers to scale ... and soon a de facto inter-gatekeeper call flow emerged. Without much prior debates in standard bodies.

As we discussed in Section 2.2.3.2, most VoIP networks today still use direct mode gatekeepers. In order to be compatible with direct mode gatekeepers, the de facto inter-gatekeeper call flow uses the **Location Request (LRQ)** RAS message. It is probably not the optimal choice: using the ARQ/DRQ message would have facilitated the correlation between the RAS messages exchanged between direct mode gatekeepers and the Q.931 messages exchanged between endpoints (SETUP, CONNECT, etc.). But this call flow is so widely deployed today that the usage of the LRQ message is not likely to change. What is happening instead is that most vendors are adding (in a proprietary way, within the H.323 extension tokens) the information that is missing in LRQ messages, notably the call identifier: all messages used in H.323 can easily be extended.

Between routed mode gatekeepers, the most efficient call flow is to simply forward Q.931 messages between gatekeepers. This is identical to the call flow used between class 4 central offices in the TDM networks. RAS messages are unnecessary, but can be used if desired: some routed mode gatekeepers will send a Q.931 SETUP message directly to the next hop gatekeepers (this assumes the prior knowledge that the next hop gatekeeper is also a routed mode gatekeeper), and some will begin by sending an LRQ message, in case the next hop gatekeeper is using direct mode only and cannot handle Q.931.

2.2.4.1   Direct call model

2.2.4.1.1   Call set-up

In the direct call model, only RAS messages are routed by the gatekeepers. Now, John wants to call his grandma using a gateway managed by a service provider, Cybercall. The service provider has its own gatekeepers. Therefore, John's terminal and the gateway will be located in different zones. John's terminal will register to his own gatekeeper and the gateway will be registered to the service provider's gatekeeper.

When John became a customer of Cybercall, his gatekeeper IP address was configured in the routing tables of Cybercall's gatekeeper, and vice versa. Therefore, these gatekeepers know about each other. Security is usually based on identifying the IP addresses of both gatekeepers, but can be enhanced by adding security tokens in LRQ messages.

The admission request is sent by John's terminal to the gatekeeper to which it has registered (Figure 2.15). This gatekeeper knows that all calls to the PSTN are handled by Cybercall. Therefore, it sends a Location Request (LRQ) to the gatekeeper of Cybercall, the LRQ message queries the Cybercall gatekeeper for a next hop IP address where the Q.931 signaling can be sent for a specific destination. Because the LRQ comes from a gatekeeper that is known, and assuming John is authorized to make the call, Cybercall's gatekeeper will accept it and returns a **Location Confirm (LCF)** to John's gatekeeper. The LCF message contains the IP address of the gateway where John's terminal should send the SETUP message. John's gatekeeper has still not replied to the initial ARQ, because it did not have enough information to do so. Now, with the IP address contained in the LCF, the gatekeeper knows where the call should be routed and sends this information to John's terminal in an ACF. If this is taking too long, the gatekeeper can send **Request In Progress (RIP)** messages to John's terminal to prevent any timeout that could cause John's terminal to reject the call or resend an ARQ.

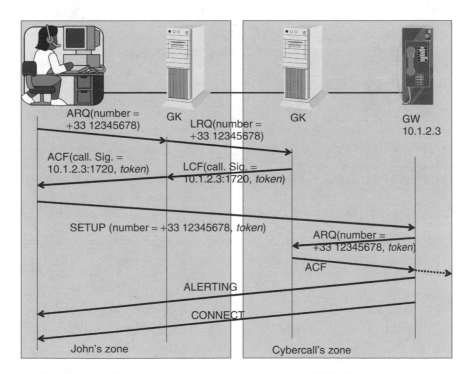

Figure 2.15 Direct call model across two domains, using LRQ/LCF messages.

Cybercall can also include a **token** in the LCF. A token is an optional parameter that consists of a 'bag of bits'. Unless it knows of this specific token, an H.323 entity should simply pass it along transparently. Here, the token serves as a secret which will be copied by John's terminal in the SETUP message. Cybercall's gatekeeper has put in this token a digital signature of some important aspects of the call, such as the destination and the current time. When it receives a SETUP message including this token, the gateway can now verify that the call has been previously authorized by the gatekeeper. However, Cybercall, in order to centralize security management, has not given the gateway enough information to decode and verify the token locally. The gateway will simply pass this token to the gatekeeper in the receive side ARQ, Cybercall's gatekeeper will check it, and return an ACF if the token is correct. Otherwise, the call would be rejected with an ARJ (Admission Reject) message, and the gateway would release the call with a Q.931 RELEASE COMPLETE message.

When it receives the ACF, the gateway will set up the call on the PSTN side and send a CONNECT message to John as soon as Grandma picks up the phone.

John then establishes the H.245 control channel to the gateway using the address and port specified by the gateway in the CONNECT message. Then, logical channels are established using OpenLogicalChannel messages, and John can talk.

2.2.4.1.2   Call tear-down

This time (Figure 2.16) if Grandma hangs up first, the gateway will send an EndSession-Command and RELEASE COMPLETE message to John's terminal, as described in the

Figure 2.16   Call released end to end at H.245 and Q.931 level, locally at RAS level.

first H.323 examples. Then, the gateway sends a DRQ message to Cybercall's gatekeeper, and John's terminal sends a DRQ message to its own gatekeeper. Note that because the LRQ message exchanged between the two direct mode gatekeepers is a stateless query message, no message is exchanged between the direct mode gatekeepers when the call is released. This illustrates the problem arising from the use of the LRQ message, as opposed to the ARQ and DRQ, for inter-gatekeeper communications.

2.2.4.2 *Gatekeeper-routed model*

There are many reasons that Cybercall would like to have finer control over John's communication. With the direct model, Cybercall doesn't know what occurs during the call (e.g., if grandma's phone is busy Cybercall's gatekeeper will see it simply as a very short call). This forces Cybercall to do all accounting at the gateway level, which may be a pain if the Cybercall domain has several dozens of gateways. Cybercall may also want to protect its domain and prevent John from potentially initiating denial-of-service attacks on the gateways; signaling ports. This is impossible to do using the direct model.

These are just a few of the reasons the gatekeeper-routed model—or a mixture of direct and gatekeeper-routed model—will be preferred in most situations where the network involves several administrative domains.

2.2.4.2.1 *Call set-up*

In the example shown in Figure 2.17, Cybercall's gatekeeper decides to route the call by putting its own IP address (10.1.2.2) in the LCF call-signaling address (as we saw in Section 2.2.2.3, this call flow can be optimized using the preGrantedARQ procedure). John's gatekeeper also decides to route the call by putting its own IP address in the ACF call-signaling address. But John's gatekeeper could also have used the direct model by copying the call-signaling address provided in Cybercall's LCF in its own ACF: in this case John's terminal would have sent the set-up message directly to Cybercall's gatekeeper, this would be a call using a mixed model. If John's gatekeeper knows in advance that Cybercall is always using the routed model, then the LRQ is unnecessary and a direct SETUP can be sent to Cybercall's gatekeeper IP address.

You probably remember that one of the most important information elements of the ALERTING or CONNECT message is the H.245 call control channel address that John's terminal must use to establish the call control channel. Here, the H.245 channel will also be routed because both Cybercall's and John's gatekeepers have put their own IP addresses in the call control transport address field of the ALERTING message. It is also possible to route the Q.931 messages but let the H.245 control channel be established directly between the endpoints.

What about the media channels? They could be routed too, but there would be very little to gain from doing so, since all the significant events of the call are signaled using H.245 or Q.931 messages.[2] But unless there is a very specific need to do so, *media channels*

---

[2] An exception could be fax, because the entire T.30 protocol is encapsulated in a media channel; therefore, the gatekeeper needs to have access to the media channel to know how many pages have been transferred.

Figure 2.17   Routed call model across two domains.

flow directly between the endpoints, even if the gatekeeper-routed model is used. Doing otherwise and routing the media channels would remove most of the scalability benefits of VoIP over TDM.

By letting media streams flow directly between endpoints, the media latency is optimized, even if call-signaling has to go through many gatekeepers, because IP shortest path routing protocols will be used to route RTP packets. This gives VoIP a very interesting 'location-independent' property, which allows customers to be served from remote gatekeepers, thereby reducing the number of points of presence required to offer the service. The 'Voice for IP VPN' service from service provider Equant, which serves over a hundred multinational companies over a VoIP network, operates from only two VoIP gatekeepers, one located in the US, one in Europe.

Some service providers are concerned about security issues that could occur using the RTP stream. Although most VoIP networks worldwide let RTP flow through transparently, we have never heard of such problems. In order to secure such a VoIP network the following protection should be configured:

- **Access Control Lists (ACLs)** on edge routers should allow VoIP signaling information only toward the routed mode gatekeeper.

- Other ACLs should allow only UDP RTP traffic to ports higher than 1024^3 (checking the proper RTP patterns in UDP packets is possible on most routers) and only to VoIP endpoints (the easiest is to allocate well-defined subnets to the gateways).

- The only possible attack is denial of service, because RTP doesn't have much logic in it! Gateways are expecting a lot of traffic on RTP ports, so bringing them down with RTP traffic requires significant bandwidth, making the attack detectable. Furthermore, gateways will accept the RTP traffic only if the logical channel has been opened properly by the routed mode gatekeeper ... in this case the identity of the attacker is known, which acts as a determent to such attacks. The last remaining possibility is a DoS attack on closed UDP ports, causing the gateway to reply with ICMP IP-level error messages. Filtering these can give an early warning of the attack; once again, the attack would require significant bandwidth as sending back an ICMP message is not CPU-intensive on the gateway.

- As in any IP network, anti-spoofing (preventing anyone from injecting in the network packets with the source IP address belonging to someone else) should be taken very seriously, as it is the only real protection against DoS attacks.

- Finally, because we are only expecting RTP traffic and know what bandwidth to expect, if per-flow traffic policing is available on the edge routers, it should be used. DoS attacks will exceed the allowed bandwidth and be rejected by the edge routers.

If you still want to relay media streams, devices exist that do just this at the edge of a network ('border session controllers'). But, by forcing RTP packets to go through these devices without care, you may significantly reduce the QoS of the VoIP network (e.g., if the device is in New York, a San Francisco to San Jose call may have its streams relayed through New York, instead of flowing directly between the two cities using IP shortest path routes).

2.2.4.2.2   Call tear-down

The call tear-down is very similar to the direct model case, except of course that Q.931 messages and optionally H.245 messages are routed through the gatekeepers.

2.2.4.2.3   More LRQ usage scenarios

When a gatekeeper is used at the interface between two administrative domains, LRQ call flows can be more complex. Gatekeepers at the edge of a domain need to manage:

- Multiple simultaneous LRQ targets.
- The sequencing of LRQ and Q.931 messages.

---

[3] Blocking ports below 1024 makes unreachable most applicative ports that could potentially be opened, and subject to attack. VoIP applications use ports higher than 1024.

Figure 2.18   Call release scenario, under the routed model.

2.2.4.2.3.1   LRQ blast

If a call to a certain destination can be sent to multiple termination networks, each with its own gatekeeper, it may be interesting to check the availability or willingness to accept the call of all partners. In order to do this, multiple LRQ messages can be sent (simultaneously, or in sequence), to all these potential termination networks. This is sometimes called an **LRQ blast**. Among the LCFs received, one will be selected by the source gatekeeper.

Note that it is tempting to do the same with SETUP messages (some SIP vendors do this with the INVITE message[4]), but only the sending of multiple SETUPs in *sequence* is allowed. A call establishment message should never be duplicated. This is because the PSTN network can send announcements before CONNECT (200 OK in SIP). If multiple calls receive network announcements, the softswitch would be unable to properly relay them to the caller.

---

[4] When using SIP, duplicating INVITE messages should be allowed *only* if the expected answer is a redirect message or the expected media is not voice. Some vendors use it for a 'simultaneous ringing' feature . . . although this is a cool demo, it simply does not work on real telephony networks. For this reason the 3GPP Group defining the UTMS 3G standard has decided not to use SIP forking for now.

2.2.4.2.3.2 Proper LRQ sequencing

When an edge routed-mode gatekeeper at the interface between several administrative domains receives an LRQ, it can choose between the following call flows:

(1) Reply immediately with an LCF, then receive the SETUP message from the calling device, then send an LRQ message to the potential termination zones, then forward the SETUP to the selected termination device.

(2) Forward the LRQ to the potential termination zones, wait for the LCF for the termination softswitches that will accept routing the call, and only then reply with an LCF with its own IP address. When the edge gatekeeper receives the SETUP, it routes it to the selected softswitch.

Both call flows seem equivalent, but they are not when *all* potential termination gatekeepers reject the call.

In call flow (1), the SETUP has already been routed to the edge gatekeeper, so the edge softswitch is responsible for finding a fallback route for the call.

In call flow (2), the edge gatekeeper can reject the call by sending back an LRJ (Location Reject) to the calling party gatekeeper. This gives the possibility of rerouting the call to the initiator.

Call flow (1) reduces call latency, but may not be appropriate if a service provider connected to the edge gatekeeper wants to keep the possibility of rerouting calls. This is the case with most clearing houses.

Call flow (2) solves this issue, but introduces more latency in the call.[5]

In any case, both call flows are useful, and a gatekeeper used as an edge device should offer the possibility of choosing between the two modes for each route.

2.2.4.2.4 Some issues with the LRQ message

In the first edition of this book, the calls flows for interdomain calls were not stable, and it was anticipated that the ARQ message, not the LRQ, would be used between gatekeepers. Unfortunately, the first implementation of the call flow by Cisco Systems used the LRQ, and then the whole industry followed. There are mainly two information elements which are missing in the LRQ message that would really be useful: a call reference identifier and a hop counter.

2.2.4.2.4.1 The missing call reference identifier

The LRQ misses a unique call reference identifier, typically the CallID. This is the main difference between an LRQ and an ARQ. The absence of this unique call reference number makes it impossible to correlate the LRQ and the subsequent SETUP message. There are many cases where this correlation would be useful. For instance, when a routed mode

---

[5] In addition, because the LRQ/LCF is stateless (no resource is reserved when replying with an LCF), it should also reinitiate an LRQ when the SETUP arrives. This doubles the number of LRQs. LCFs could be cached for a short time, but this is a violation of the standard.

gatekeeper is used as an edge element between multiple domains (clearing house function), then the owner of the clearing house would like to be able to easily identify each connected domain in the CDRs generated by the gatekeeper. The CDRs generated from the SETUP messages will include the source IP address of the device that initiated the SETUP to the edge gatekeeper and destination IP addresses of the SETUP message sent from the edge gatekeeper. If each connected domain also uses a routed mode gatekeeper, then these IP addresses will be the IP addresses of each gatekeeper, and they allow easy identification of the administrative domains. But many VoIP service providers still use the direct model. In this case the IP addresses will be those of the PSTN gateways. It is very time-consuming to keep track of all these IP addresses and correlate them to a service provider. Since the direct mode gatekeeper of each service provider will send an LRQ before each call, it would be nice to include the LRQ source IP address in the CDRs. Unfortunately, because the LRQ cannot be correlated to the SETUP message, this is impossible. Another case where the presence of a call identifier in an LRQ would be useful was described in Section 2.2.4.2.3.2. If the edge gatekeeper is required to completely proxy the LRQ message before accepting the SETUP message, then two LRQ messages will be generated for each call, because the LRQ is stateless. Correlating the LRQ to a specific call would make it easier to keep the LCF response of edge domains in cache, knowing that these edge domains can now reserve resources for the coming call.

2.2.4.2.4.2   The missing hop counter. Discussion of call loops

The second element that would be useful in an LRQ message is a hop counter, to prevent loops in the VoIP domain. Note, however, that this is nothing more than a useful tool, because it would still be possible to loop calls using SETUP messages without RAS and also because call loops can include a PSTN hop that would reset the counter. The only way to completely prevent loops in VoIP networks is to not only include a counter in LRQ but also SETUP messages, and to take into account the SS7 ISUP hop counter if there is a PSTN hop. This is possible if the edge gateways support the H.246 encapsulation of ISUP information, or H.323 annex M2, or H.323v5, which adds such a hop counter to standard SETUP messages. Cisco Systems also proposed a mechanism called **Global Transparency Descriptor (GTD)**, where ISUP national information elements are passed and stored in a uniform way within a data structure in H.323 Q.931 messages. GTD is much more powerful that H.246 (or its SIP equivalent SIP-T) because it proposes a uniform coding of the ISUP information, as opposed to transporting national ISUP 'flavors' as is. If the proposal becomes a standard it will certainly be the best way to address the loop problem, among many other interworking issues. Even with these improvements, call loops remain possible if edge devices connected through user interfaces (analog, ISDN) are allowed to loop calls back to the network, because in this case the hop counter is reset. This is one of the reasons the call forward of external calls back to the PSTN is usually forbidden as part of the certification program of edge devices.

2.3 OPTIMIZING AND ENHANCING H.323

2.3.1 Issues in H.323v1

2.3.1.1 Call set-up time

One of the major weakness of H.323 in its version 1 was the time required to actually establish the media channels for a new call. Even in the simple cases, H.323v1 procedures involve:

- One message round trip for the ARQ/ACF sequence.
- One message round trip for the SETUP-CONNECT sequence.
- One message round trip for the H.245 capabilities exchange.
- One message round trip for the H.245 master slave procedure.
- One message round trip for the set-up of each logical channel.

This looks bad enough, but the real situation is even worse since the Q.931 and H.245 channels use TCP connections which must also be set up.

Each TCP connection needs an extra round trip to synchronize TCP window sequence numbers (Figure 2.19). In a WAN environment where each round trip may take several hundred milliseconds, this can lead to unacceptably long set-up delays, especially when using the gatekeeper routed-call model where a TCP connection needs to be established between each gatekeeper.

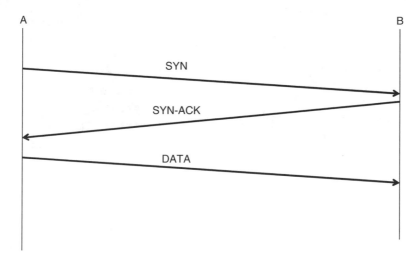

Figure 2.19 TCP connection three-way handshake.

2.3.1.2 The TCP slow-start issue

The use of TCP causes at least one unnecessary round trip due to the SYN/ACK hand-shake. In fact, the situation can be worse if the SETUP message is larger than the maximum transmission unit (MTU) or if the first segment is lost.

If the SETUP message is larger than the MTU, the sender must send the SETUP in two or more TCP segments. The problem is that most TCP implementations are designed to be friendly to the network and follow RFC 2001, which mandates a slow-start procedure (Figure 2.20). In our case, after sending the first segment, the sender must wait until it has received an ACKNOWLEDGE before transmitting the next segment. Only then it can increase the window size and send two segments at once.

Because of this, large SETUP messages may cause one additional round trip. A good practice is to try to limit the size of the SETUP message below 576 octets, as this is the minimal IPv4 MTU, or at least below 1,500 octets (the Ethernet MTU), but this may not always be possible.

During an active TCP connection, the TCP stack dynamically estimates the round trip time (RFC 793) and uses this value to detect lost packets. But, for the first segment, TCP starts by using a worst case estimate. For the initial connection, the timeout value is normally set to the average round trip time A (initialized to 0 seconds), plus twice the deviation D (initialized to 3 s). If the first segment is lost, most TCP implementations

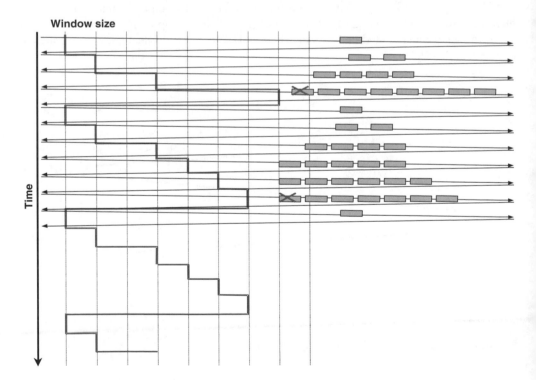

Figure 2.20 TCP slow-start and back-off after packet loss.

will therefore wait for up to 6 s (retransmission timeout $= A + 2D$) before retransmitting the first segment. Some operating systems (e.g., Microsoft Windows®) only allow this value to increase (this is used for satellite connections). Once the first segment has been lost, things get even worse. The RTO (retransmission timeout) after the first segment is calculated as $A + 4D$ (12 s with the default values), and after each segment loss the RTO is doubled. Therefore, if the first segment is lost and retransmitted, the timeout value for this second segment will be 24 s! If the first two attempts fail, there will be 30 s between the first of the call set-up attempts, and the third retransmission.

Gatekeepers, gateways, and IP phones should tune the TCP stack settings:

- To send signaling TCP packets on a higher quality of service (e.g., using DiffServ, see the companion book, *Beyond VoIP Protocols*) in order to minimize packet loss.

- To aggressively retransmit TCP segments. This can be achieved by lowering the initial setting for D for the connection establishment and lowering the 500-ms granularity of the RTO used on many operating systems. The maximum value of the RTO before the connection is lost—64 s—should also be lowered. On Linux, the number of retransmissions can be controlled with the/proc/sys/net/ipv4/tcp_syn_retries and/proc/sys/net/ipv4/tcp_synack_retries parameters.

- To increase the initial window size of the TCP slow-start mechanism up to two to four segments instead of one.

- To disable any buffering of information before packets are sent. This is Nagle's algorithm, which was designed to optimize transmissions for keyboard input; it delays transmission of a packet until a sufficiently large transmission buffer is accumulated, or 200 ms have elapsed. This can be done by setting the TCP_NODELAY socket option.

- To reduce the number of failures that are necessary to notify the application that 'something is wrong', or to clear the faulty socket. On Linux, this is configured by setting the/proc/sys/net/ipv4/tcp_retries1 and/proc/sys/net/ipv4/tcp_retries2 parameters, respectively.

- To reduce the time the gatekeeper waits for client confirmation when it closes a socket. This avoids accumulating sockets in the half-closed state. Most operating systems will wait up to 7 RTOs before closing the socket, which can exceed 15 min! On Linux, this is the/proc/sys/net/ipv4/tcp_orphan_retries and/proc/sys/net/ipv4/tcp_fin_timeout parameters. This setting is extremely important for VoIP networks with PC based softphones, due to the number of occurrences of abrunt disconnections of PCs (crashes, physical disconnections, loss of modem connections due to a call-waiting signal, etc.). Each 'orphan' connection also uses memory, which can lead to relatively simple denial-of-service attacks (the number of orphan connections in Linux can be controlled by the/proc/sys/net/ipv4/tcp_max_orphans parameter).

- To reduce the amount of time the gatekeeper waits for an acknowledgment of sent data if the socket has been closed (typical problem when sending the RELEASE COMPLETE message). This requires using the SO_Linger option (disabling linger or using a small linger timeout): after the timeout, if the ACK of the sent data has not been received (graceful close), the socket is forcibly closed with a TCP RST packet.

- Try to use the selective acknowledge feature of some TCP stacks. This option (SACK) is negotiated during the three-way handshake. On Linux, this is the/proc/sys/net/ipv4/ tcp_sack option. This helps to speed up recovery of lost fragments and avoids retransmitting segments that have not been lost.[6]

This tuning makes the performance of TCP comparable with the performance of UDP-based retransmission schemes on most VoIP networks.

If TCP tuning is not enough in specific cases, H.323v3 introduced a new transport mode allowing the use of UDP signaling instead of (or simultaneously with) TCP signaling. This is in annex E of the standard. In H.323v4, a new possibility was introduced of keeping the TCP channel open and reusing it for the H.225 signaling of multiple calls. This option was designed to facilitate the design of routed mode gatekeepers and large-scale gateways on operating systems that have a low limit to the maximum number of TCP connections, or have scalability problems on the 'poll()' function required to detect incoming events on multiple sockets (this is a well-known issue with Linux, but it can be solved by using a modified version of the poll function). This option is not widely implemented and should be considered with great care because it causes head-of-line blocking (i.e., if one call is blocked for any reason, no further event related to other calls will be transmitted). Overall, the option of using multiple TCP sockets is much more robust and should be preferred.

The best solution is to use the new SCTP, which is optimized for telecom applications and offers the best of UDP and TCP simultaneously . . . this option has just been introduced in H.323v5!

2.3.1.3   Network-generated prompts

Another problem was discovered by network experts after H.323v1 had been standardized. In the switched circuit network there are situations in which a message is played to the caller before it receives a connect:

- When the SCN is congested and the call cannot be established, you can get a prompt saying 'Due to congestion, your call cannot be connected, please call later'. This prompt is generated by the network itself at the local exchange, and, because it does not originate from the called endpoint, no CONNECT message is ever sent.

- In some applications the Intelligent Network can also generate network messages (e.g., for televoting applications: you dial a phone number and you get back a message saying 'Thank you for voting YES, the current status is 34% YES, 66% NO'). Similar pre-connect prompts are also used in many countries to implement prepaid calling card services in the network: the destination number and the PIN code are requested before connect, and the Q.931 CONNECT message is sent only when the call connects to the final destination.

---

[6] the default TCP ACKNOWLEDGE is cumulative, which means all the packets since the last acknowledged packet will be retransmitted even if only one segment has been lost.

With H.323v1, it is impossible to send a voice message to the calling party before sending a CONNECT message, because the media channels are not yet established.

2.3.2 The 'early H.245' procedure

The early H.245 procedure is used when the H.323 SETUP message contains an H.245 address, which is available to the called party if it wants to start connecting to the H.245 channel immediately. Alternatively, if the calling party has not proposed an H.245 connection address, the called party can make its own proposal by including an H.245 address in the call control messages sent before the CONNECT message (CALL PROCEEDING, ALERTING). The early H.245 helps the H.245 procedure to start as early as possible in the call, sometimes even before it actually connects. In most cases, unless the call connects right away, it makes the inherent delays due to the multiple message exchanges required by H.245 invisible to the call participants. It also solves the network prompt before connect issue explained in Section 2.3.1.3. All this while preserving full compatibility with all H.323 features, including out-of-band DTMF transmission, third-party call control using the TCS = 0 procedure (see more details see Section 2.7.1.2.2), and sophisticated video and conference control procedures.

The early H.245 procedure is so useful that it is should be one of the most important criteria for the selection of any H.323 equipment. This call flow should be used whenever possible.

2.3.3 The 'fast-connect' procedure

The fast-connect procedure was introduced in H.323v2 to enable unidirectional or bidirectional media channels to be established immediately after the Q.931 SETUP message and eliminate any post-connect delay in the audio path.

The usefulness of the fast-connect procedure is questionable, as the early H.245 procedure (described in Section 2.3.2) also solves these issues. In the early days of H.323 there was still some confusion on which was the best method to solve the delay issue, and all possible solutions were welcome. Fast connect has one little advantage over early H.245: it removes any post-connect audio delays even in the case of an immediate call connection. It also has major drawbacks compared with the early H.245 procedure: for instance, it does not allow the out-of-band transmission of DTMF information[7] and it does not provide a third-party call control feature before H.323v5 (this version adds this possibility in the H.460.6 extension). It is tempting to use both early H.245 and `fastStart` (see next paragraph) at the same time, which in fact many vendors are currently doing. Since H.323v4, this is officially possible, but the actual H.245 communication should not transmit anything but the endpoint capabilities and master/slave information before the completion of the initial fast-connect negotiation.

---

[7] RFC 2833 may be used, but does not allow feature servers to act on DTMF commands without also relaying the media stream.

An endpoint which decides to use the fast-connect procedure will include a new parameter, called `fastStart`, in the SETUP user-to-user information element. This parameter includes a description of all the media channels that the endpoint is prepared to receive and all the media channels that the endpoint offers to send. This description includes the codecs used, the receiving ports, etc.

If the called endpoint cannot or doesn't want to use the fast-connect procedure, it will not return the `fastStart` element in subsequent Q.931 messages. In this case the normal procedure involving H.245 will take place.

If the called endpoint supports the fast-connect procedure, then it will return, in the CALL PROCEEDING, PROGRESS, ALERTING or CONNECT Q.931 message, a `fastStart` element that selects from among the media offered by the caller.

The `fastStart` element is always inserted in the H323-UU-PDU of the user-to-user element (its use with the SETUP message is shown in bold):

- Protocol discriminator field (08H).
- Call Reference Value (CRV).
- A Message Type (SETUP).
- ...
- Called party number and subaddress.
- Calling party number and subaddress.
- The H.323 user-to-user element which contains the SETUP user-to-user information element in which we find:
  - The protocol identifier.
  - ...
  - The sender's aliases.
  - The destination address.
  - The CID and CallID.
  - `fastStart`: used only in the fast-connect procedure, `fastStart` is a sequence of **OpenLogicalChannel** structures. Each **OpenLogicalChannel** structure (Figures 2.21) describes *One* media channel that the caller wants to send (**forwardLogicalChannelParameters** within the **OpenLogicalChannel** structure) or receive (**reverseLogicalChannelParameters**). All proposed **OpenLogicalChannels** can be selected simultaneously, unless they share a common **sessionID** value in the **H2250-LogicalChannelParameters** of the **OpenLogicalChannel** structure, in which case they are considered alternative options for the same channel.
  - The **mediaWaitForConnect** boolean.

The calling terminal can select one or more acceptable **OpenLogicalChannel** structures within the offered `fastStart` parameter and return them in a `fastStart` parameter within an H.225 CALL PROCEEDING, PROGRESS, ALERTING, or CONNECT message. The selected logical channels are considered open after this.

```
OpenLogicalChannel ::=SEQUENCE
{
        forwardLogicalChannelNumber    LogicalChannelNumber,
        forwardLogicalChannelParameters SEQUENCE
        {
                portNumber  INTEGER (0..65535) OPTIONAL,
                dataType    DataType,
                multiplexParameters  CHOICE
                {
                        h2250LogicalChannelParameters H2250LogicalChannelParameters,
                        none        NULL
                },
                ...,
                forwardLogicalChannelDependency    LogicalChannelNumber OPTIONAL,
                replacementFor        LogicalChannelNumber OPTIONAL
        },

        reverseLogicalChannelParameters  SEQUENCE
        {
                dataType    DataType,
                multiplexParameters    CHOICE
                {
                        h2250LogicalChannelParameters H2250LogicalChannelParameters
                } OPTIONAL, -- Not present for H.222
                ...,
                reverseLogicalChannelDependency  LogicalChannelNumber OPTIONAL,
                replacementFor        LogicalChannelNumber OPTIONAL

        } OPTIONAL, --Not present foruni-directional channel request

        ...,
        separateStack        NetworkAccessParameters OPTIONAL,
        encryptionSync       EncryptionSync OPTIONAL -- used only by Master

}
```

Figure 2.21 OpenLogicalChannel ASN.1 structure.

Note that the network can send media to *any* of the receiving channels mentioned in the SETUP message of the caller, immediately after the calling terminal has sent this message, unless MediaWaitForConnect is true. Therefore, even if the calling terminal plans to use only one of these channels for regular conversation, as indicated in the fastStart response, it *must* be prepared to receive media on any one of these channels (before the response). Although most H.323 vendors have implemented the fastStart procedure, many of them actually do not support this requirement and are not able to receive audio before the remote endpoint has selected a media channel proposal in its own fastStart element. This is because most implementations do not have enough memory to load multiple voice coders at the same time, and the vendor selects the right coder to load when it receives the remote fastStart. In the best implementations that fully support fastStart, the first RTP packet that is received can be used to load the codec, without waiting for the remote fastStart element. An example of a case where supporting the full fastStart requirement is important is network-based redirection announcements: multiple announcement servers may have to send audio to the calling endpoint, before it connects to the party redirected to. In this case the fastStart element will be sent only by the last endpoint, but the announcement servers still need to stream audio toward the calling party before this happens.[8]

---

[8] Some vendors can receive multiple fastStart elements in sequence and always take into account the last one received. This greater flexibility makes it possible to activate media reception

Usually, in a normal ISDN call the called endpoint does not send media before the CONNECT message has been sent. It is possible to force this behaviour with the `fast-Start` procedure by setting the `mediaWaitForConnect` element of the Q.931 SETUP to true.

As shown in Figure 2.22, the fast-connect procedure dramatically improves the number of round trips required to set up the conversation, and eliminates all post-connect audio delays.

Since the fast-connect procedure solves a major flaw of H.323v1 regarding interworking with the traditional TDM networks, it has been made a core feature in the H.323 profile of the ETSI TIPHON project.

Fast-connect procedure usage becomes a bit subtle in certain circumstances (e.g., when H.323 is used to provide class 5 services). The interactions between the call forward on no answer service and the use of `fastStart` are extremely complex, and not well studied by the standard (see Chapter 5). In addition the `mediaWaitForConnect`Boolean is a bit too simple to fully account for the variety of call flows found in the PSTN. The sending of audio information before connect is controlled by the in-band audio indicator in Q.931 messages, including the PROGRESS message. Some call flows can become extremely complex, as in the following example where ring-back tones alternate with redirection prompts.

Such call flows today are possible (the previous scenario is currently used in Milan, Italy), but need prior vendor agreement on the exact handling of **in-band audio indicators**. Also, some call flows really would require the sending of multiple `fastStart` elements to update the RTP logical channel information before CONNECT. H.460.6 has refined the `fastStart` procedure and allow the refreshing of `fastStart` elements before CONNECT.

Figure 2.22   Audio path delays with the fast-connect procedure.

---

only after having received a `fastStart` response, without restricting the feasible services. Unfortunately, this way of interpreting the `fastStart` response, which creates a form of third-party media control, is not yet taken into account in the standard.

Using just the fast-connect procedure, DTMF transmission is not possible as H.245 UserInputIndication is not available (no H.245 channel is opened). Because of this limitation, early H.245 (establishing the H.245 channel before CONNECT) should be used in conjunction with `fastStart` in most cases, as it is officially allowed since H.323v4. Many types of calls will require DTMF information before CONNECT. For instance, this is what would happen typically when calling an IN-based card telephony service. An IVR server requests the card code before sending a CONNECT Q.931 message because the call is not yet charged. The CONNECT message is sent only after the code has been checked and the destination party has answered the call (as shown in Figure 2.23).

Caveat: The addition of fast-connect mode to H.323 has made it possible to manufacture a simpler, yet H.323-compliant terminal. In fact, this was one of the goals of the initial `fastStart` proposal, at a time when SIP began to claim simplicity compared with H.323. By supporting only H.323 in fast-connect mode, developers can avoid implementing H.245 in simple appliances like IP phones in this way. However, most of the potential of H.323 comes from the conferencing and third-party call control features which are enabled only by H.245. Simple H.323 terminals without H.245 will not be able to participate in such conferences. Moreover, DTMF is normally carried using H.245: simple terminals will have to use in-band DTMF coding which does not work as soon as complex services like prepaid servers or contact centres are implemented. Overall, such 'simplified' terminals would not meet even the basic requirements for telephony and would make the design of services on networks with such endpoints extremely challenging. Interestingly, facing the same issues, SIP has become significantly more complex over time, notably adding third-party call control and out-of-band DTMF capabilities to the basic call, and is now virtually identical to the H.323 protocol, with `fastStart` and H.245 tunneling enabled.

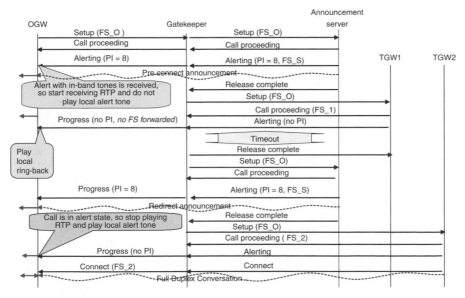

Figure 2.23 Complex call flow with pre-connect audio. Note: there is no DTMF capability before connect.

Figure 2.24 Use of a PROGRESS message for pre-connect audio.

2.3.4 H.245 tunneling

Most H.323 devices today use two separate TCP connections for each call: one for the Q.931 messages (SETUP, ALERTING, CONNECT, etc.), and one for the H.245 messages (OpenLogicalChannel, TerminalCapabilitySet, etc.). This may become a problem for some gatekeeper or gateway implementations that run on operating systems with low limits on TCP connections and that do not use distributed designs. It also unnecessarily doubles the issues associated with the use of TCP.

H.323v2 offers a way of using a single TCP connection by encapsulating H.245 messages in Q.931 messages: this is called H.245 tunneling.

An endpoint which wants to use H.245 tunneling must set the h245Tunneling element of the SETUP message and all subsequent Q.931 messages to TRUE. A called endpoint also indicates its willingness to accept H.245 tunneling by setting this same element to TRUE in all Q.931 messages.

The calling endpoint simply encapsulates one or more H.245 messages in the h245Control element of any Q.931 message. If the called endpoint is also capable of receiving it, all H.245 messages can be exchanged in this way and there is no need to open a separate TCP connection for the H.245 channel. Otherwise, if the called endpoint has not set the h245Tunneling to TRUE in the first Q.931 message it sends back (it could be CALL PROCEEDING, PROGRESS, ALERTING, or CONNECT), the calling endpoint knows this is not supported and the normal procedure for opening an H.245 channel is followed. Q.931 messages are modified as shown in bold:

- Protocol discriminator field (08H).
- Call Reference Value (CRV).
- A Message Type (SETUP, ALERTING, . . .).
- . . .
- The H.323 user-to-user element (H323-UU-PDU) now contains:
  - The **H245Tunneling** boolean.
  - **H245Control**: a sequence of ASN.1 octet strings representing ASN.1 PER-encoded H.245 PDUs. The H323-UU-PDU type can be the usual SETUP, CALL PRO-CEEDING, CONNECT, ALERTING, USER INFORMATION, RELEASE COM-PLETE, FACILITY, or PROGRESS, or a new NULL value called *empty*, which is explained later.

When using H.245 tunneling the Q.931 channel needs to remain open during the entire duration of the call. If an H.245 message needs to be sent when no Q.931 message is pending to be sent, then the H.245 message will be encapsulated in a Q.931 FACILITY message. In this case the Q.931 FACILITY message is sent, but the H.323 user-to-user element only contains the H245Tunneling boolean and the H.245 PDUs encoded in H245Control: in the next paragraph the H323-UU-PDU type isn't the usual FACILITY type, but is set to the new 'empty' value. Such a need to use FACILITY messages may also occur in the gatekeeper-routed model (Figure 2.25).

Figure 2.25 Use of FACILITY messages for H.245 tunneling.

A terminal can signal that it wants to use H.245 tunneling in a Q.931 message which already contains a fastStart element with OpenLogicalChannels parameters by setting H245Tunneling to TRUE. In H.323 prior to version 4, it was forbidden to encapsulate H.245 messages in the same Q.931 messages as the fast-connect procedure (as they would supersede the indications found in the OpenLogicalChannels parameter). This was clarified in version 4, where capability messages and master/slave messages can be sent during the fast-connect negotiation. All other H.245 messages must wait until the fast-connect exchange has occurred.

2.3.5   Reverting to normal operation

In some cases a terminal using fast-connect and/or H.245 tunneling may need to use a separate H.245 control channel in the middle of an established call (e.g., when a terminal that has opened an audio connection in fast-connect mode needs to open a new media channel). In this case the terminal can send a FACILITY message to the other terminal indicating it wishes to establish a separate H.245 channel and proposing a transport address for it. The terminal which receives the facility message must establish a new TCP connection for the H.245 channel using this transport address. Once the new connection is established, the terminals must stop using the H.245 tunnel.

2.3.6   Using RAS properly and only when required

Most tutorials on H.323 initially introduce the direct call model using RAS. This tends to lead people to believe that RAS messages are a necessary overhead. This is not always the case, especially when using routed mode gatekeepers. It is important to understand the exact role of RAS messages and when they are redundant.

2.3.6.1   Uses of ARQ and LRQ messages

2.3.6.1.1   ARQ and LAN access permission

The ARQ is used to request an access to the network. Once the connection is open, the terminal may use a BRQ to request more network resources when it opens new logical channels. It does the same to accept new logical channels. H.225.0 states: 'As part of the process of opening the channel, before sending the open logical channel acknowledgment, the endpoint uses the ARQ/ACF or BRQ/BCF sequence to ensure that sufficient bandwidth is available for the new channel (unless sufficient bandwidth is available from a previous ARQ/ACF or BRQ/BCF sequence).'

Since the ARQ is used to request access to the network, a calling endpoint, once it has sent an ARQ, is expected to send a SETUP. The CRV parameter can be used to link the two messages (e.g., within a gatekeeper routing the call). H.225.0 states for the SETUP message: 'If an ARQ was previously sent, the CRV used here shall be the same.'

2.3.6.1.2   Address resolution

Both the ARQ and the LRQ messages can be used for the address resolution function (i.e., obtaining a destination IP address from a destination). The name 'location request' seems to imply that the LRQ message is the right message to use, but when comparing the semantics of the LRQ and the ARQ, it becomes obvious that the ARQ can also serve this purpose. In fact, an ARQ is a superset of an LRQ, requesting not only an address translation but also LAN access. This is good because, otherwise, terminals might need to first do an LRQ, then an ARQ, and finally a SETUP! Note that this address resolution can be:

- A pure alias to transport address resolution (e.g., from an email alias the gatekeeper indicates the IP address to send the SETUP).
- An alias translation, if the terminal specifies it, can map aliases (canMapAlias parameter of the ARQ); in this case the gatekeeper can also force the terminal to change the destination alias in the SETUP.

2.3.6.1.3   Conclusion

Usually, a terminal needs to do an address resolution just before it sends a SETUP. In this case the ARQ can be used to request LAN access as well as an address resolution, and thus the LRQ is redundant. This is reflected by the fact that H.225 mandates the support of ARQs by H.323 endpoints, whereas the support of LRQs is optional. Overall, the LRQ message is not very useful, except that it is now the de facto standard for inter-gatekeeper call flows (the ARQ would have been more appropriate here too). The most legitimate use of the LRQ message is when a call control gatekeeper needs to query an access gatekeeper for the current location of an endpoint. Such a 'split gatekeeper' architecture is used in networks with hundreds of thousands of endpoints, in order to distribute the registration function (see Section 2.3.6.1.1 for more details).

2.3.6.2   Disabling the ARQ/ACF sequence

If the gatekeeper is used in routed mode, it has the possibility of authorizing or blocking the call when it receives the SETUP message, and, since it has access to all aspects of call signaling, keeps complete control over the call for the entire duration of the conversation. In this case, the ARQ message is really unnecessary and only adds an extra round trip to the call SETUP delay. The gatekeeper can instruct an endpoint to send a SETUP directly, without a prior ARQ/ACF, by using the **preGrantedARQ** parameter that is contained in the RCF (registration confirm) message.

If preGrantedARQ is not configured, the terminal is required to send an ARQ to the gatekeeper before each call SETUP.

If preGrantedARQ is configured in the RCF, the gatekeeper can give one or more of the following privileges to the terminal:

- Initiate a call without first sending an ARQ.
- Initiate a call without first sending an ARQ, but *only* if it sends the SETUP to the gatekeeper.

- Answer a call without first requesting permission from the gatekeeper.
- Answer a call without first requesting permission from the gatekeeper *only* when the SETUP message comes from the gatekeeper.

Using preGrantedARQ is an excellent way to optimize call SETUP time. In most end-points, the ARQ can also be disabled manually using the configuration tool—sometimes indirectly (e.g., on some endpoints the routed mode gatekeeper should be declared as a 'gateway').

2.4   CONFERENCING WITH H.323

2.4.1   The MCU conference bridge, MC and MP subsystems

There are two distinct functions that may be present in any conference. The first one is the control function, which decides who is allowed to participate or not, how new participants are introduced in existing conferences, how the participants synchronize on a common mode of operation, who is allowed to broadcast media, etc. This role is assumed in H.323 by a functional entity called the **multipoint controller (MC)**.

When many people talk in an audio conference, they might simply multicast their audio, and all terminals can do the mixing of individual media streams themselves. In most cases, however, individual terminals will have limited capabilities, or it might be impractical to multicast all media streams (especially in the case of video). If multicast cannot be used, an entity in the network needs to do the mixing or switching of incoming media streams, and send only the resulting processed outgoing stream to each terminal. In the case of video it can be the image from the last active speaker, in the case of audio each terminal will receive a stream resulting from the addition of all streams from other speakers in the conference (plus some of its own, but attenuated). In H.323, this functional entity is called the **multipoint processor (MP)**.

A dedicated callable endpoint, which contains an MC and optionally one or more MPs, is called a **multipoint control unit (MCU)**. It is not just MCUs that have the MC functionality, however, a terminal or gatekeeper with sufficient resources can also have the capability to act as an MC and may be able to do some media mixing locally. However, the MC functional entity in a terminal or a gatekeeper cannot be called directly, but will be included in the call when it becomes multiparty.

A conference is called a **centralized conference** when a central MP is used to mix or switch all media streams for participating endpoints. When each terminal sends its media streams to all other participating terminals (in multicast or multi-unicast), it is called a decentralized conference.

2.4.2   Creating or joining a conference

2.4.2.1   Using an MCU directly

Most of the time, people will decide to create a conference and name it, for instance, myconference@conferencerooms.com. So, the participants know from the beginning that it is going to be a conference call. The easiest way to create such a conference is simply to call an MCU and send a SETUP message with the `conferenceGoal` parameter set to `Create` and a globally unique CID. It may also include the alias of the conference (myconference@conferencerooms.com). So far, nothing differs from a regular call.

If the MCU decides to accept the call (this can be based on previous reservation done through a website), it replies with a CONNECT.

The endpoint and the MCU exchange their TerminalCapabilitySets. Then the master–slave procedure begins, the MCU always wins and becomes *the* active MC of the conference. The MCU indicates it is the active MC of the conference by sending a MCLocationIndication message to the calling endpoint. It can also assign an 8-bit number to the terminal with a terminalNumberAssign message (the terminal must copy those 8 bits in the low 8 bits of the SSRC field of all its RTP datagrams).

2.4.2.1.1 Inviting new participants

Once a terminal is in a conference, it may invite others (e.g., terminal C) to participate by sending a SETUP message to the active MC with a new CRV, the CID of the conference and the `conferenceGoal` parameter set to `invite`. The destination address and, optionally, the destination call-signaling address of the SETUP message must be those of terminal C.

When it receives this message (Figure 2.26), the MCU will send a SETUP message to terminal C with the CID of the conference and `conferenceGoal=invite`. Terminal C accepts by sending a CONNECT message. At this point the MCU sends a RELEASE COMPLETE to the inviting terminal. The active MC establishes an H.245 control channel with terminal C using the transport address provided in the connect. They exchange their TerminalCapabilitySets. The MC signals during the master–slave procedure that it is already the active MC and may send an MCLocationIndication message. When this is done, the MC sends a `multipointConference` message to the inviting and invited terminals. If there were already other terminals in the call, the MCU will send them a `terminalJoinedConference` H.245 message to make them aware of the new entrant.

Because the incoming terminal might have capabilities which are incompatible with the existing media channels in place in the conference, the MCU must send a `communicationModeCommand` to all terminals specifying the new set of allowed transmitting modes for each stream. All media channels that happen not to conform must be closed.

At this point the MC can begin to send OpenLogicalChannel messages to the endpoints. The endpoints should also wait, when they have received a `multipointConference` message, until they receive a `communicationModeCommand` message to open logical channels. All endpoints must send the openLogicalChannel messages to the MC.

The MCU can also initiate the invitation (e.g., if the invitation is not done from an H.323 terminal, but from a Web interface of the MCU). In current H.323 commercial products, this is the most widely used model, because it does not require support for any specific H.323 call flow on the participating endpoints.

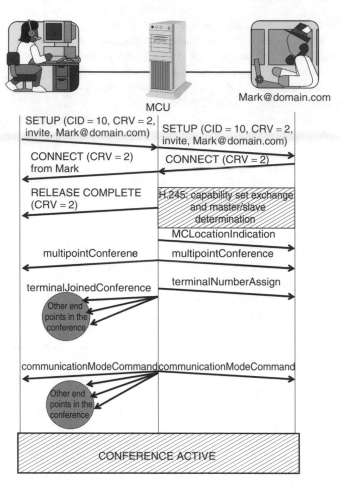

Figure 2.26 Inviting participants in an MCU-controlled conference.

2.4.2.1.2 *Joining an existing conference*

A terminal can easily join an existing conference by sending a SETUP message to the MCU with the CID at the conference and `conferenceGoal=join`. If the terminal only knows the Alias of the conference, it must provide it and leave the CID at 0. Most commercial MCUs use this simpler model, where all participants calling the same number are automatically bridged into the same conference. This model does not suppose any support for the H.323 conferencing features at the endpoint. In order to secure the conference and prevent any random user from calling the bridge directly, the call can be routed though a routed mode gatekeeper, which decides on the fly if the participant is allowed and may translate the initial called party number of the SETUP message into a conference number (this is the solution used by the France Telecom eVisio IP videoconferencing service). Other vendors have an embedded interactive voice response server in the MCU, which authenticates users by prompting for a PIN code.

2.4.2.1.3 Browsing existing conferences

The MCU can in theory provide a list of existing conferences that a terminal could join by sending a conferenceListChoice H.245 message to a terminal. This can be used, for instance, when an alias that has been used in the SETUP message is in fact the name of a group of conferences (e.g., H323support@conferences.com might be a group name for Q931support, H245support, and RASsupport@conferences.com).

Again most commercial MCUs use a simpler Web-based administrator interface to browse for ongoing conferences.

2.4.2.2 Ad hoc conferences

When two endpoints (John and Mark) have started a call as a point-to-point call, they still might want to include someone else in the conversation. Someone might call one of the two parties, or suddenly they might need to talk to someone else to solve a problem. In these cases the call has not been set up directly using a MCU because John or Mark had no idea, when first placing the call, that it would become a conference call. This type of conference is called *ad hoc*.

2.4.2.2.1 John invites Mary

During the discussion, John and Mark decide to go to the cinema, but they need to talk to John's wife, Mary, so that she can choose the movie.

2.4.2.2.1.1 If John and Mark are using the direct call model

In this case, either John or Mark's terminal needs to have an MC functionality and the MC will basically behave as the MCU in Section 2.4.2.2.1. For instance, if Mark's terminal has a MC, John will send to Mark's terminal a SETUP message with a new CRV (not the one used for the point-to-point call between John and Mark), `conferenceGoal=invite`, and the alias of Mary. The rest is exactly as in Section 2.4.2.2.1.

If neither John nor Mark have an MC-capable terminal, they must clear the call and call Mary again via an MCU ... not very user-friendly! Fortunately, many H.323 phones now include three-way conferencing capabilities.

2.4.2.2.1.2 If John and Mark are using the gatekeeper-routed model

Now, with the gatekeeper-routed model, if no terminal has an MC, John and Mark can still invite Mary if the gatekeeper has an MC capability. In this case the gatekeeper behaves as a MCU in the INVITE example above. In some cases the gatekeeper will not have an MC in the same box, but it can easily redirect all conference-related messages to an external MC entity since it routes all Q.931 and H.245 messages.

2.4.2.2.2   Mary calls John

When Mary calls John, she has no reason to know that he is already in a call, so she sends a regular SETUP message.

2.4.2.2.2.1   If John and Mark are using the direct call mode

When John's terminal receives this SETUP message, it will probably propose a menu to John asking whether he wants to:

(1) Reject the call.
(2) Put the call with Mark on hold and talk with Mary.
(3) Include Mary in the call with Mark.

If either John's or Mark's terminals have an MC capability, John can choose (3). For our example we consider that only Mark's terminal has an MC capability.

Mary's terminal will receive a FACILITY message indicating that the call should be routed to John's terminal (a parameter `routeCallToMC` will be present in the facility message) which is the MC-capable terminal. This message also indicates the existing CID of the call with Mark. Mary's terminal releases the call with John (RELEASE COMPLETE) and sets up a new call with Mark.

Now, the SETUP message sent by Mary's terminal contains the same CID as the ongoing call between Mark and John, and the parameter `conferenceGoal=Join`. Then, the call flow continues as if Mark was the MCU in the JOIN example of Section 2.4.2.2.2.

This example, which follows the theory of H.323 conferences, is very unlikely to occur in practice. First, most H.323 endpoints designed for use with a direct mode gatekeeper have an internal MC function. Another reason is that source-based redirections (initiated by a message sent to the terminal that should redirect his call) only work on private networks: on public networks, the dialing convention of the redirecting terminal and the redirected terminal may be different, and, therefore, the number given by John to Mary in the FACILITY message may be unusable by Mary. There are many other reasons that make such a redirection scenario impossible to use over public networks (see Chapter 5).

2.4.2.2.2.2   If John and Mark are using the gatekeeper routed model

This is very similar to the direct call case, except that now the FACILITY message sent by John's terminal will contain the address of the gatekeeper, which will be responsible for the MC function, directly or by invoking an external server. Since the call from Mary arrived through the gatekeeper, the FACILITY message can be intercepted by the gatekeeper and the gatekeeper can use the third-party rerouting procedure (see (2.7.1.2.2)), instead of propagating the FACILITY message all the way to Mary's endpoint. This eliminates the problem of incompatible dialing conventions and all other issues associated with public networks.

2.4.2.3 Conferences and RASs

In most cases, terminals will only know the alias of the conference they want to join. Therefore, the initial ARQ will contain only the conference alias in the `destination-Info` parameter of the ARQ. The CID parameter will be set to 0, which means it is unknown. The callIdentifier must be set by the caller as usual. The gatekeeper will return the Q.931 transport address of the endpoint containing the MC (the MCU, or an endpoint with MC capability) in the ACF.

In theory, as soon as the caller knows what is the exact value of the CID (after receiving a connect from the MC), it must inform the gatekeeper using an IRR RAS message; but, in practice, we don't know of any endpoint doing this.

2.4.3 H.332

The conference model described in Section 2.4.2.3 is **tightly coupled**: all the participants maintain a full H.245 control connection with the MCU. This is very resource-intensive and this model breaks when the number of participants increases beyond a few dozen.

Conferences with a large number of participants tend to be organized with a panel of several speakers (less than 10, typically) and a large audience that listens most of the time and speaks only when requested by the moderator. II.332 describes the electronic equivalent of a panel conference (Figure 2.27), called a **loosely coupled conference**, and is designed to scale to thousands of participants. H.332 is a mix of a usual tightly coupled conference (used by permanent speakers) and a multicast RTP/RTCP conference (as known on the mBone) for passive listeners. The RTP/RTCP-only listeners must know which codec is used and other details (UDP ports, ...). H.332 uses the syntax of the IETF **Session Description Protocol (SDP)** to encode the value of those parameters. A new SDP type (a = type:H332) is defined to let the RTP listener know that this is an H.332 conference. The information can be conveyed using the IETF **Session Announcement Protocol (SAP)** or a simple file on a webpage or sent by email.

Due to the large number of participants, the highly coupled conference among panel members is subject to several constraints: the codecs used should remain stable. If a new member forces a new capability negotiation and triggers a change of codecs, a new SDP announcement must be created. Spreading this information using SAP or otherwise takes time, and most RTP listeners are left out until they have been notified of the new announcement.

The difficult part is to allow the panel to invite a listener to talk, to let listeners request and be granted the right to ask a question. In order to join or be invited by the panel, the RTP listener must also have some H.323 capabilities. Simple RTP/RTCP terminals can only listen. In order to join the panel, a listener must use the regular H.323 conference join and must know the address of the MC that is provided in the SDP.

Similarly, the panel needs to know the callable address of the terminals to be able to invite them to the conference. This is possible because conference listeners periodically

Figure 2.27 H.332 conference.

transmit information elements, such as their name on email address, as RTCP SDES packets (see Chapter 1). A new information element, RTCP SDES item 'H323-CADDR', conveys the H.323 callable address of the terminal. Since bandwidth reserved for RTCP traffic is limited, it takes some time to build a complete list of listeners, and therefore a listener may become callable only some time after he has joined the conference.

2.5   DIRECTORIES AND NUMBERING

2.5.1   Introduction

In the early days of IP telephony (this wasn't so long ago!), one of the major problems was to call someone using a dial-up connection, since the IP address of such users was allocated dynamically. Early solutions all used the same scheme: when the IP telephony software was started, it immediately contacted a central server on a preconfigured IP address and sent a message with the name of the phone user and the current IP address. There were many implementations, ranging from **Microsoft ILS/ULS** to solutions running on top of IRC servers . . .

H.323 makes these solutions completely obsolete. A terminal implementing RAS properly has to register to a gatekeeper, and the RAS message contains all the necessary information—in particular, the current IP address—needed to contact the terminal from a known alias.

2.5.2 Contacting an email alias with H.323 and the DNS

2.5.2.1 *Resolution algorithm*

The gatekeeper can be queried for the current network location (as a call-signaling transport address) of the aliases within its zone, using an LRQ or an ARQ if a call follows immediately. Therefore, when trying to reach an alias, the first step is to find the gatekeeper responsible for this alias. A possibility in small environments is to multicast the LRQ until someone answers, but this is obviously not scalable.

If the alias used is an email alias, like someone@domain.org, then a much better strategy is described in the informational annex of H.225.0. Much information about a domain name can be found by using the Internet **Domain Name System (DNS)**.

The DNS was originally invented to help resolve names into IP addresses. When your computer needs to communicate with another computer named othercomputer.domain.org, it uses the DNS to find its IP address. This involves several steps:

- First, the computer asks a well-known master server which second-level DNS server has the information about domains ending with. org. In fact, this information is likely to be cached locally, and this step is probably skipped.
- Then, the computer queries the appropriate second-level DNS about domain.org, and the reply tells it the IP address of the third-level DNS that stores the information about all names within the domain.org name space.
- Finally, your computer queries this third-level DNS server and obtains the IP address of othercomputer.domain.org. (There are more steps to resolve host names that have more than three hierarchical levels.)

In fact, the DNS holds much more information about each domain, such as the name of the administrator, the address of the mail server for this domain, etc. All this information is stored in 'DNS **resource records (RRs)**'. For instance, the DNS record:

Othercomputer   IN   A   10.0.1.1

means that the computer named 'othercomputer' can be reached at the IP address 10.0.1.1. The DNS record:

Domain.org   IN MX   10   10.0.1.2

means that the mail server (Mail eXchange) for domain.org can be reached at 10.0.1.2.

How can we use this? A special DNS record of type TXT can hold any text. So, we can use it to store the location of the gatekeeper handling the alias resolution for the entire domain. The syntax used is the following:

Domain.org IN TXT

ras[<gk id> @] <domain name>[: <portno>][<priority>]

<domain name> can be the DNS name of the host running the GK software or its IP address. The other fields are optional, <portno> specifies a non-standard RAS port,

<gk id> can be used if multiple logical GKs are running on the same host and, therefore, the name of the host cannot be used as the GK ID. Priority can be used if there are multiple gatekeepers in this domain, smaller numbers have precedence. For instance, valid strings for gatekeeper TXT records could be:

ras 10.0.1.3

ras gatekeeper.mydomain.com : 1234   10

Now, when trying to call someone@domain.org, a computer can first locate the appropriate DNS for domain 'domain.org' as explained above, and then retrieve all the TXT records for domain.org. If a TXT record begins with 'ras', then the IP address or server name that follows is the name of the gatekeeper for this domain. There can be several RAS records; therefore, a zone can be served by several gatekeepers and a domain can be used by several H.323 zones.

Once the gatekeeper has been found, the caller can learn the transport address to send the SETUP by sending an LRQ or an ARQ. This call flow is supported by most commercial gatekeepers, making it simple to organize the VoIP network routing for email aliases.

2.5.2.2   H.323 URL

In H.323v4, the syntax for an H.323 URL was added. The URL should begin with 'h323:' and be followed by a user name and optionally a server name, separated by the '@' sign. The server name can be an IP address, but in general it will be a DNS name. The procedure described above will be used to locate the relevant gatekeeper.

The H.323 URL can be located within a web page to cause a browser to make a call to the indicated address, if a properly configured H.323 VoIP softphone has been installed on the PC. Note that NetMeeting® does not support H.323v4 and will not react to such an URL. Microsoft uses a proprietary URL scheme to trigger NetMeeting calls.

2.5.3   E164 numbers and IP telephony

2.5.3.1   A country code for the Internet

If IP telephony becomes a successful technology, more and more people will have an IP phone or IP telephony software running on a computer. How can they be called from the PSTN (Public Switched Telephone Network)? It is not possible to use email aliases.

Obviously, it is possible to call a gateway with an interactive voice response system that will ask which person must be called on the IP network. For instance, it could ask for a subscriber identifier and then place a VoIP call. This is not very practical. Many VoIP service providers buy large blocks of numbers from traditional carriers and allocate these numbers to VoIP subscribers.

The problem with all these solutions is that there is no way to know in advance that these numbers are reachable over IP; therefore, incoming calls are routed over the PSTN all the way to the network owning these numbers, then to a VoIP gateway. This may not be the optimal route, in many cases it would be much better to route the call sooner over IP.

The problem is that no specific numbering resource has been allocated to IP telephony: in 1997 discussions began within ETSI TIPHON about the numbering resource that would be most appropriate for IP telephony. Many solutions were proposed:

- A special prefix to be chosen in each country (e.g., in Japan all numbers beginning with 050 followed by 8 digits are allocated for IP telephony endpoints).
- A global service code for IP telephony. An example of such a global service code is 800 for free phone calls. From most countries in the world, it is possible to dial the international access code, followed by 800 and the service number.
- A global network code. Providers offering services in several countries can request to have a 5-digit network code allocated to them (e.g., a satellite phone company could be allocated the network number 99999 and you would reach its subscribers by dialing the international access code + 99999 + the subscriber number).
- A country code for IP telephony.

The first solution can be implemented very easily in each country if the local carrier wishes to and if local regulations allow it. This solution, adopted in several countries (e.g., Japan and Norway), may allow some countries to implement VoIP on a large scale easily; but, it has several drawbacks:

- The chosen prefix will be different in each country, and make IP phone numbers less easily recognizable.
- Although these IP phone numbers are now grouped under a single prefix, as opposed to being allocated at random according to each specific VoIP carrier request, this is still a large number of prefixes to manage on a global scale. It makes it difficult for telecom carriers to detect whether these calls should be routed over IP as soon as possible: in most cases the calls will still be routed to the user's country via regular telephone lines. If the call is not recognized as a VoIP call, which is very sensitive to media transcoding (tandeming) and delays, it may also get routed through DCME equipment or satellite links, thereby reducing call quality significantly.

The three last solutions are technically identical, the difference lies in the ITU rules for allocating each type of global code. IP telephony falls between these rules:

- It is not a geographic country.
- It is not a private network.
- It is not a specific service.

There were even proposals to present the need for an 'Internet country code' as an interworking requirement.

Still, the need is there and a global code would present the advantage of enabling VoIP users to connect anywhere in the world and always receive calls over the shortest IP path, as opposed to receiving their calls through their home country. A global code for VoIP is coherent with the global nature of the Internet. For instance, when receiving a call that originated in France, a US citizen temporarily working in France would receive a call that was routed over the French IP network, as opposed to a call routed first to the US over PSTN, then back to France over IP. This would justify lower prices for communications to IP phones behind such a country code. But, there are also some regulatory issues linked with such a solution, such as number portability (is it possible to port a VoIP number to PSTN and vice versa?) and legal interception (should the US CALEA apply to a US citizen working in France ... and therefore do we want all these calls routed through the US?).

Although some political discussion is likely to prevent the introduction of a country code for VoIP, many standard bodies have started to work on the issue. But, ITU Study Group 2 will make the final decision (ITU has a draft on the subject called E.IP, which is not very advanced at the moment). Some countries have already started to allocate a special prefix for IP telephony: on 14 December 1998, Norway allotted prefix 850 to Telenor Nextel for its VoIP service 'Interfon PC' (http://www.totaltele.com/view.asp?ArticleID=20742), and Japan allocated one million telephone numbers behind the '050' prefix.

Meanwhile, several technical proposals have been made to support the address resolution of a telephone number to a call-signaling IP transport address.

All proposals use a database (flat or hierarchical) to find the home gatekeeper handling the resolution of the phone number into a call-signaling address that can be used to send the SETUP message.

The number is not resolved directly into the IP call-signaling address of the endpoint because this IP address may change very often: for a dial-up user, a new address will be dynamically allocated each time he connects to the Internet. In addition, supplementary services based on the gatekeeper may redirect the call to different terminals based on time-based or other rules.

2.5.3.2 UPT

The ITU already has a framework for a service called Universal Personal Telephony (UPT). UPT is based on a special access code and presents many similarities to the solutions presented above. UPT calls are routed to a 'serving exchange' which resolves the original number into an E.164 number. UPT includes several models:

- Model 3a is a flat numbering scheme behind access code +878.
- Model 3b is a numbering scheme behind access code +878 that substructures the numbering space with country codes.

In order to extend the scope of UPT beyond classical voice, the UPT model would only need a modification enabling the UPT information to include not only the address of one or more localization servers, but also the technology to use in order to reach them (classical voice, VoIP, protocol information, etc.).

The reader will not be surprised to learn that UPT 3a faces similar problems to the IP country code requests. There is no agreement on how to administer the flat space, and many political implications are far more complex to solve than all the technical problems put together.

Some ongoing UPT VoIP trials use UPT numbers that are first routed to a national switch of the operator owning the UPT number, then handed off to a VoIP system: the calls are not routed to a VoIP network at the source.

2.5.3.3 DNS-based number resolution

2.5.3.3.1 History

This proposal was first presented to ETSI TIPHON. It uses the network part of the IP address as the first part of the phone number. For instance, a user managed by a gatekeeper having an IP address allocated from class C 192.190.132.xxx could have +999192190132678 as a telephone number, where 999 is the country code allocated to IP telephony (only one thing is certain for the moment: 999 cannot be allocated for IP telephony!).

Readers familiar with IP addressing will probably be shocked by the last three digits, which clearly are outside the 1–254 range. We chose these numbers on purpose, because they do not need to respect the IP addressing rules.

According to this proposal, when a gatekeeper routes a call to +999192190132678, it will decide by analysing the first digits whether the network part of the phone number is class C (see Companion book, Beyond VoIP Protocols multicast chapter for more details on IP addresses classes). Therefore, the part of the number that is an IP network identifier is 192.190.132.

Then, the gatekeeper will locate the DNS that has information about this network by doing an operation called a REVERSE LOOK-UP. During this operation, the network address is mapped to the DNS name 132.190.192.in-addr.arpa.

Once the proper DNS is located using the regular hierarchical DNS procedure, the gatekeeper queries the DNS server for information on 678.132.190.192.in-addr.arpa. At this stage 678 is just a name for the local DNS server, so there is no need to follow the rules for IP addresses. There should be an SRV record or a TXT record with the IP address or DNS name of a 'home gatekeeper'.

Once it has obtained this information, the gatekeeper can route the call to the home gatekeeper, or query this gatekeeper using an LRQ.

At first glance, this proposal was attractive: it was simple and enabled the use of distributed databases, solving the 'who owns the database' issue. However, it also had a number of showstopper issues that made it impossible to adopt.

(1) It is a hierarchical dialing plan. It is impossible to assign blocks of numbers of arbitrary sizes, which leads to rapid exhaustion of the numbering space.
(2) It is unfair. Some US universities have a class A of their own ($255*999*999*999$ potential phone numbers) while entire countries have only a few class C networks (999 numbers). This is the unfortunate result of careless IP address allocation in the

past, but it is unlikely that people would accept this historical artifact to be replicated for telephony.

Finally, although one of the key features of this proposal is the ability to distribute the management of the database, considering features like number portability, and taking into account the fact that the very concept of network classes is now obsolete (a single 'class' is distributed over many owners in most IP networks[9]), leads to the conclusion that the whole database should be centrally managed.

2.5.3.3.2 ENUM

The key idea of the previous technique, using the DNS to resolve a telephone number into a call control resource information, has been refined and expanded by the Telephone Number Mapping working group of the IETF, resulting in RFC 2916bis. This procedure is more commonly called 'ENUM'.

ENUM decomposes any telephone number into a pseudo host name. The number is first written in international format including the country code (e.g. +46-8-9761234 for a number in Sweden), then all non-digit characters (−, .) are removed, and finally the number is written in reverse order (to respect the right to left hierarchical nature of the DNS system, to form a pseudo host name in the new domain name E164.arpa. For instance, the previous number becomes:

```
4.3.2.1.6.7.9.8.6.4.e164.arpa
```

When a system needs to locate the appropriate resource to reach +46-8-9761234, it must query the DNS for the **Naming Authority Pointer Record** (**NAPTR**, DNS-type code 35, defined in RFC 2168 and RFC 2915) corresponding to the pseudo host name `4.3.2.1.6.7.9.8.6.4.e164.arpa`.

The NAPTR record is used to attach a rewrite rule, based on a regular expression, to the DNS domain name. Once rewritten, the resulting string can be interpreted as a new domain name for further queries, or a URI (Uniform Resource Identifier) which can be used to delegate the name look-up. The syntax of the NAPTR RR is as follows:

```
Domain TTL Class Type Order Preference Flags Service Regexp
   Replacement
```

Domain, TTL, and Class are standard DNS fields; Type is set to 35 in the case of the NAPTR. The order and preference field specifies the order in which records must be processed when multiple NAPTR records are returned in response to a single query. The ordering is lexicographic, order is used first, then preference.

The 'S', 'A', 'P', and 'U' flags indicate how the next query should be processed. The next query should request SRV records for flag 'S', A records for flag 'A'. Flag

---

[9] IP addresses are now allocated in blocks of arbitrary size (as long as it a power of 2). The mechanism of reverse DNS look-up can be adapted for classless IP addresses, but most DNS servers do not support it and very few of those that do have been properly configured.

'P' indicates a protocol-specific algorithm, in this case the 'replacement' field will be used as the new name to fetch the corresponding resource record. If the flag is 'U', the regular expression (an expression composed of a series of symbols each defining a specific modification to a string, and defined in POSIX) specified in the RegExp field should be applied to the domain name to get an absolute URI. This last option is used by ENUM.

The service field defines the protocol that will be used for the next step of the resolution (H323, LDAP, SIP, TEL), and the type of service that will be provided ('E2U' in the case of ENUM).

The input of the regular expression is the E.164-encoded telephone number (international format), with a leading '+' sign, and only digits (e.g., +4689761234).

For instance, the following record transforms +4689761234 into sip:info@company.se and mailto:info@company.se, respectively. The preferred method is H.323, then SIP, then the Simple Mail Transfer Protocol (SMTP):

```
$ORIGIN 4.3.2.1.6.7.9.8.6.4.e164.arpa.
  IN NAPTR 100 10 "u" "h323+E2U"    "!^.*$!h323:info@company.se!" .
  IN NAPTR 100 20 "u" "sip+E2U"     "!^.*$!sip:info@company.se!"
  IN NAPTR 102 10 "u" "mailto+E2U" "!^.*$!mailto:info@company.se!" .
```

LDAP could also be used to continue the query:

```
$ORIGIN 6.4.e164.arpa.
 *IN NAPTR 100 10 "u" "ldap+E2U" "!^+46(.*)$!ldap://ldap.se/cn=01!".
```

At present, ENUM is still very much theoretical, as it raises complex administrative and technical issues. The delegation of the e164.arpa domain is a very sensitive issue, being discussed between the Internet Architecture Board and the ITU Study Group 2 in charge of numbering issues.

The technical issues are many, prominent among which is latency, as the DNS resolution can be very low. This is a characteristic of any database relying on hierarchy to scale. Another issue is the timing of record updates: the caching mechanism used by DNS times cache records out after the time-to-live period, but this is not sufficiently precise for telephony use, notably for number portability which would require resource record updates simultaneously by all service providers.

Another major concern is security, because many denial-of-service attacks or, even worse, phone number-hacking schemes are made possible in ENUM. DNSSEC (RFC 2535) solves some of the security issues, at the expense of a more complex administration and possibly increased resolution delays.

In order to be used in operational networks, ENUM will probably require further work. It seems likely that the question of which protocol to use to request a phone number resolution and which back-end database to use will need to be more clearly separated.

2.5.3.4 Dialing plan distribution

The previous discussions apply mainly to the issue of reaching an IP terminal from its phone number allocated from a E.164 numbering space. This is the problem of the 'last hop'.

But there will probably be several transit IP networks offered to reach a particular VoIP alias, and similarly many IP telephony providers will offer outgoing gateways to call regular phones. Therefore, another interesting problem is the distribution of reachability information and any related data (prices, QoS levels, etc.) that may help a service provider to choose among several possibilities when routing a call to a regular PSTN number or an H.323/SIP alias.

In H.323v3, H.225 annex G ('Communications between administrative domains'), was introduced as a way to exchange reachability and pricing information between administrative domains. Essentially, the protocol is very similar to a simple routing protocol, like RIP in IP networks, but distributes reachable phone numbers instead of reachable IP networks. Annex G really takes a simplistic approach to exchanging reachability and cost information, and shares many limitations and drawbacks also found on simpler IP routing protocols. There is a big difference though, VoIP calls are charged and any error, even temporary, can cost millions or ruin the reputation of a carrier. The potential issues that can be introduced by annex G, from call loops to denial of service (by announcing low prices on fake routes) have caused all carriers so far to back off. Therefore, there hasn't been any operational deployment of annex G by any significant VoIP carrier.

Exchange of route and price information between carriers is still done in the old-fashioned fax-and-paper way, or using file transfers, in most cases, and routing information is entered in the central routing systems under human supervision. Essentially, nothing has changed from the days of traditional carriers.

2.5.3.5 Conclusion

The dream of many engineers who worked on DNS usage for phone number resolution or automated phone-routing protocols was to make the phone network as informal as the Internet is today, where a new service provider can get connected in minutes just by listening to the routing protocol advertisements of neighbouring service providers. The introduction of a country code also aimed at facilitating easy identification of calls to IP phones and promoting lower prices to such destinations, justified by the fact that calls would be routed over IP directly from the source.

It is likely that no global code will be allocated to VoIP, service providers will instead negotiate bilateral agreements to exchange traffic over IP. This happens more and more because many traditional carriers now exchange minutes over VoIP. When routing a call to a VoIP service provider, a carrier with a VoIP-capable transit network will certainly send the call directly over IP, and not through the PSTN. Therefore, calls will be routed over IP straight from the source, but charged as regular telephone calls since the absence of a global code identifies these calls as calls to IP phones.

The use of DNS-like mechanisms for telephone networks is also proving more difficult than anticipated. The key difference between IP networks and telephone networks is number portability. In most countries, it is no longer possible to know the service provider owning a number from a prefix in the number. In fact, most telephone service providers already have to carry out phone number resolution today, not to find a home gatekeeper, but to find the carrier owning the destination phone number. Today, number resolution schemes in the PSTN roughly fall in three categories:

- Onward routing: every service provider is responsible for resolving a block of numbers, even if these numbers get ported out. Call routing in the network remains hierarchical, calls are always routed first to the owner of the block identified by a well-known prefix (e.g., 123xxxxxx is always routed to Acme Telecom). The technique used by each service provider for number resolution of incoming calls is arbitrary, as no external query protocol is used: the call is simply rerouted to the new owner based on a local resolution. For instance, Acme Telecom finds that 123555555 has been ported out to Competitor Telecom and reroutes the call to this carrier, usually using a special routing prefix (e.g., D000991234555555). This technique can be used by VoIP 'as is'; in fact, it already works in several networks today (Italy, Germany, France). The major advantage of VoIP is that media tromboning is avoided. This technique does not require ENUM or anything else, since each service provider is free to choose its local resolution mechanism (many use LDAP).

- All-call query: a central database is queried by all service providers for each call in order to obtain a new routing number, identifying the new owner of the number. This scheme can also be extended easily to include technology information (this call should be H.323, SIP, . . .). All-call query is used in the US and several other countries.

- Synchronized databases: each service provider is required to have a locally synchronized database of all numbers and their current owner. National authorities define the database synchronization primitives, which are rather complex in order to take into account number portability (numbers can have a portability request pending, be technically ready to be ported out, technically ready to be ported in, etc.). This technique is used in Denmark. The 'synchronized database' approach would require a protocol that has nothing to do with the current DNS query protocol.

The DNS query protocol of ENUM could be used to replace the current 'all-call query' mechanism (based on SS7 TCAP), with the same or even better functionality, but this will in many cases require a centralized database, as opposed to the distributed DNS database foreseen by ENUM. This is because the DNS database is hierarchical, and the DNS delegation model supposes that you delegate 'blocks' of numbers to a service provider, which is no longer true with number portability. In order to use the DNS delegation model, the original owner of a number would need to be required to manage the associated resource record indefinitely, even if the number is ported out (this is the equivalent of the onward routing method for DNS requests). If this is not acceptable, then the DNS delegation model can be used only at the country code level, but all national resolution servers would need to be owned and managed by a single entity. In addition, ENUM would also need to clarify the use of caching mechanisms in the context of number portability (DNS caching with a TTL value of T creates an interval of time of length T where servers can respond to queries with the new or the old record) and would need to take into account non-E164 numbers (e.g., national 800 numbers and short service numbers that do not have an international form).

Overall, VoIP will not change much the way service providers work today:

- Network interconnections and call-routing decisions, including the choice of transport technology and protocol, are negotiated on a bilateral basis, not via a routing protocol.

- The number resolution mechanism will depend on the country. Number resolution performed by onward routing may be enhanced by requiring all carriers to maintain an ENUM database for their blocks of numbers. Alternatively, a national database can be used, queried via SS7/TCAP (an IP-based query protocol), the ENUM DNS query protocol, or anything else.

2.6 H.323 SECURITY

2.6.1 Typical deployment cases

Building a telephony trial based on IP technology is easy; it is much more difficult to build a production-grade VoIP network, providing scalable and secure operations despite the vast heterogeneity of connected devices and networks.

A security and authentication issue exists whenever the VoIP core network is connected to untrusted IP networks or untrusted VoIP devices.

This section presents typical deployment cases, and the combination of softswitch and network features that can be used to solve security and authentication issues.

2.6.1.1 Carrier-to-carrier connections

Interconnection between **Internet Telephony Service Providers (ITSPs)** is required in order to:

- Send traffic to another ITSP for least cost routing application.
- Terminate traffic in the local VoIP network for another ITSP.
- Route traffic from ITSPs to other ITSPs according to a least cost routing policy (arbitrage, clearing house).

These call flows present multiple security challenges.

2.6.1.1.1 Third-party dependence for security

Many ITSPs currently run their internal call routing using direct mode softswitches (also called light class 4). These softswitches are in fact used as simple directories, and reply to VoIP gateway queries with the IP address of the target VoIP gateway used for each call. In the direct call model, call-signaling is set up directly between gateways, in a peer-to-peer model. This model presents a number of issues even in a closed VoIP network (e.g., no ability to reroute calls on PSTN congestion, no ability to connect untrusted devices such as IP-PBXs, gateway-based billing, etc.), but presents an unacceptable security risk when used across different VoIP domains. For each A to B call, the peer-to-peer model supposes that gateways from domain A will be given (by the direct mode gatekeeper of domain B) the IP address of the gateway in domain B where the call SETUP message should be sent.

This implies that any gateway in domain A should be able to reach the signaling ports of any gateway in domain B. In H.323 this means opening TCP port 1720 (Q.931), and even all TCP ports over 1024 if H.245 tunneling is not used, to all gateways in network A. Since the IP addresses of gateways are not known or frequently changing, this means opening access to all gateways in network B from the entire IP network A. If network A security is compromised, network B security becomes compromised as well. As the amount of network peering grows, the security of the overall network becomes weaker.

The VoIP gateways that signal ports should always be protected from the outside. Most VoIP gateways cannot sustain more than a few calls per second and, therefore, are very easy to break using relatively light denial-of-service attacks. Even worse, some VoIP gateways, which normally are required to be authorized by the direct mode gatekeeper before placing an outbound call, have presented security leaks in the past which allowed the placing of unauthorized outbound calls.

Whenever possible, the routed call model, which does not authorize call control messages to be exchanged directly between gateways, should be used. In router access control lists (ACLs), only the call control communications between the gateways and the routed mode softswitch (e.g., TCP port 1720 for H.323) should be opened. The softswitch sitting between the ITSPs acts as a 'fuse' between networks: any attack may bring down the edge softswitch, isolating the two networks, but will not compromise the protected network.

2.6.1.1.2 *Authentication*

Arbitrage requires the ITSP to generate bills for each connected ITSP and, therefore, to be able to determine which calls are coming from which ITSP. This authentication must be done in a way that is as independent as possible from the IP telephony vendor used in the connected networks, because ITSPs use a wide variety of gateways and softswitches, many implementing a number of proprietary extensions. Some methods using cryptography have been proposed (e.g., the **Open Settlement Protocol** on **OSP** which uses SSL HTTP requests), but they cannot in general be deployed in real networks due to the interoperability challenges caused by the cryptographic token formats used, the variety of VoIP protocol flavours, and the performance issue of TCP-encrypted links. The most efficient and flexible way of authenticating calls coming from a given ITSP is to validate the source IP address of the signaling against the IP network allocated to that ITSP. When the routed mode is used the IP address of the ITSP gatekeeper can be preconfigured. IP address forging is impossible, since VoIP requires messages to be sent to and received by each ITSP. Therefore, ITSPs that could forge their source IP addresses, even if this wasn't blocked by the router ACLs, would not be able to place calls. Having a routed mode softswitch which supports source IP address validation and reports the source IP address in its CDRs, combined with IP spoofing protection at the router level, ensures that inter-carrier billing is both reliable and scalable, and can be deployed in a multi-vendor, multi-protocol context.

2.6.1.1.3   Confidentiality

ITSPs running an arbitrage or a clearing house business may not want to expose the IP addresses of their partners' VoIP gateways, for fear that third parties may discover the names of their termination partners and establish direct business relationships bypassing the clearing house. This issue is highly theoretical since termination tariffs are generally negotiated by clearing houses based on the aggregate call volume of all their customers and would not be accessible to individual customers anyway. Clearing houses are really protected by their purchasing power more than anything else. Still, exposing termination partners' IP addresses is a concern for some clearing houses.

Most routed mode softswitches natively include IP address-hiding features in all the messages that are relayed by the softswitch. If these options are activated, all signaling-related IP addresses of the network hidden by the routed mode softswitch are replaced by the IP address of the softswitch. However, the softswitch does not see the media stream and, therefore, cannot by itself hide the IP addresses of media streams. In most cases, hiding the IP address of signaling messages is deemed sufficient, as it will make it impossible for the connected ITSP to see the hidden IP addresses in all of its common reporting tools: CDRs and traffic reports will only indicate the IP address of the softswitch. Learning media-level IP addresses would imply using network sniffers on a regular basis, which is unlikely in any serious deployment environment. If media-level IP address hiding is necessary, it will be necessary to use RTP relays, which have in some cases a significant impact on network engineering, quality of service and scalability.

2.6.1.1.4   Scalability and quality of service

Considering the potential problems exposed above, some ITSPs have decided to peer using only traditional TDM switches connected to gateways. Ignoring the cost, this poses two issues: audio delays will be at least doubled, probably exceeding acceptable limits for most users, because both gateways will have audio jitter delays and coding delays. In addition, the gateways on both sides of the TDM switch may use different coders for each portion of the call (e.g., GSM on one side and G723.1 on the other, converted to G.711 in the middle because the TDM switch can only process this codec: this creates codec-tandeming issues further degrading the audio quality). Another solution is to use an IP/IP VoIP gateway, which terminates all signaling and RTP media flows, thereby hiding the entire network behind its IP address. Such a solution significantly improves audio quality compared with the previous technique, because the RTP stream does not need to be decoded and recoded, but still it prevents the RTP stream from being routed along the shortest path between the calling party and the called party, and obviously introduces additional delay compared with solutions which route only the signaling data.

Note that it is generally a bad idea to use a general purpose firewall to relay RTP media streams, because RTP packets are significantly smaller than average data packets, leading to unusually high packet-per-second rates. Most firewalls will at least require hardware add-ons to properly relay RTP streams without introducing packet loss or jitter on carrier size links.

2.6.1.2 Class 5 residential networks

Residential telephony services over an IP network are becoming increasingly popular. Next-generation service providers want to provide bundled video data and telephony services that can only be provided over IP, and consumers may also be willing to use the software IP phones provided with the Microsoft Windows® OS. One of the key challenges posed by residential VoIP services is the heterogeneity of endpoints that can be used:

- Analogue telephone adaptors, with or without modems.
- Voice-enabled cable modems.
- Softphones.
- IP phones.

Most of the time these endpoints are registered to an access gatekeeper or SIP registrar (using RAS messages in H.323, Register in SIP), while call-related signaling is handled by a routed mode softswitch as shown in Figure 2.28.

Class 5 application adds the security requirements given in Sections 2.6.1.2.1–2.6.1.2.5.

2.6.1.2.1 Authentication

Residential phones or gateways need to be authenticated, and the authentication method needs to be as independent as possible of the phone manufacturer.

Figure 2.28 Large-scale residential network combining a set of access gatekeepers and one or more routed mode call control gatekeepers.

Unfortunately, strong authentication methods implemented in VoIP edge devices are still mostly proprietary at this time. VoIP standards define frameworks for implementing security (e.g., H.235 for H.323), but fail to describe the exact implementation of security mechanism, or propose too many options. In order to implement strong security mechanisms in a class 5 context, the separation of access functions and call control functions in the class 5 architecture allows access gatekeepers or registrars from the same manufacturer as the CPEs to be used, while using a third-party softswitch for centralized call control. This enables the use of multiple proprietary authentication schemes at the edge, while maintaining a unique call control device.

In H.323 this can work as follows:

- If all the CPEs and trunk gateways are from the same manufacturer, then when the CPE is cleared to make the call by the access gatekeeper, it receives in the ACF message a cryptographic token which is copied in the SETUP message. The authentication and authorization policy can be used by the access gatekeeper. The token is then carried transparently by the class 5 call control switch, and when it arrives at the trunk gateway the trunk gateway will validate this token again using a receive side ARQ sent to an access gatekeeper. In the case the token is not present or valid, the call is rejected.

- If the CPEs and network gateways are from different manufacturers, then a hierarchical authentication mechanism can be used. Once the access gatekeeper has validated the call attempt from the CPE, it sends an LRQ message to the class 5 call control gatekeeper. The class 5 gatekeeper signs the information contained in the LRQ and puts this electronic signature in the LCF in a token. Then, the access gatekeeper returns this token in the ACF to the CPE, which will include it in the SETUP sent to the class 5 gatekeeper. The call will be accepted by the class 5 gatekeeper only if there is a valid security token. This scheme preserves the independence of each vendor to design their own security mechanisms (or 'flavors' of H.235) at the access level, while keeping a vendor-independent routing core and centralized access control.

Some other simple security mechanisms also work with all protocols and all vendors (e.g., the endpoints can be registered, not with their phone number, but with a secure ID containing a hash code). The security level of this method is identical to the level provided by a static password.

2.6.1.2.2   Denial-of-service attacks

Denial-of-service attacks are perhaps the most serious threat in residential telephony deployments. For all applications subject to denial-of-service attacks, the most efficient prevention is to make IP address spoofing impossible in the residential network. If DHCP dynamic address allocation is used, a trace should be kept of all IP address allocations to subscribers in order to be able to trace the originator of any attack. A further prevention made possible if the IP access layer supports it (e.g., if sophisticated IP-aware Ethernet switches are used in Ethernet-to-the-building or ETTB deployment) is to configure token bucket rate control at the edge on all signaling flows. For instance any data flow going to

TCP port 1720 of the gatekeeper from a residential H.323 subscriber should be allocated a very low average bitrate.

Any routed mode softswitch by definition routes all signaling, and therefore acts as a fuse in case a DoS attack should occur. Therefore, an independent routed mode softswitch should be placed at network boundaries potentially subject to DoS attacks, notably those where anti-spoofing is not in place. When such an attack occurs, the CDRs provide immediate information on the IP address of the calling party (except for the distributed DoS attacks, against which there is no known prevention today for any application). Some softswitches have built-in protection against DoS: preservation of a minimal level of service, immediate detection of the attack, and automatic recovery as soon as the attack stops.

2.6.1.2.3 Billing

Customer premises' devices can never be trusted, therefore billing records should never depend on them (e.g., Radius records generated from a CPE gateway cannot be trusted). Again, the solution is to use a routed mode softswitch, which processes all H.323/SIP call control signaling:

- No call can be made to the network unless the softswitch allows it.
- All call signaling events, including call establishment, call termination, abnormal abort, are known to the softswitch and taken into account in the billing records.

Prepaid communications are a special case, as they require a dynamic call cut-off capability. The routed mode softswitch may support this dynamic call cut-off feature by injecting call release messages in the call-signaling path: the two half-calls (from the caller and to the callee) are released by the softswitch, causing the network-side device (e.g., a VoIP gateway) to close all signaling and media ports associated with the call. Even if the residential endpoint ignores the call release message and continues to send audio packets, the device at the other end will not transmit these audio packets and will reply with ICMP error packets.

This architecture ensures that no call lasts longer than actually measured by CDRs; but, the opposite problem exists as well, some calls may be shorter than indicated in the CDRs.

In all VoIP networks, but even more importantly when software IP phones or residential devices are used, it is very difficult to get a reliable indication of call termination. Such devices may suddenly stop responding without properly releasing calls, due to a failure or a network connectivity problem. This can create very significant business issues if customers get billed for minutes they did not use. The softswitch should implement a dead endpoint detection algorithm ensuring that calls from/to any non-responding endpoint will be cleared within seconds. The transport layers used by traditional network protocols include a 'keep-alive' mechanism that has been forgotten in most VoIP protocols and must be recreated. There are several ways of doing this, either at the transport level (e.g., in H.323 sending malformed TCP segments from time to time and checking that the other end rejects them as expected, this is called '**TCP keep-alive**'), or at the protocol level by sending asynchrous queries from the softswitch to the endpoint (e.g., SIP OPTIONS message or MGCP AUEP). The use of the new **Stream Control Transmission Protocol** (**SCTP**, RFC 2960) protocol for VoIP would help solve this issue in a more systematic way, because it includes a built-in keep-alive mechanism.

2.6.1.2.4 Regulatory features

National regulations require the availability of the following services for all IP to PSTN, PSTN to IP, and IP to IP calls:

- Legal call interception.
- Malicious call identification.
- **Calling Line Identity Presentation (CLIP).**
- **Calling Line Identity Restriction (CLIR).**

Many VoIP networks today support these features only through PSTN switches: they are supported only for IP to PSTN or PSTN to IP calls, not for IP to IP calls. This is not acceptable for deployments, as regulatory agencies require these features for all calls regardless of the technology used. The softswitch should implement all of these features in the network and for all call flows, including IP to IP calls. The legal intercept feature in particular is always a bit tricky to implement in VoIP networks, as it should not introduce any noticeable delay in the audio path. It is impossible to use traditional conference bridges (which add delays due to RTP decoding and jitter buffers): dedicated devices must be designed which duplicate RTP packets on the fly without decoding them.

2.6.1.2.5 Is media encryption required?

Sometimes, the availability of technologies creates the need. This is what happened with media encryption for telephony. It is often heard that IP telephony is not secure if media streams are not encrypted. Such a statement is exaggerated. Indeed, in traditional residential or business networks the call can be intercepted at the following places:

- On the phone wire, simply by connecting a loudspeaker (or a passive sniffer for ISDN and digital phones).
- On the link between the user site and the service provider, simply by connecting a loudspeaker for analogue lines and T1/E1 network sniffers for digital lines.
- In the service provider network, where it is necessary to be on the path on the call and to use the ISUP SS7 information to find the TDM time slots and circuits used to transport the call voice stream. This is close to impossible without the collaboration of the service provider.

So far this level of security has been sufficient for most uses. Standard VoIP offers a similar or superior level of security:

- If LAN switches (not hubs) are used in the customer premises, the call can be wiretapped only by using a network sniffer with VoIP capabilities, *and placing it on the Ethernet wire directly connected to the IP phone*. This is because switches do not broadcast packets, which therefore are routed only on the shortest wire path between the source and the destination. The level of security is therefore comparable with that of standard home or business telephony networks. There is no way you can listen to your neighbour's conversation; you do not even have access to the IP packets.

- On the link between the service provider and the company/user the call can be wiretapped by using a WAN network sniffer. If IPSec is used between the customer router and the network router, this becomes impossible without government-level encryption-cracking technology.

- In the carrier network the H.323/SIP signaling of the call must be correlated with the IP addresses of the end-to-end media stream, then with the IP routing information, to locate the path of call IP packets. This is a lot more complex than in traditional TDM networks.

In short, standard unencrypted VoIP is more secure than traditional technology. Even governments are having a hard time trying to break at down, even with softswitch vendors and service providers' active co-operation in implementing wiretapping. Media-level encryption at the phone level should be reserved only for niche markets (e.g., for military use—VoIP is very popular in the armed forces, as it significantly simplifies and accelerates wiring in battlefields).

Even VoIP over wireless LAN will not require any specific development, as the air link is already secured by layer 2 mechanisms.

2.6.2   H.235

H.235 aims at providing application-level privacy (no eavesdropping) and authentication (assuring that people are really who they pretend to be) to H.323 communications and more generally all protocols using H.245. Because H.323 can be used on the open Internet, it inherits the reputation of the Internet; some say that it is less secure than regular telephony. In fact, even without H.235, it is much more difficult to listen to an H.323 phone call than to wiretap a phone line, because you need to implement not only sophisticated network-sniffing tools, but you also need to be on the path of IP packets and to implement the proper voice codec algorithm. Legal call interception, for instance, is much more complex with voice over IP than traditional phone networks. With H.235, IP telephony becomes *much* more secure than regular telephony. Security based on H.235 can be implemented at several levels. With the strictest options, it becomes virtually impossible, even for someone having free access to the IP network, to listen to any conversation that has been secured by H.235 on even to know the number that is being called. In most cases though, H.235 will be used only to make sure that users do not forge their identities. For other aspects the security level provided by standard H.323, comparable with or better than the security level of the TDM networks, is deemed sufficient.

2.6.2.1 A short introduction to cryptography

This chapter has been written purposefully to avoid any reference to the complex notions of algebra, but unfortunately we cannot avoid it altogether. It is possible to read H.235 by considering each encryption function as a black box, but then many parameters, random numbers here and there, remain obscure. We have chosen to describe the cryptographic algorithms used by H.235 in a simple way, but it doesn't mean that they are simple. The real complexity of cryptography is in the detail: How do you choose a random number? How do you calculate a large prime? And so on. So, this chapter will probably seem crude to cryptography experts, but we hope it will help those just wanting to have an overview.

2.6.2.1.1 Common terms

Cryptography is a set of techniques and mathematical algorithms which address one or several of the following needs:

- Privacy: the need to keep the content of a piece of information unknown to anybody except a controlled set of individuals.
- Authentication: the need to check and verify identities.
- Non-repudiation: the ability to attribute with certainty a document, a call, or any piece of information to an author.
- Integrity: the need to preserve the original content of a document from any modification or falsification.

2.6.2.1.2 Cryptographic techniques

Two main techniques are in use today:

- The first one (called symmetric cryptography, shared key cryptography or secret key cryptography) is probably as old as civilization. Caesar was already using it to send messages to Rome.
- The second one (called asymmetric cryptography or public key cryptography) is much more complex and is based on elaborate mathematical algorithms.

2.6.2.1.2.1 Secret key cryptography

(a) Simple algorithms

Secret key cryptography relies on a shared secret between the sender of a message and the receiver. The shared secret can be the algorithm used to encode the message (e.g., a given permutation of letters), or a 'key' used as a parameter in a well-known algorithm.

Simple algorithms, such as letter permutation, are very weak unless the permutation changes frequently: a message can usually be cracked by examining only about 40 letters of the cryptogram when a fixed permutation is used.

A refinement of this algorithm, called **one time pad** was described by Vernam in 1926: if a message is encrypted by adding a completely random key of the same length (e.g., doing an XOR with a random bitstream), then the cryptogram contains absolutely no information for anyone not knowing the random key. Nothing can distinguish it from a random message. In other words, the security of this system is perfect and mathematically proven, provided of course the random string is *really* random and unknown to anyone except the two parties exchanging information. A pseudo-random string can be used instead, but then the security of the system then depends completely on the quality of the pseudo-random generator.

One time pad has, however, a serious drawback: it needs to send an extremely long random key in advance to the recipient of the messages in a secure way. The advent of the CD-ROM has made this relatively easy, and this system is still used today for military or diplomatic communications.

(b) DES and its successors

The most widely used secret key algorithm in use today is the **Data Encryption Standard** (**DES**). DES is a consequence of a consultation by the US Commerce Department in 1971 asking for a secure, yet easily implementable encryption algorithm. They requested a publishable algorithm (i.e., security could not rely on the fact that the algorithm was unknown).

It was only in 1974 that an appropriate proposal was submitted: IBM's Lucifer algorithm. The proposed algorithm was modified and finally resulted in the algorithm that was standardized in 1976 as DES. The current standard is Federal Information Processing Standards Publication 46-2 (**FIPS PUB 46-2**) of 1993.

DES is a block algorithm that can code a message of 64 bits into a cryptogram of 64 bits using a 56-bit key (the actual key has 64 bits, but 8 are used just for error detection). Regarding patents, IBM grants under certain conditions free licenses for devices using DES.

Coding 64 bits is not extremely useful, and an additional standard (FIPS PUB 81) describes how to extend the use of DES to data of arbitrary size:

- **Electronic code Book** (**ECB**) is the direct application of DES on a message split into 64-bit chunks using the same key repeatedly. It is not very secure because similar sequences in the initial message will also appear in the coded message, leading to several potential attacks.

- **Cipher Block Chaining** (**CBC**) avoids this weakness of ECB by using the result of the encryption of a block n to perform an XOR (eXclusive OR) with block $n + 1$ before encrypting it. A transmission error on one block of a CBC-encoded file will prevent the proper decoding of both this block and the next one.

- **Cipher Feedback** (**CFB**) is more appropriate for coding sequences of less than 64 bits.

- **Output Feedback** (**OFB**) uses DES to generate a pseudo random bit sequence that is added (XOR) to the message to be encoded. OFB can in theory code small messages of less than 64 bits but is considered more secure when coding messages over 64 bits. OFB is not subject to error propagation and for this reason is quite appropriate for coding audio or video. (The coding used in the GSM cellular standard is derived from OFB.)

Because CBC, CFB, and OFB all use chaining, the sender and the receiver must be provided with a common initialization vector, in addition to the key. When using H.235, the key is carried in the **EncryptionSync** parameter. H.235 also describes how to construct an initialization vector for CBC, CFB, and OFB.

With the power of computers ever increasing, the safety provided by DES has been questioned. Triple DES simply chains three individual DES blocks using different keys, which raises the complexity of the algorithm from a 2^{57} equivalent codebook to 2^{112} (because of some theoretical reasons, the exponent is increased only by a factor 2, not 3). The justification for three stages instead of two is quite involved, but in short it has been proven that using just two stages does not increase the security of basic DES.

There are many other efficient algorithms using a shared secret key:

- RC2 codes blocks of 64 bits with a key of 40, 56, or 128 bits.
- RC4 is flow-oriented and uses a 40- or 128-bit key.
- The brand new Rijdael algorithm, which will be the successor of DES.

The RC2 and RC4 algorithms have been implemented by many US companies, as they once were easier to export than DES-based solutions.

2.6.2.1.2.2 Asymmetric cryptography

Asymmetric cryptography is based on a new pragmatic way to consider the security of information: a piece of information is not secure only if you don't know how to extract the information; it is also secure when you do know how to extract the information but you cannot practically do it because it would require too much time to run the extraction algorithm even for the fastest computer.

(a)   One-way functions, 'hash' functions

Assymetric cryptography uses many so-called 'one-way' functions. A function F is a one-way function when it is extremely difficult to find x knowing $F(x)$. The idea is that if we have such a function mapping a set of messages M to another set of messages C, it is possible to code a message m from M by using $F(m) = c$ as a cryptogram. For instance, the following function is 'one way':

$$Z/pZ \rightarrow Z/pZ$$

$$x \rightarrow q^x \bmod p$$

where p is a very large prime number (typically with over 100 digits). It is possible to demonstrate that in this case there is at least one number q that is 'primitive' (i.e., for any element E of Z/pZ it is possible to find an element x such as $q^x = E$. If q is chosen to be primitive, there is a one-to-one mapping between the initial message m and the cryptogram $F(m)$. There are classic methods to find a primitive element of Z/pZ once p has been fabricated to have some 'good' properties.

In theory, it is possible to find the initial message from the cryptogram by calculating $F(x)$ for each possible x and compare the result with the cryptogram. But, there are about 10^{100} possible values for x and each calculation is very costly—just imagine how long it takes to calculate, say $12343323223^{654654654654}$! Even being smart and trying to optimize the calculation (e.g., trying to calculate $((((((q^2)^2)^2)^2)^2)^2 \dots)$ there will still be about $\log(654654654654)$ multiplications. But, why not 'just use the inverse function'. The fact is we don't know how to invert this function efficiently. There is no other way than to try each possible solution until a match is found!

A straightforward application of one-way functions is password storage. Instead of storing the clear form of a password p, we store $F(p)$ on the authentication server. This way the password file cannot be used to find the original passwords. When a user logs on with a password p', the system can simply check the validity of the password by verifying that $F(p')$ equals the stored value $F(p)$. In this case the function does not need to provide one-to-one mapping: we can tolerate having several passwords mapping to the same code, if of course the number of possible codes remain very high; this is called a hash function. Hash functions are also frequently used to 'summarize' and electronically sign information (e.g., in H.323 the information contained in H.235 ClearTokens (Section 2.6.2.2.1.1) can include a hash code parameter, which is the result of F(token information, secret)). This prevents anyone from easily modifying the token information, because they cannot recalculate the proper hash code without the secret.

(b)   How to negotiate a shared secret with the Diffie and Hellman algorithm

This algorithm allows two persons, say Bob and Mary, to negotiate a common secret over a *public* link. First, Bob and Mary need to agree on a large prime p and an integer q. It is not a problem if other people know this choice as well. Then, Bob and Mary execute the following steps:

(1) Bob secretly chooses an element of Z/pZ: a. Mary secretly chooses an element of Z/pZ: b.
(2) Bob sends $q^a \bmod p$ to Mary. Mary sends $q^b \bmod p$ to Bob.
(3) Bob and Mary choose $S = q^{ab} \bmod p$ as a common secret. They can calculate it easily because $S = (q^a \bmod p)^b = q^{ab} \bmod p = q^{ba} \bmod p = (q^b \bmod p)^a$!

There is no known way to calculate S knowing only q^a or q^b! Bob and Mary have managed to negotiate a common secret on a public link and can use any symmetric cryptography method using this secret to exchange messages. The Diffie–Hellman algorithm is in the public domain.

(c)   Public key encryption with the El Gamal algorithm

The public key encryption system presented below was authored by El Gamal and derives directly from the Diffie–Hellman method. Again, we use the discrete logarithm function $F : x- > q^x \bmod p$, in which q and p are known to both the sender B and the receiver A and possibly other persons as well. In this system recipient A has a public key $P_a = F(a)$ built from his secret a.

B wants to send a secret message M to A, and of course B wants to be sure that only A can decipher the message. For simplicity, let's assume for a moment that the message is a number in Z/pZ.

B chooses a random number k and sends q^k mod p and $M * P_a^k$ mod p to A. Note that in this system the cryptogram is twice as long as the original message.

Anyone intercepting the coded message needs to know the value P_a^k to find M. P_a is widely known, so all that is needed is k. But, as we have seen above, it would take an enormous amount of calculations to find k from q^k mod p if p is large enough. So, unless a government agency with a large budget is really determined to discover M, B can be pretty safe.

For A, it is very easy to find M from the information sent by B. First, we have to remark that $Pa^k = (q^a)^k = (q^k)^a$. A knows q^k and a, so can easily calculate the value of Pa^k and deduce M immediately.

Public key encryption is very CPU-intensive, and should never be used when not strictly necessary. It is much more efficient to use it until such a shared secret has been determined and then use a secret key algorithm, such as DES.

(d)   RSA

The **RSA algorithm** is based on the difficulty of decomposing a large number into its prime factors when some of these factors are very large primes. The principle is to encrypt a message m using m^e mod n as the cryptogram. e is a prime (e.g., $e = 3$). n, for instance 15, is the *public key* of the recipient of the message. The public key n is not just a random number, it is also the product of two large primes p and q. In our oversimplified example $n = p * q = 3 * 5$. Both p and q are kept secret.

For instance, if '7' is the message to transmit securely, $C(7) = 7^3$ mod $15 = 343$ mod $15 = 13$. Therefore, '13' is the corresponding cryptogram.

In order to decipher the message, the recipient seeks a number d with the property $m^{ed} = m$. This is equivalent to saying that $m^{(ed-1)} = 1$ mod m.

If the greatest common denominator of m and n is 1 ($\gcd(m, n) = 1$) we know from Euler's generalization of the Fermat theorem that $m^{\phi(n)} = 1$ mod m, where $\phi(n)$ is the cardinal of the set of numbers having no common divisors with n. When n is a product of primes it is easy to calculate $\phi(n) = (p - 1) * (q - 1)$. Of course, there could be cases where $\gcd(m, n) \neq 1$, but the probability $(p + q - 1)/pq$ is negligible for large primes. Calculating $\phi(n)$ is straightforward when you know p and q. But, if you know only the public key $n = pq$, you cannot find p and q in a reasonable period of time.

So in our example the recipient is seeking d such that $ed = 1$ mod $\phi(n)$, with $\phi(n) = (p - 1)(q - 1) = 8$. The problem now reduces to finding d and k in $3d + k8 = 1$. We know that a solution can be found with the Euclide an algorithm because $\gcd(3, 8) = 1$ since $e = 3$ is a prime. Here, the solution is $d = 11 (3 * 11 - 8 * 4 = 1)$. $d = 11$ is the *private key* of the recipient and can be used to decipher the message. $M = C^d$ mod $n = 13^{11}$ mod $15 = 1792160394037$ mod $15 = 7$.

Note that the roles of the private key d and the public key e are completely symmetric. So, it is possible to encrypt a message using the private key and then decrypt it with the public key. This is used for digital signatures.

(e)   Digital signatures

There are many ways to cryptographically sign a document. In this section we present only one of these methods, based on the ability to cipher with a private key and decipher with a public key in the RSA algorithm.

Figure 2.29 Principle of a digital signature.

First, a hash of the document to be signed is calculated (see Figure 2.29). A hash is a function that takes a long document as input and produces a small string as output. If the initial document changes slightly (e.g., only one bit changes), a good hash function should lead to a completely different and unpredictable result. One of the most popular hash algorithms is called MD5 (Message Digest 5).

Then, the result of the hash function is encrypted with the private key of the person who signs the document; this becomes the signature.

It is easy to check whether a document is original (has not been modified since the signature) and really comes from the alleged author:

- First, a hash of the document is calculated with the same algorithm.
- Then, the signature is deciphered using the public key of the alleged author. If the alleged author is really the author of the document, we obtain the hash code of the original document.
- Finally, both digests are compared, if the document has been modified in any way since the signature they will differ.

2.6.2.1.2.3 Certificates

Digital signatures are very useful, but only if you are sure that the public keys used by receivers to check the authenticity of the message are associated with the correct identity. The public key can be distributed using a secure method, but this is not very practical. Certificates are a much more efficient way of ensuring that a public key is not a fake.

Certificates usually contain the public key of the presenter, along with some identity information (name and address, corporation name, etc.) and a validity period. In order to avoid any falsification, all the information contained in the certificate is digitally signed by an authority.

An authority is someone owning a widely known public key—so widely known that no one can fake it (e.g., it can be included by default in the operating system, or configured by an administrator). When the authority signs a message with its private key, everyone can verify the signature using the public key. If the authority is known to create certificates only after adequate verification of the alleged identity of the certificate owner, then the certificate is a secure association between a public key and an identity.

If there were only one root authority R, it would rapidly get difficult to handle the workload associated with checking the identity of people or organizations requesting a digital certificate. But, having many root authorities is also difficult because the new authority needs to make its public key widely known (e.g., by approaching Microsoft and asking that they include it in the default configuration of their Internet browser).

Fortunately, there is a solution enabling a root authority to delegate this certificate creation task to intermediary authorized agencies. In order to do this, the root authority creates certificates for each intermediary agency by signing a document containing their public key, name and possibly other elements (see Figure 2.30). An intermediary authority A, when requested to create a certificate C by signing a message which contains the name of the customer, his public key, etc. can just sign this message with its public key Pa.

The new certificate C is returned to the client, *with the certificate of the intermediary authority A signed by R*.

How does it work ? When someone needs to verify the validity of certificate C, he first checks that the signature of the certificate document by A is valid. He can do this because the public key of A is included in the certificate of the intermediary authority A signed by R.

But the public key of A is not well known, it could just be an untrusted local agency. So, it is also necessary to check that the certificate of the intermediary authority A has been properly signed by R, certifying that A can be trusted. This is easy, because the public key of R is well known.

It is relatively easy to check the identity of someone once you are in possession of his certificate. You need to check that the person you are communicating with is the

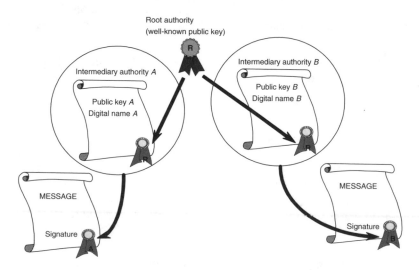

Figure 2.30   Delegation of authority using a hierarchy of certificates.

legitimate owner of this certificate (i.e., is the person owning the private key corresponding to the public key contained in the certificate). This is not trivial because the presenter of the certificate cannot simply show his private key, as it would become compromised! Fortunately, there are other ways to perform this verification without compromising the secret key: for instance, one can encrypt a random string with the public key, send it to the presenter of the certificate and ask him to decrypt it. If the presenter of the certificate has managed to decrypt the string, then he has access to the private key and must be the owner of the certificate.

2.6.2.2 *Securing H.323 with H.235*

During the ITU meetings that led to H.235, because of the general perception that the Internet was not secure and that it would be easy to listen to calls, the initial focus was on securing the media channels. The H.245 procedures were extended to support the encryption of media channels by adding security parameters in the **OpenLogicalChannel** message. If the H.245 channel itself is secure in the first place, then the parameters within the **OpenLogicalChannel** message need no specific protection. This was the main motivation for securing the H.245 channel, but other reasons are equally important (e.g., protecting the DTMF information carried in H.245 **UserInputIndication** messages which may contain sensitive credit card data or passwords).

With a little more experience it became clear that on most IP networks the media channels were hard to access, and therefore already secure enough for the average user. The key requirement was in fact coming from service providers who need to avoid charging the wrong account for a call and also need to be able to prove that someone placed a call in case of a disrupted call. So, the call-signaling channel must be authenticated and optionally encrypted. In the context of H.235, if signaling encryption is used, any server in the network which needs to know the contents of the H.225.0 or H.245 messages needs to be trusted by the communicating endpoints, because it will have access to all confidential information elements: DTMF digits, encryption keys of the media channels, etc. These servers include the gatekeepers in the gatekeeper routed model, the MCUs and gateways otherwise.

Today, apart from of niche markets (military or financial applications), H.235 is only used to make sure that access to public H.323 networks is restricted to subscribers and to authenticate the calling party in order to prevent any misuse of network call-accounting functions. In many networks, notably when customer premises equipment are owned and managed by the service provider, these goals can be achieved without H.235, using data layer security or source IP address validation. For instance, in a VoIP network providing voice VPN capabilities between corporate PBXs, the VoIP gateway connected to the PBX is probably already using some form of IP or transport-level security, and can be considered a trusted element of the network even without H.235.

2.6.2.2.1 *H.235 tools*

2.6.2.2.1.1 *Tokens*

Tokens are parameters transmitted within H.323 messages that are opaque for H.323 itself but can be used by higher level protocols. H.235 uses two types of tokens:

- A **ClearToken** is an ASN-1 sequence of optional parameters, such as timestamp, password, Diffie–Helmann parameters, challenge, random number, certificate, ... ClearTokens are appropriate whenever the need is to ensure only the integrity of the transported information.

- A **CryptoToken** contains an object identifier of the encrypted token, followed by a cryptographic algorithm identifier, some parameters used by the algorithm (e.g., initialization vector), and the cryptographic data itself. CryptoTokens can be used to convey hidden tokens, signed token, or hash values. The cryptographic algorithm needs a key of a specific size N. For symmetric key algorithms the key is derived from a secret shared between the communicating parties. If the secret is shorter than the required key, the secret is simply padded with zeros, if it is longer than the key then the secret is split into blocks of size N octets (or less for the last chunk) which are XORed. The resulting value is used as the key. When the shared secret is not configured in advance a method to negotiate a common secret is required (Section 2.6.2.2.1.2).

As stated above, the most frequent requirement in real deployments is to make sure the devices registering to the VoIP network cannot forge their identities. For this purpose, it is possible to include in the SETUP message a ClearToken containing a call ID, a timestamp, and a hash value (computed from the call ID and the timestamp—the letter prevents replay attacks). This coupled with the calling and called party information is usually enough.

2.6.2.2.1.2 Generating a shared secret with Diffie–Helmann

Many H.235 procedures require a shared secret. If the communicating endpoints do not already share a secret, they must create one common secret, beginning with a communication that someone can potentially intercept.

The Diffie–Hellman key can be negotiated as described in 2.6.2.1.2.2 by using H.235 tokens.

In Figure 2.31, the *DhA* parameter contains p, q and q^a, the *DhB* parameter contains p, q, and q^b. The random value passed in B's reply is used for XORing parameters for further exchanges to prevent replay attacks. The CryptoToken is optional and can be used to digitally sign some parameters in order to prove the identity of the sender.

2.6.2.2.2 Securing RAS

RAS messages are exchanged between an endpoint and a gatekeeper prior to any other communication. H.235 does not provide a way to ensure privacy on the RAS link, but it does provide authentication and integrity. If the security mechanism to be used is not known in advance, two parameters are present in the Gatekeeper Request (GRQ) message that allow negotiation of the right mode and algorithms: **authenticationcapability** indicates the authentication mechanism that can be used, and **algorithmOIDs** contains the list of algorithms supported (DES CBC, DES ECB, RC2, ...). Of course, in real networks the service provider will usually preconfigure all equipment with the selected algorithm, so this is a bit theoretical.

Figure 2.31 Diffie–Helmann parameters encapsulated in ClearToken and CryptoToken structures.

There are two modes of operation depending on whether the gatekeeper and the endpoint share a secret or not.

If there has been no previous relationship and no shared secret between the gatekeeper and the endpoint, they need to negotiate one. For this purpose a Diffie–Helmann negotiation occurs during the GRQ, GCF phase using a ClearToken as described in Section 2.6.2.2.1.2. After this, the gatekeeper and the endpoint share a common secret. This secret can be used to authenticate any subsequent RAS message between the gatekeeper and the endpoint, in particular the RRQ and URQ. This is done by including in those messages a CryptoToken (encrypted using the DH secret) containing an XORed combination of the GatekeeperIdentifier, the sequence number of the request, and the last random value received from the gatekeeper (in the RCF or an xCF message). The key used to code the CryptoToken is derived from the Diffie–Hellman secret as described above. The gatekeeper provides new random values in each xCF in a ClearToken.

When the gatekeeper and the endpoint share a common secret, defined at subscription time, then the easiest procedure is to include in each RAS message a ClearToken with a timestamp and a hash code computed on the calling and called party numbers, the call ID and the timestamp. But there are more complex options; for instance, the following procedure can be used:

- The terminal sends a GRQ with **authenticationcapability** set to **pwdSymEnc** (other modes can be used besides **pwdSymEnc**, such as hash-based or certificate-based authentication, with a similar procedure) and a choice of algorithms in **algorithmOIDs**.

- The GK replies with a GCF containing a ClearToken with a challenge string and a timestamp to prevent replay attacks, **authenticationmode** set to **pwdSymEnc** and **algorithmOID** set to the chosen algorithm (e.g., 56-bit DES in CBC mode).

- At this point, the endpoint may have received more than one answer from several gatekeepers. It chooses one gatekeeper and registers it by sending an RRQ. This RRQ should contain a CryptoToken (using the algorithm chosen by the GK, here 56-bit DES CBC) with the encrypted challenge. The gatekeeper can check the validity of this answer by encrypting the challenge locally with the key associated with the endpoint alias (known from the GRQ), and comparing the result with the endpoint-provided encrypted challenge.

- After this, other RAS messages can be authenticated by including a CryptoToken with the XORed combination of the GatekeeperIdentifier, sequence number, and GK random values provided in xCF messages.

Using one of these methods, the gatekeeper can authenticate the RAS messages of each terminal in its zone.

2.6.2.2.3   Securing the call-signaling channel (H.225) and the call control channel (H.245)

The call-signaling channel can be secured using transport-level mechanisms like TLS or IPsec. An endpoint knows that it needs to secure a channel using TLS if it receives the call on port 1300. This is the more advanced option of H.235, providing confidentiality in addition of integrity and authentication. In practice, such complexity is not required, because only integrity and authentication are required. Most commercial H.323 endpoints implement one of the following methods:

- The SETUP message includes a token with a timestamp and a hash value computed from the most important parameters of the call (at least the calling party number, called party number, and callID). The gatekeeper, in routed mode, can verify this token with the shared secret, ensuring integrity and authentication.

- The RAS Admission Request (ARQ) message includes a token with a timestamp and a hash value computed from the most important parameters of the call (at least the calling party number, called party number, and callID). The gatekeeper performing the RAS function can verify this token from the shared secret. If the token is verified, it sends back in the ACF a token that should be included in the subsequent SETUP message. This token will have to be verified by the call control server receiving the SETUP message. The usefulness of this hierarchical authentication is clearer when considering the scalability of authentication in a large H.323 network (see Section 2.6.2.3.3).

In the SETUP, the caller will indicate which security schemes it supports for the H.245 channel in the **h245SecurityCapability** data structure. **h245SecurityCapability** includes a specific object identifier for each cryptographic algorithm, 56-bit DES CBC, and 56-bit DES OFB (e.g., each has its own identifier). The callee chooses one in the **h245Security-Mode** data structure carried by one of the Q.931 response messages (e.g., CONNECT).

If no common security mode can be found, the callee can release the call with the reason code set to **SecurityDenied**. The necessary messages needed to secure the H.245 channel are exchanged before any other H.245 message.

Different methods can be used to initiate the secure channel, depending on whether the communicating endpoints share a secret or not. These procedures are very similar to those described for the RAS channel. Again, if a shared secret does not exist, it can be created using a Diffie–Hellman procedure. To our knowledge there is no commercial implementation of this yet; so, we will not detail the procedure further.

2.6.2.2.4 Encryption of media channels

Once the H.245 channel is secured, the terminals need to know which security modes can be used for the media channels. This is part of the capabilities exchange (e.g., terminals can signal that they support GSM capability, and/or encrypted GSM capability). A new capability has to be defined for each combination of codec and encryption mode. Since encryption algorithms can use a significant portion of the CPU, it is possible to signal such capabilities as plain GSM + H.263 video *or* Triple DES-encrypted GSM. H.323 is very powerful when it comes to expressing capabilities.

When a new logical channel is opened, selected security mode is specified (chosen by the source) and the key that will be used for logical channel encryption is *provided by the master* (as determined in the master–slave negotiation, see Section 2.2.1.2.2) either in the **OpenLogicalChannel** or in the **OpenLogicalChannelAck** using the **encryptionSync** field. The key is associated with a dynamic payload type, so a receiver which has just received a new key in the **encryptionSync** field will know it must use it as soon as the payload type of the RTP packets it receives matches the payload type associated with the key. The key can be refreshed at any time using the dedicated H.245 commands, **EncryptionUpdateRequest** and **EncryptionUpdate**. If the master decides to update the key (using the H.245 **EncryptionSync** message), then the payload type of the RTP stream must change for the RTP packets that use the new key.

Key negotiation can be made inherently secure using certificate exchange, or can be secured by first securing the H.245 channel. If the H.245 channel is not encrypted for some reason, then H.235 has provisions to open a separate specific LogicalChannel of type **h235Control** to negotiate key parameters for the logical channels. Again, there is no commercial implementation of this.

For multipoint communication, the secured H.245 channel is established with the MCU, and therefore the MCU must be trusted. New endpoints arriving in the conference can retrieve other endpoints' certificates, through **ConferenceRequest/ConferenceResponse** messages. However, they must trust the MCU to check whether the endpoints actually own those certificates.

As already mentioned in Chapter 1, many popular algorithms, such as DES ECB or CBC, are block-oriented. They are designed to code data aligned on the block size (64 bits for DES). The most simple way to cope with this is the RTP padding method described in RFC 1889 (see Chapter 1 for more details). When it is used, the P bit of the RTP header is set. However, there are other techniques that can be used with DES. In addition, to regular RTP padding, H.325 mandates that all implementations support ciphertext stealing for ECB and CBC, and zero pad for CFB and OFB. These techniques are modifications of the regular ECB/CBC/CFB/OFB chain-coding process for the incomplete data block and its predecessor, leading to a cryptogram exactly as long as the original message. When payload length is not a multiple of block size and the P bit is not set, then the decoder must assume that one of these methods is used.

In all cases, when an initialization vector is needed, it is constructed from picking as many octets as needed from the concatenated sequence number and timestamp octets, repeated if needed.

2.6.2.3 Scalable and secure H.323 deployments in a multi-vendor environment

2.6.2.3.1 Split gatekeeper architecture

In very large deployments of H.323 networks, from 50,000 endpoints up to several hundred thousand, the RAS and routed mode call control functions cannot be performed by the same gatekeeper. A typical network of 100,000 endpoints will generate about 500,000 calls a day, but will generate at least 144,000,000 registration requests during the same time (assuming one refresh registration request per minute per endpoint). This translates to about 1,700 RRQ messages per second. Assuming a single computer were powerful enough to handle this load in case of a temporary network failure, and assuming all endpoints randomly send the first RRQ after failure over a period of 20 s, the average load during network restarts may exceed 5,000 RRQs per second. A single softswitch is unlikely to be able to reliably handle such a load, in addition to the call control functions.

The right solution is to separate the access function, handled by regional 'access gate-keepers', and the call control function, handled by a central call control gatekeeper (Figure 2.32). There are typically several access gatekeepers for each call control gate-keeper in the network. A router-based access gatekeeper can be expected to handle about 30 to 50 registrations per second, representing about 5,000 endpoints. Call control gate-keepers see only active calls, and some scale up to about 15,000 simultaneous calls, sufficient for 300,000 to 500,000 residential users. Note that most vendors publish the calls-per-second limitation of the call control softswitch; this shows 500 active residential calls translate to only about 5 calls per second (the average call duration is about 3 min, and 30% of calls are dropped due to a busy or no answer condition). A softswitch with 15,000 active calls will receive only about 150 calls per second.

2.6.2.3.2 Securing the network edge

The security of the registration of an endpoint to its access gatekeeper uses one of the H.235 methods explained above. Access gatekeepers may check the identity and pass-word of endpoints (most of the time, the shared password methods of H.235 are used) with Radius or LDAP interfaces to a central database. Only the first RRQ is checked, the other keep-alive RRQs do not need an external password database access. A significant advantage of the 'split gatekeeper' architecture described above, besides scalability, is that it facilitates the deployment of multi-vendor solutions. Because of the number of options in H.235, most vendors' security implementations do not interoperate with one

Figure 2.32 Split gatekeeper architecture.

another, because the exact information provided in each security token is slightly different, the hash code calculation method is different, etc. With the split architecture it is possible to group endpoints of the same brand together and point them to an access gatekeeper that can understand secure tokens from this brand while keeping call control centralized.

Providing security for the call admission function is more complex. Even if the ARQ/ACF RAS exchange with the access gatekeeper is secure, it is still necessary in most cases to secure the Q.931 call control channel. There is one exception: if the network is from a single vendor and if the PSTN access gateways support the same format of security tokens as the edge endpoints; in this case, it becomes possible to secure the network at the edge. There are several large-scale instances of such networks today.

In the simple case illustrated in Figure 2.33, a token is provided by the access gatekeeper back to the edge endpoint in the admission confirm (ACF) message. The endpoint is required to copy this token in the SETUP message. The call control gatekeeper ignores this token and assumes this call can be forwarded to the destination. The destination is located with a Location Request (LRQ) to the proper access gatekeeper, and the SETUP is forwarded to the destination with its security token. The destination is required to get an authorization from its own access gatekeeper before it can continue processing the call. The token is passed to the access gatekeeper in the Admission Request, and the access gatekeeper verifies it by controlling that the hash code is correct for the given timestamp, callID, and call destination. If it is correct, the call is accepted with an ACF; otherwise, the call is rejected (ARJ) and released by the gateway.

Figure 2.33 Call secured at the edge by tokens.

This simple single-vendor case is completely secure for calls to the PSTN, because the gateways are the property of the network provider and can be trusted. It is slightly less secure for calls to endpoints, because potentially an endpoint could be hacked to not do any admission request before it proceeds with a received call. However, this requires both the calling and called endpoints to be hacked, a fairly limited probability.

2.6.2.3.3   Security at the access level and at the call control level

If the network involves multiple vendors, or if the PSTN gateways do not belong to the same service provider as the calling endpoints (e.g., if a residential network sends international calls to a clearing house), then the simple solution above is not sufficient and the call control gatekeeper needs to enforce the security policy. The simplest way to do this is to have the call control gatekeeper simply validate the token passed in the SETUP message, which was copied from the ACF. But, even this is not trivial. Most vendors simply send a hash code based on the endpoint password and call parameters. This forces the call control gatekeeper to know the password of each endpoint to validate the token, which is time- and memory-consuming for 500,000 users. A more scalable way of doing this is to have the access gatekeeper return a new token, signed with the access gatekeeper password, once the call has been authorized. This makes things much easier for the call control gatekeeper, because now it only needs to know the password of the access gatekeepers, not of the end users.

The assumption for this solution to work is that the access gatekeeper and the call control gatekeeper support the same format for security tokens. In a multi-vendor environment, this is unlikely. A more involved call flow is necessary to implement call control security in this environment.

The call flow of Figure 2.34 also leverages the idea of hierarchical authentication. Once the access gatekeeper has authorized the call, it requests an authorization token from the call control gatekeeper by sending a location request message (LRQ). The call control gatekeeper does not need to perform any specific authentication work, as it only accepts LRQ messages from access gatekeepers, as long as these access gatekeepers have not sent an LRQ if edge authentication failed. For each LRQ received, the call control gatekeeper returns an authorization token (tokenCC) in the LCF, which is copied by the access gatekeeper in the Admission Confirm, and then by the edge device in the SETUP message.

When it receives the SETUP message, the call control gatekeeper simply validates the call control token (tokenCC); it does not even need to look at the access token (tokenA). In fact, the access token is really unnecessary in this case.

This method works even with multiple vendors of CPEs and access gatekeepers, since the call control gatekeeper is not even attempting to understand the format of the security token of the edge device.

Figure 2.34 Hierarchical security with token assigned by the call control GK.

2.7 SUPPLEMENTARY SERVICES

2.7.1 Supplementary services using H.450

H.323v2 introduced the new H.450 standard series, supplementary services for H.323. H.450 is based on the QSIG extensions of ISDN for use by PBXs, and therefore is targeted at private installations. H.450 should be used with caution on public networks, as only a few of the H.450 services (e.g., H.450.7 for message-waiting indication) can be deployed safely in a public environment. H.450 caused a lot of confusion as many perceived it as 'the' way of doing supplementary services in H.323. For clarification, H.323v4 added the following note:

> *Within the H.323 environment, there are several different methods by which services can be provided: the H.450 series of Recommendations, H.248[12] in association with its packages, stimulus signalling and Annex K. Although there is commonality of certain design goals for each of these solutions, the emphasis varies and each is more appropriate for certain circumstances. [...] The H.450 series of Recommendations is designed for interoperability of services at a functional level. Its derivation from QSIG ensures interworking with many private networking systems. Services are defined for peer–peer relationships, with feature intelligence typically resident in the endpoint. An H.450 based service must normally be explicitly supported by each affected endpoint in the system.*

H.450.1 defines the general framework for exchanging supplementary service commands and responses for use by supplementary services, H.450.2 defines a **call transfer** procedure[11] (**blind call transfer** and **call transfer with consultation**), and H.450.3 defines the **call diversion** procedures (**call forwarding unconditional**, **call forwarding on no answer**, **call forwarding on busy** and **call deflection**).

H.323v3 added a few services, notably call hold (H.450.4), call park and call pickup (H.450.5), call waiting (H.450.6), and message waiting indication (MWI, H.450.7). Of these, only the MWI, H.450.7, is widely supported today by IP phones and residential gateways.

H.323v4 further expanded the H.450 series with H.450.8: (Name Identification Service), H.450.9: (call completion), H.450.10: (call offer), and H.450.11:(call intrusion).

2.7.1.1 H.450.1

H.450.1 defines the 'generic functional protocol for the support of supplementary services in H.323'. This recommendation is based on Application Protocol Data Units (APDUs) carried in the call-signaling messages (ALERTING, CALL PROCEEDING, CONNECT, SETUP, RELEASE COMPLETE, PROGRESS) or in FACILITY messages.

---

[10] H.248 is a stimulus protocol very similar to MGCP.

[11] Another possibility is the use of call control tromboning and redirection of media streams using the Null Capability Set sequence, or TCS=0, as described in Section 2.7.1.2.2.

H.450.1 can be used to convey call-related instructions (e.g., redirecting a call) or call-independent instructions (e.g., program call screening). In the latter case, a special SETUP message with specific bearer capability and conferenceGoal information elements is used to transport the APDU. H.450.1 APDUs have the following structure:

- Optional **Network Facility Extension (NFE)** with the source entity type (endpoint or anyEntity) and address, and the destination entity type and address.
- A description of what to do with unrecognized messages (discard, clear call, . . .).
- A structure with the actual operation invoked.

The NFE part of the APDU provides a way to route supplementary service messages. The network entity receiving a SETUP message with an H.450.1 APDU may not be the intended recipient of the instructions contained in the APDU. It may have to relay it, or choose to intercept it in the case of a gatekeeper. All H.450 services are built on top of H.450.1.

2.7.1.2   H.450.2 (call transfer)

This recommendation provides a way of transferring calls between H.323 endpoints once the initial call is established (the callee has answered).

2.7.1.2.1   Call transfer between H.450.2-aware endpoints

The scenario shown in Figure 2.35 is an example of call transfer between endpoints. The call could be routed through a gatekeeper, but the gatekeeper would simply relay all H.450.2 APDUs:

- User B calls user A (the transferring user). This is the primary call.
- User A answers the call and uses H.450.2 to transfer the call to user C. Uses A may previously establish a separate call (secondary call) with user C to announce the transfer, for instance. If this secondary call exists, endpoint A notifies C of the pending call transfer, C returns a temporary identifier I for this secondary call if it can participate in the transfer. Otherwise, the attempt aborts here.
- Endpoint A sends an H.450.2 request to user B to call C (if there is an A–C secondary call, the temporary identifier I is mentioned). The endpoint may handle this directly if it is H.450.2-capable, or A's gatekeeper may choose to do it.
- When the new call request initiated by B arrives at C, C releases the secondary call if it existed. Then, if C answers the call the primary call is also released. B and C can now talk.

In this scenario, A could have called B in the first place, or C could have called A. The next steps of the call transfer would remain the same. The invoke and result APDUs are carried in normal Q.931 messages whenever possible, in FACILITY messages otherwise, as shown in Figure 2.36.

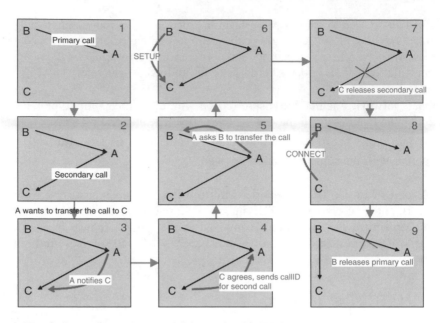

Figure 2.35   Call transfer with consultation using H.450.2.

A note on FACILITY REDIRECT: H.323 mentions another simple way of supporting call transfer. An endpoint can simply send to the transferred endpoint a FACILITY message with the address of the endpoint transferred to. When it receives such a FACILITY message, a terminal should release the current call and restart a new call to the address specified in the FACILITY message. This is a simple way of transferring a call without consultation, but to our knowledge only very few endpoints and gateways support it.

2.7.1.2.2   Transfer using the gatekeeper

H.450.2 is not very easy to implement, and is unusable end to end in public network environments (see Chapter 5 for more details). The flow chart (Figure 2.36) describes only the normal case, but it would get much more complex if it took into account the many options of H.450.2 and the error conditions. Because of this complexity, many H.323 endpoints, such as stand alone IP phones with stringent memory constraints, may not implement H.450.2.

However, in Section 2.7.1.1 we emphasized the fact that all H.450 APDUs could be routed (using the origin and destination addresses found in the NFE) or intercepted by a gatekeeper. Therefore, if the terminals involved in the primary call were using the gatekeeper routed model (all Q.931 and H.245 messages get relayed by a gatekeeper), then the intermediary gatekeeper can intercept and act on H.450.2 APDUs on behalf of endpoints B and C. This allows H.450.2 to be used even if only terminal A is H.450.2-aware. Terminal A could be a sophisticated secretary terminal, while endpoints B and C could be ordinary simple IP phones. If we go one step further, endpoint A itself could be a simple IP phone, but the gatekeeper would have a web interface allowing the user of

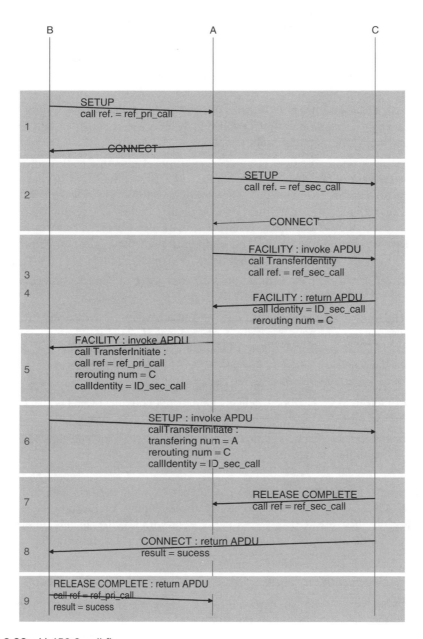

Figure 2.36 H.450.2 call flow.

terminal A to ask the gatekeeper to initiate the call transfer. This logic has been adopted by stimulus mode endpoints (see chapter 4 for more details).

The task of the gatekeeper is more complex than what we have seen in the end-to-end H.450.2 case. In the previous case endpoint B initiated a *new* call to terminal C, and the normal H.323 procedure was used. Now, the gatekeeper must find a way to cause endpoint B (connected to A) to transfer the call to C without ever releasing the ongoing call.

Fortunately, this operation, called third-party rerouting, has been taken into account in H.323. It is done as follows (case of a blind call transfer):

- As soon as it knows it needs to transfer the call to C, the gatekeeper calls endpoint C: it send a SETUP and receives a CONNECT. If it receives a RELEASE COMPLETE (busy terminal, . . .), then it aborts the operation.

- The gatekeeper sends an empty terminal capability set to endpoint B, endpoint A, and endpoint C. This is possible because it relays the H.245 messages between A and B and therefore can 'insert' messages. The empty capability sets indicate that the remote terminal has no receive capabilities, and, logically, this causes terminals A and B to close all active logical channels. Terminal C will also not attempt to open a logical channel to the gatekeeper. Further, all endpoints reset their H.245 state machine and go back to a state where they are waiting to receive capabilities.

- The gatekeeper can close the connection with A now (H.245 end session command and Q.931 release complete), or wait until the transfer is completed.

- During H.245 channel establishment with C, the gatekeeper has received the capability set of terminal C, and now forwards it to terminal B. This will cause terminal B to restart the H.245 state machine just after the capability set exchange, and B will start a master–slave determination exchange.

- Then B sends an openLogicalChannel command over the H.245 channel, the gatekeeper relays it to terminal C. The openLogicalChannelAck contains the RTP/RTCP addresses of terminal C, so B will now establish the logical channels with C.

- C also opens logical channels with B through the gatekeeper, and this completes the transfer: B communicates with C.

If the endpoints are H.450.2-aware, the gatekeeper can still perform call redirection. In this case H.450.2 APDUs are used to notify the endpoints of the progress of the call transfer. For instance, a FACILITY message is sent with a CallTransferComplete invoke APDU to endpoint B to inform endpoint B that it has been transferred to C.

2.7.1.2.3 Blind transfer, secure transfer, transfer with consultation

To sum up what we have learned in Section 2.7.1.2, here is how an H.450.2-aware terminal can perform the classic types of call transfers:

- Blind transfer: in this type of transfer, A doesn't want to check if C is available before disconnecting from B. As A and B are still in an active call, A sends a FACILITY message to B, the FacilityReason field is of type *callTransfer* and contains a Call-TransfertInvoke (CallTransferInitiate) invoke APDU informing B of the address of C. Then, A terminates its conversation with B using the regular H.323 procedure. When it receives the FACILITY message, B initiates a call with C.

- Secure transfer: now, if B cannot connect to C, A wants to remain in conversation with B. A sends the CallTransferInvoke to B but does not disconnect from B immediately. Instead, it waits until it receives the FACILITY message from B with the result of the call transfer (CallTransferResult=success on failure). Depending on the result, A releases the conversation with B (B might also send the RELEASE COMPLETE) or keeps it active.

- Transfer with consultation: now, A might be a secretary who needs to check whether C is available or in a meeting. In a regular private phone system, A can put B on hold. But for some reason the Q.931 message HOLD is forbidden in H.323. So, A can simply stop sending media to B (or send prerecorded music).

- Another solution would be to use the H.450.4 hold supplementary service. Now, A can establish a new call with C. Once A has been allowed to perform the transfer, A can use either the blind transfer or the secure transfer procedure. The procedure shown in the flow chart (Figure 2.36) is a transfer with consultation using secure transfer after the consultation.

2.7.1.3   H.450.3: call diversion. Introduction to H.323 annex K

Recommendation H.450.3 is focused on the redirection of calls between H.323 endpoints before the call is established. This includes call forwarding on busy, call forwarding on no reply, call forwarding unconditional and call deflection. The diversion might be performed by a gatekeeper, or by the endpoint itself. H.450.3 allows the number of successive call diversions to be controlled/limited.

When activating the call forwarding unconditional (CFU) supplementary service for a particular address A, a user can still originate calls, but all calls to A will be redirected to another address. The user can activate/deactivate this service directly on the endpoint associated with A, or remotely on a gatekeeper. H.450.3 also provides ways to interrogate an entity to ascertain whether the supplementary service is activated or not, and for which addresses. The endpoint receiving the diverted call is notified that the call has been diverted, and also where the last diversion point was. The calling endpoint may also optionally be notified that the call has been diverted, with or without the new call destination address.

When activating the call forwarding on busy service, the same operations as in CFU will occur if the line of the user is busy. There might also be more specific conditions (diversion if more than N calls are waiting, . . .).

The call forward on no reply is similar, but occurs if the called user using this supplementary service has not answered after a programmable period of time.

The call deflection supplementary service can be invoked by a called user dynamically before the user answers a call. It causes the call to be diverted to the address entered by the called user.

Many types of call redirection can easily be performed by a routed mode gatekeeper. For the call forwarding unconditional service it can simply change the called party information element of the SETUP message and forward it to the new destination. For call forward on busy it will first forward the SETUP to the original destination address, then if it receives a RELEASE COMPLETE (cause busy), send a new modified SETUP to the next destination. Call forwarding on no answer is more involved, because the gatekeeper needs to make sure that audio channels are not established before redirecting the call. Chapter 5 gives more details and call flow examples on call redirections.

The activation of all these H.450.3 supplementary services is performed by exchanging H.450.1 APDUs. For instance, the call deflection supplementary service needs to be triggered by the called endpoint: this is done using the callRerouting invoke APDU. Even this simple task requires many information elements:

- The `reroutingReason`, if needed.
- The new `calledAddress` to use for the redirected call.
- A `diversionCounter` that is useful to avoid loops.
- The `lastReroutingNr` with the address of the last endpoint that performed the rerouting.
- `subscriptionOptions`: does the terminal want to inform the calling party?
- The original `callingNumber` (note that the 'number' here can be any H.323 address).
- More textual information in fields, such as `callingInfo`, `redirectingInfo`, ...

At this stage the reader probably wonders whether it is necessary to introduce so much complexity for such a simple feature. This is true of many supplementary services. In fact, when the services are activated by a gatekeeper, a lot can already be done with the ubiquitous web interface for user interaction if you have a web phone and you want to program call forward on busy; this can be done by filling in an HTML form. With a little imagination it is even possible to let the user customize his call control with much more flexibility with a web interface: something like 'if-my-boss-is-calling-then-ring-my-desk-phone-and-try-my-cellular,-otherwise-go-to-the-answering machine,-or-if-it's-my-banker-calling-again-to-say-my-account-is-low-then-sound-as-if-I-wasn't-here ...' H.323 annex K describes how an IP phone with a proper Web-capable user interface can use it for the control of supplementary services, which makes it possible to present any user interface for any feature with no impact for the IP phone. But this annex has had no success so far, probably because it provides a lot less functionality than its MGCP equivalent (business phone package, see Section 4.2.2.2.3 for more details).

2.7.2   Proper use of H.450 supplementary services, future directions for implementation of supplementary services

H.450, the VoIP version of QSIG, is an appropriate way to convey supplementary services in a private network of PBXs, but end-to-end call transfer supplementary services should not be used in public networks. In Chapter 5 we will describe in more detail the specific requirements of public networks (i.e., networks interconnecting multiple enterprises or residential users, where the service provider must bill for calls originated from each connected device and cannot trust connected devices). Clearly, because of security, call-routing, and accounting issues, only a handful of the H.450 standards can be reasonably deployed in a public network. Just like QSIG is not used today on public networks, H.450 will probably remain only a PBX interconnection protocol.

H.450 can also be implemented to provide supplementary services on business phones, but it has a serious competitor in MGCP. Today, virtually all PBXs use a stimulus protocol to control PBX business phones ('Unistim' for Notel, 'ABC' for Alcatel, etc.). This simplifies the phone design and gives a lot more control to the PBX. For instance, many PBXs offer services as soon as the phone is off-hook, even without dialling any number: you can have notification of voicemail, warnings if the line has been forwarded, etc. Such services cannot be implemented with an H.323 or SIP phone, because these phones do not send any notification to the network when they are off-hook. The rigidity of the H.323 and H.450 standardization process is also a great obstacle to its use in business systems, where the race for new features and differentiation leaves no room for endless discussions in standard bodies. Most PBXs today offer over a hundred features. H.450 took over two years to sort out the first dozen!

MGCP, with its new extensions (business phone event package, BTXML from Cisco, etc.), now has the ability to control virtually any device, see MGCP, section 4.2.4 chapter including feature buttons, screens, loudspeaker modes, etc. For the first time a standard protocol gives access to the same power as proprietary stimulus protocols. Many implementations of IP-PBX and Centrex services already exist for MGCP, which goes far beyond the reach of H.450. Automatic off-hook, CTI calls, paging calls are already available! Therefore the prospect of H.450 for business phones also seems very limited, probably restricted to the same niche markets as ISDN business phones today.

2.8 FUTURE WORK ON H.323

H.323 is now carrying billions of VoIP minutes per year. Most networks are running H.323v 2 with some version 3 and 4 extensions. The current version of the protocol benefits from the experience accumulated by VoIP vendors in hundreds of VoIP networks. However the telephone network is a lot more complex than anticipated in the early days of voice over IP, and the H.323 protocol, despite its maturity, is still in need of improvement and extensions to cover the very specific call flows found in PSTN networks. Among the call flows that are still problematic today and require proprietary extensions or specific vendor to vendor tuning, one can cite:

- Pre-connect announcements. Although the basic pre-connect announcement is covered by early H.245 and Fast Connect, the H.323v4 protocol is still not flexible enough to allow multiple media servers to stream media to the calling endpoint (dynamic update of the Fast Connect media information). H.323v5 solves this problem with H.460.6.

- Call release scenarios using Q.931 messages 'forgotten' by the H.323 standard (e.g., the DISCONNECT message). In some instances media is being played while the call is still in a half-released state: one of the most common cases is in-band network announcements played when dialing the wrong number. A consequence is that it is relatively difficult to implement **Advice of Charge (AOC)** on H.323. In ISDN, AOC is sent at the end of the call in one of the messages releasing the call. Since the release messages use three messages, there is always one that is sent by the network to the endpoint. In H.323, when the endpoint releases the call first with a RELEASE COMPLETE message, the network has no chance of sending the AOC information to the endpoint.

- Precise rules on how to transport and use the 'progress indicator', which specifies whether in-band information is present or not (it should be ignored when present). Some complex call flows can be found where locally generated ring-back tones alternate with in-band tones. In H.323, vendors need to be careful to take the progress indicator information into account, as opposed to blindly playing the media they receive from the other party.

- Precise rules on how to interwork with the ISDN network, and in particular the handling of media-type information (3.1-kHz audio, fax, . . .). Simply carrying this information transparently causes errors in the ISDN network because it advertises capabilities not yet present in the VoIP network.

- H.323v4 does not handle call loop detection in a robust way (i.e., with hop counters in every message and mapping rules with SS7 ISUP messages). H.323v5 solves this problem.

- More experience and understanding of the implications of call redirection in VoIP networks (see Chapter 5).

- In general, a much better way of mapping SS7 ISUP messages to H.323 messages without loss of information, H.246 and H.323 annex M works only if the ISUP flavor on both sides is the same, because the ISUP information is considered a 'black box' (SIP-T uses a similar approach, and has the same problem). The Global Transparency Descriptor (GTD), a work in progress authored by Cisco Systems and based on a complete decoding and mapping of the ISUP information to a network-independent format, is a significant improvement for both H.323 and SIP. It is not yet standardized though.

Despite all the improvements that are still required, H.323 is, with MGCP, the most mature of VoIP protocols today. H.323 is not limited to trials, single-vendor networks, or niche markets, such as PC to phone, any more. Most international traffic clearing houses use H.323 to exchange calls, and even incumbent carriers increasingly make use of VoIP and H.323 when terminating international traffic. There are H.323 class 4 transit networks in production today that serve millions of end users, and even much more complex class 5 residential networks with multiple vendors, serving over 250,000 users (FastWeb, Italy), and providing all regulatory features, such as emergency calls, local number portability, lawful call interception, in addition of traditional class 5 services (call forward, call hold, three-way conferencing, etc.).

H.323 has also entered the 3G space: the H.324-M standard used for videoconferencing of recent 3G handsets also uses H.245 for session control and, therefore, interworks seamlessly with H.323. H.323, sometimes misrepresented as a 'legacy' protocol with no future, does seem to keep a significant momentum not just for the size of the installed base, but even in 3G!

3

The Session Initiation Protocol (SIP)

3.1 THE ORIGIN AND PURPOSE OF SIP

The concept of a **session** was first introduced in RFC 2327 (the **Session Description Protocol** or **SDP**) as a set of data streams carrying multiple types of media between senders and receivers. A session can be a phone call, a videoconference, a user taking remote control of a PC, or two users sharing data, chatting, or exchanging instant messages.

The **Session Initiation Protocol (SIP)** was originally defined in RFC 2543 by the **MMUSIC (Multiparty Multimedia Session Control)** working group of the IETF. The MMUSIC working group is focused on loosely coupled conferences as they exist today on the **mBone** (see the companion book, *Beyond VoIP Protocols* Chapter 6 for additional details on the mBone) and is working on a complete multimedia framework based on the following protocols:

- The Session Description Protocol (SDP, RFC 2327) and the **Session Announcement Protocol** (**SAP**, RFC 2974).
- The **Real-Time Stream Protocol** (**RTSP**, RFC 2326) to control real-time data servers.
- SIP.

These protocols complement existing IETF protocols, such as **Real-Time Transport Protocol** (**RTP**, RFC 1889) from the **AVT** (**Audio/Video transport**) working group, used for the transfer of isochronous data, or RSVP from the **IntServ** (**Integrated Services**) working group for bandwidth allocation.

SIP now has its own working group within the IETF which maintains close co-ordination with the MMUSIC group (mainly because the MMUSIC group is still working on improving the Session Description Protocol which is used extensively in SIP).

IP Telephony   O. Hersent, J.P. Petit, D. Gurle
© 2005 John Wiley & Sons, Ltd   ISBN: 0-470-02359-7

One of the initial goals of SIP was to remain simple, and to this purpose 'classic' tele-com protocol design principles, such as protocol layer isolation or complete separation of functional blocks (e.g., message syntax, message encoding and serialization, retransmis-sion), were initially left behind as unnecessary heaviness. The initial SIP RFC aimed at defining in a single 150-page document all the technical details required for session man-agement, covering message reliability, transport, security, and a set of generic primitives for the following functions:

- User location: determination of the technical parameters (IP address, etc.) required to reach an end system to be used for communication, and association of end users with end systems.

- User availability: determination of the reachability of an end user, and the willingness of the called party to communicate.

- Endpoint capabilities: determination of the media types, media parameters and end system functions that can be used.

- Session set-up: 'ringing' a remote device, establishment of media session parameters at both the called and calling parties.

- Session management: including transfer and termination of sessions, modifying ses-sion parameters, and invoking services. The scope of SIP has been restricted to loose multiparty conferences (i.e., functions such as chair control are out of the scope of the current SIP specification). These conference control functions are left to extensions that can be carried within SIP messages.

It took just about a year for SIP to become surprisingly popular for a telecom protocol, but this can be understood from the context. Just like its contemporaries WAP or UMTS, the development of the Session Initiation Protocol occurred at the peak of the Internet bubble, and many start-up companies spent an inordinate amount of marketing resources to promote SIP to omnipotent status. Just as the 'new economy' was being praised as a simple new paradigm vastly superior to the 'old economy', burdened by obsolete conventions and processes, the word began to spread that SIP was a new simple way of designing telecom systems, and that the old public network was unnecessarily complex and inefficient. Surprisingly, even the H.323 protocol, only a couple of years older than SIP, was caught in this wave and began to be criticized for its heaviness and traditional telecom heritage.[1]

After the explosion of the Internet Bubble, the marketing clouds slowly began to dissi-pate, and after a few years of experience the real strengths and weaknesses of SIP are now easier to assess. One strength of the protocol is that the authors constantly try to abstract it from any specific use. For instance, most of the time, SIP primitives will be used to carry 'opaque' objects required for a specific application or media, and not understood

---

[1] Indeed, H.323 is based on the Q.931 protocol used in current telecom networks, and uses the most recent software modeling tool, the **Specification and Description Language** (SDL), capable of automatic test case generation. H.323 defines and separates many functional software modules, uses an abstract syntax (ASN.1) to build its messages, and automatic generation of parser/serializer protocol data units from the abstract syntax.

by the SIP protocol stack.[2] This did stimulate the imagination of developers, and led to interesting ideas (e.g., the use of SIP for instant messaging).

The simplicity of Figure 3.1 also explains much of the initial enthusiasm for SIP.[3] From this simple example, we can see that SIP is very efficient: the callee to caller media channel can be set in exactly one round trip, and the caller to callee media channel can be set up in one and a half round trips. This is much better than the many round trips that were required by the bootstrap nature of H.323v1. That being said, H.323v2 is as efficient as SIP if 'fast-connect' call set-up is used; in fact, the call flow is almost identical.

But the weaknesses of the protocol are also many, and the SIP community is now working hard to solve or improve them:

- Because SIP 'can potentially' be expanded, it is often believed and touted that SIP 'does' everything. This is the well-known 'it's just software' syndrome. Year after year there has been an accumulation of proprietary extensions of SIP, sometimes described in draft documents, sometimes not even documented; but, the lack of a well-defined standardization process has prevented convergence of implementations to occur. The reality, despite claims of the contrary in 'sponsored' interoperability events, is that

Figure 3.1 Simple phone call scenario with SIP.

---

[2] This ability to transport opaque parameters is also present in most other protocols, notably H.323 using the 'non-standard parameters' that can be freely defined within the framework of the standard. Note also that in SIP the size of opaque parameters is restricted by the fact that no segmentation mechanism has been defined for SIP over UDP.

[3] This figure does not use the offer–answer model introduced later by RFC 3261, see Section 3.3.2.3.2.

only the most trivial call flows work across vendors, and they are too trivial to be used in any real-world application. Too often, SIP is still only a nice name hiding a proprietary protocol. As a result, operational SIP networks today are built mostly with infrastructure equipment from a single vendor.

- The PSTN appeared to be a lot more complex than originally anticipated, and therefore SIP lacked many of the features required for proper interworking with the PSTN. H.323v1 had also missed quite a few details, but its Q.931 heritage made it easier to fix the issues quickly in a standard way across vendors. The result is that the vast majority of VoIP networks interworking with the PSTN are today using H.323, not SIP. The few SIP networks interworking with the PSTN required many proprietary extensions to the protocol.

- The increased complexity of the protocol required by the PSTN interworking and other fixes in the initial RFC is becoming hard to manage with the original 'informal all-in-one design' approach. The 'old' way of layering protocols and defining clean functional modules aimed at managing complexity and ensuring consistent quality as software evolved. The latest SIP specifications clearly head back to this modular approach, but the original design and the lack of formal methodology makes this very difficult, and the latest RFCs are still burdened with exceptions and shortcuts between software layers that make the protocol difficult to implement and test. SIP is certainly not 'simple' any more.

In November 2002, the VASA consortium (BellSouth, Chunghwa Telecom, Equant, France Telecom, SBC, Sprint PCS, Telecom Italia Lab, VeriSign, Verizon, WorldCom) published an independent study of 'SIP in Carriers Networks' which emphasized that *'some network operators have experienced significant difficulties in interworking different vendors' products in laboratory trials of both SIP to SIP and SIP to legacy network element interoperability. In contrast with the initial objectives of SIP, operators are driven towards single vendor solutions,'* and concluded: *"For existing networks, the arguments against immediate migration from TDM or H.323 to SIP outweigh the potential benefits."*

SIP is becoming more complex and implements traditional telephony features, while H.323 is implementing some ideas originally from SIP (e.g., H.323 can be used for any communication between named users—there is even a specification for instant messaging[4]). The bottom line is that for use in interactive communications, SIP and H.323 are becoming virtually identical in features and complexity. This chapter will describe the most common PSTN interworking scenarios, which work without extensions of SIP, and will list the major cases where extensions are still required. When available, the documented extensions of major SIP vendors will be discussed.

One of the applications that has emerged out of the multiple theoretical possibilities of the protocol is instant messaging. With the adoption of SIP for instant messaging by Microsoft, it is likely that all vendors will converge on the Microsoft implementation, considered as a de facto standard (and formalized in draft-ietf-simple-presence—version

---

[4] T.140.

07 in November 2002). So, at least the major aspects of instant messaging will stabilize.[5] This section will describe this implementation of instant messaging, which conforms to the main guidelines of RFC 3265. Unfortunately, the details are still not thoroughly documented and may evolve over time.

3.2 FROM RFC 2543 TO RFC 3261

SIP remained a draft document for a long time before if was first published as an RFC in March 1999 (RFC 2543). The first published version of the protocol was SIP 2.0. Unfortunately, this first version of the RFC was trying to embrace too much, contained many errors, and was too vague and ambiguous to be a real specification document. It was more a sort of technical brainstorming document, and was taken as such by the many start-up companies that began to implement SIP products. The first trial SIP networks all used their "flavor" of SIP, with their own corrections and expansions to the original SIP specification, and used only the simplest call flows defined by the RFC.

As the first useful feedback was gathered from these trials, the RFC was updated with nine 'bis' versions, and finally all changes were merged in June 2002 in a new RFC, RFC 3261. Important aspects of the initial specification are now split in separate RFCs. Although RFC 3261 does not update the SIP version number, which remains SIP 2.0, it not only corrects errors and clarifies ambiguities, but really makes major changes to RFC 2543. The protocol is now more robust and more clearly documented, although the RFC is still a bit verbose and vague, with expressions like 'modest level of backwards compatibility' or 'almost identical' that can be misleading. RFC 3261 is really a major new version of SIP, and is not backwards-compatible with RFC 2543, although most simple call flows will work across the two RFC versions.

In the process, the SIP protocol lost the apparent simplicity of its early days, and the size of the main RFC nearly doubled with 270 pages. The new RFC is an umbrella document that points to other RFCs for specific details or applications. The complete documentation includes over 400 pages. Among the most important documents are:

- RFC 3262: Reliability of Provisional Responses in Session Initiation Protocol (SIP). This RFC is required in all cases where SIP needs to interwork with a telephone network.
- RFC 3263: Session Initiation Protocol (SIP): Locating SIP Servers. The location of SIP servers is really an independent module in a SIP implementation and is now documented separately from the main SIP RFC.
- RFC 3264: An Offer/Answer Model with Session Description Protocol (SDP). This was one of the most necessary clarifications of the original SIP RFC, where the exact use of the SDP syntax was ambiguous and led most vendors to implement a single

---

[5] There is still one major competitor of SIP for instant messaging: the JABBER open-source protocol, used for instance, by France Telecom.

codec or their own interpretation of codec negotiation. The new RFC is much clearer and implements a mechanism that is almost identical to the H.323 FastStart and the tunneled H.245 logical channel management.

- RFC 3265: Session Initiation Protocol (SIP)-specific Event Notification. This RFC is used by some vendors to transport DTMF tones, but its main use is for instant messaging. Only the main methods are specified, the content of events is still not standardized.

- RFC 3266: Support for IPv6 in Session Description Protocol (SDP). While IPv6 was taken into account since the beginning in H.323, it required some syntax extensions in SIP, now covered by this new RFC. The use of IPv6 for IP telephony remains very questionable, however, because VoIP is the most sensitive application to IP protocol overheads, which are going from bad to worse in IPv6. In addition, the IP address depletion issue which is the main motivation for the introduction of IPv6 can be overcome by use of application-level proxies and intelligent use of private IP addresses (see Chapter 6 for details. The handling of quality of service is also virtually identical in IPv4 and IPv6.

3.3   OVERVIEW OF A SIMPLE SIP CALL

3.3.1   Basic call scenario

In this section we assume that the initiator of the call knows the IP address of the called endpoint. For instance, the caller might be calling the following SIP address, also called a **Uniform Resource Locator (URL)** or a **Uniform Resource Identifier (URI)**:

sip:john@192.190.132.31

In Section 3.4.1.2 we will see that there are many other types of SIP addresses, this one is just a simple case where the IP address of the called endpoint is directly specified. Note that the syntax looks similar to web addresses (e.g., http://www.netcentrex.net). This is because they both use the URI syntax, defined in RFC 2396, which begins with a **scheme** portion, before the colon (':'). The scheme portion indicates how the rest of the string is to be interpreted, and which syntax to expect. If the scheme is 'http', then the syntax for an http resource is expected after the colon: a double slash ('//'), then a host name, then an optional path.

SIP uses two scheme names: 'sip' for communications over non-secure transport protocols, and 'sips' for communications over secure transport protocols. SIP can also be used with the 'tel' URI scheme defined in RFC 2806. The minimal expected syntax after 'sip:' or 'sips:' is a **host name**. A host name can either be described directly by its IP address in dotted form (e.g., 10.11.10.13), or using its name in the Domain Name System (e.g., hostname.subdomainname.domainname.rootdomain). The DNS system was described in Section 2.5.3.3. Usually, a SIP call is to a specific person, and therefore the SIP URL syntax allows more optional parameters to be used, the generic syntax is:

sip:user:password@host:port;uri-parameters?headers

Our example uses only the user and host portions.

SIP entities communicate using **transactions**. SIP calls a transaction a request requiring a specific action (e.g., the INVITE request below) and the response(s) it triggers (200 OK in our example) up to a final response (see the definition below, all 2xx, 3xx, 4xx, 5xx and 6xx responses are final). The initiator of a SIP request is called a **SIP client**, and the responding entity is called a **SIP server** for this transaction.

Most communications using SIP need several transactions. If the caller is Mark and the callee is John, then the end systems used by Mark and John (the SIP RFC calls them **user agents**) will play the role of the client or the server for each transaction, depending on which user agent initiates each transaction (Figure 3.2).

SIP uses several types of request methods: REGISTER, INVITE, ACK, CANCEL, BYE, OPTIONS (defined in the main RFC), PRACK (added in RFC 3262, and required for interoperability with the PSTN), SUBSCRIBE, NOTIFY (both used for instant messaging, these were added in RFC 3265).

The simple communication scenario in Figure 3.2 uses the INVITE, ACK, and BYE request methods. A SIP client calls another SIP endpoint by sending an INVITE request message. The INVITE message usually contains enough information to allow the called terminal to immediately establish the requested media connection to the calling endpoint. The called endpoint needs to indicate that it is accepting the request. This is the purpose of the 200 OK response message. Since the request was an invitation, the 200 OK

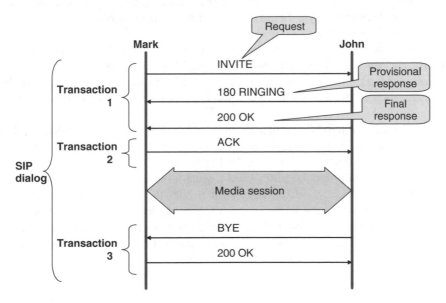

Figure 3.2 SIP dialog and SIP transactions.

response usually also contains the media capabilities of the called endpoint, and where it is expecting to receive the media data.

The other messages are explained in detail below, the ACK is a simple acknowledge of the 200 OK final response, and BYE is the request to 'drop the call'.

This exchange of SIP transactions between two user agents during a communication is called a SIP **dialog**. The combination of the To tag, From tag, and Call-ID completely defines a SIP dialog (these information elements are defined below).

3.3.1.1   Call initiation details

The messages exchanged by a SIP client and a SIP server are independent of the under-lying transport protocol except for some details (e.g., the size of non-TCP messages is limited). The original SIP RFC only required SIP systems to support the UDP transport protocol. RFC 3261 now requires SIP systems to support both the UDP and TCP trans-port protocols, but UDP is still the most widely used protocol, because it provides better control on retransmission and latency. The only issue of UDP is that it cannot transport large amounts of information without causing packet fragmentation (SIP does not define any application-level fragmentation mechanism). RFC 3261 recommends using TCP or any other reliable congestion-controlled protocol (defined in RFC 2914), if the request size is within 200 bytes of the path MTU, or if it is larger than 1,300 bytes and the path is unknown. All implementations must still be able to handle packets up to 65,535 bytes (including IP and UDP headers).

Connections over UDP or TCP require a port number. If no port is specified in the SIP URI, the connection is made to port 5060 for the transport protocols UDP, TCP, and **SCTP (Stream Control Transmission Protocol**, RFC 2960). If the transport protocol used is **TLS (Transport Layer Security**, a secure transport protocol using TCP, defined in RFC 2246), the default port is 5061.

When using TCP, the same connection can be used for all SIP requests and responses of a dialog, or a new TCP connection can be used for each transaction. If UDP is used the address and port to use for the answers to SIP requests is contained in the Via header parameter of the SIP request. Replies must not be sent to the IP address of the client.

Since no protocol or port is specified in our sample SIP URI (sip:john@192.190.132.31), Mark's user agent defaults to UDP, and will send its first SIP INVITE message over UDP to IP address 192.190.132.31, port 5060. When UDP is used, only one SIP mes-sage per UDP datagram may be sent; when a stream-oriented protocol like TCP is used, the message framing uses the Content-Length header to determine the end of the mes-sage and the beginning of the next. Therefore, this header must be present when SIP is used over stream-oriented protocol. This violation of the strict transport and presentation protocol layering avoids specifying a separate transport layer-framing layer for stream protocols.[6]

---

[6] H.323 uses a stricter protocol layering and uses RFC 1006 for the purpose of transporting messages over a stream-oriented protocol.

The dialog is initiated by the following INVITE message:

```
INVITE sip:john@192.190.132.31 SIP/2.0
Via: SIP/2.0/UDP 10.11.12.13;branch=z9hG4bK776asdhds
Max-Forwards: 70
To: ''John'' <sip:john@192.190.132.31>
From: :''Mark'' <sip:mark@10.11.12.13>;tag=1928301774
Call-ID: a84b4c76e66710@10.11.12.13
CSeq: 314159 INVITE
Content-Type: application/sdp
Content-Length: 228
```

$$v = 0$$
$$o = \text{mark } 114414141 \ 12214 \ \text{IN IP4 } 10.11.12.13$$
$$s = -$$
$$c = \text{IN IP4 } 10.11.12.13$$
$$t = 0 \ 0$$
$$m = \text{audio } 49170 \ \text{RTP/AVP } 0$$
$$a = \text{rtpmap:0 PCMU/8000}$$
$$m = \text{video } 51372 \ \text{RTP/AVP } 31$$
$$a = \text{rtpmap:31 H261/90000}$$
$$m = \text{video } 53000 \ \text{RTP/AVP } 32$$
$$a = \text{rtpmap:32 MPV/90000}$$

The last part of the message, after the blank line, is the media description using the Session Description Protocol, which will be described in detail in section 3.3.2.3.2.1. We will focus first on the Request start line and the most important headers which are mandatory in any SIP request (in bold font). As noted above, SIP is not strictly independent of the transport protocol, and the Content-Length header is mandatory only over stream transport protocols, even if there is no payload in the SIP message.

3.3.1.1.1 Start line

The start line indicates the request method (INVITE), followed by the request-URI which indicates the user or service to which this request is being addressed. For all requests except REGISTER, it is set to the same value as the URI in the To header. This is no longer true in the specific **strict routing** mode detailed in Section 3.4.2.3. The last element of the start line is the SIP protocol version, which is SIP/2.0 for endpoints implementing RFC 2543 or 3261.

3.3.1.1.2 Via

The only way to distinguish the two versions of SIP (RFC 2543 or RFC 3261) is by looking at the Via header, which in RFC 3261 must contain a branch parameter beginning with 'z9hG4bK'. The primary purpose of the Via header is of course different. If UDP is used the address and port to use for the answers to SIP requests is contained in the Via header parameter of the SIP request. Replies must not be sent to the IP address of the client.

In our case the SIP call is established directly between the caller and the callee, but when the call is routed through several SIP proxies, each proxy adds its own Via header to the received request before forwarding it. This allows the route of the SIP request to be traced, and is useful for several purposes, including loop detection. We will study this more complex call flow in Section 3.4.2 when introducing SIP proxies. The strange role of the z9hG4bK 'magic cookie' is an example of the few remaining violations of functional block isolation within the design of the protocol in RFC 3261.

3.3.1.1.3 Max-Forwards

This mandatory header is used to prevent routing loops when the call is routed by proxies in the network. Each proxy decrements the counter by one as it forwards the request, and responds with an error if it receives a request with the counter set to 0. This is a very useful feature, which has been added only in version 5 of H.323.

3.3.1.1.4 From

The From header identifies the caller. The syntax of the header places the display name between double quotes, unless the character set used in the display is not creating parsing issues, and the SIP URI of the caller between '<' and '>'. This syntax is defined in RFC 2822 and already used for email applications. If the caller wishes to remain anonymous, he can use the keyword 'Anonymous', without quotes, instead of a display name (e.g., From: Anonymous< ... >).[7] If no display name is used, no brackets are required for the URI (e.g., From: sip:mark@10.11.12.13;tag=1928301774). In the new SIP RFC, the tag parameter is mandatory; it is one of the key elements used to uniquely identify a dialogue. This parameter was optional in RFC 2543; therefore, parsing and comparing the URI fields in an incoming mid-dialog request message was necessary to identify a relevant existing dialogue. The use of tags makes this identification more formal and efficient. RFC 3261 considers the absence of a tag equivalent to tag $= 0$.

3.3.1.1.5 To

The To header identifies the target user or the target resource of the SIP request. It uses the same syntax as the From header, except it lacks the tag parameter in the initial INVITE, but this will be added by the callee in the response to complete the unique identification of the dialog.

3.3.1.1.6 Call-ID

The Call-ID header contains a globally unique identifier for this call. The easiest way to form one is to generate a locally unique identifier on the user agent, and concatenate

---

[7] This method is not sufficient to comply with the requirements of the Calling Line Identity Restriction service which is mandatory in public telephony applications. In ISDN and H.323, a specific information element (octet 3a) is used to tell the network whether the number should be presented to the called party or not. The number cannot simply be erased because it must be provided in the case of emergency calls or for intercepted calls. In very simple SIP networks with a single proxy, the proxy could restore the identity on the fly, but it would require two separate fields, as in ISDN, to convey the identity information *and* the CLIR information in larger, multi-proxy networks.

a globally unique identifier for the user agent (e.g., its public IP address[8] or its DNS host name).

The combination of the To tag, From tag, the Via branch parameter, and Call-ID completely defines a SIP dialog.

3.3.1.1.7 CSeq

The CSeq header field is an identifier which serves to match the request and responses of an active transaction. It is formed of a method name which must match that of the request, and a sequence number which can start at an arbitrary value. The responses to a request are required to have a CSeq header identical to the CSeq header of the request.

Each client is required to increase the CSeq number by one for each request they send during a dialogue (except for ACK and CANCEL requests, where the sequence number is used to identify the target of the action requested). Since provisional responses are not transmitted reliably (see 3.3.1.1.8.3), it is still possible to receive non-consecutive Cseq numbers from a server.

3.3.1.1.8 Dialog and transaction identifiers, message retransmission

The SIP dialogue is identified by the From tag, To tag, Call-ID, and Via branch combination (Figure 3.3). Once the dialogue is established, all requests and responses of the dialogue must include these header values. This makes it easy for a user agent to identify the relevant dialog on receiving a SIP message.

Each transaction is identified by the common value of its CSeq header field (both the method name and sequence number must be identical). The value of the CSeq header must be distinct for distinct transactions within the dialog. The only exceptions are the ACK after a non-2xx response (see Section 3.3.1.1.8.1 for details) and CANCEL transactions. The CANCEL transaction uses the same CSeq identifier as the transaction it is attempting to cancel, but the method name is set to CANCEL. The ACK transaction is used only in relation with a prior INVITE transaction, which uses a three-way handshake, while all other types of transactions use only a two-way handshake.

3.3.1.1.8.1 Non-invite transactions

When used over unreliable transport protocols, the reliability of non-invite transactions relies on message retransmission.

The sender of a request will first retransmit the message if it does not receive a provisional or final response (see 3.3.2.2) within 500 ms[9] (or a better estimation of the network round trip time if the user agent calculates one using the timestamps in SIP messages). It will keep retransmitting the request until it receives a response, doubling the retransmit interval at each occurrence up to an interval of 4 s. If a provisional response is received,

---

[8] The address used in our example (10.11.12.13) is private; so, in order to really generate a unique Call-ID we should use a DNS name for the call ID, not this address. We kept it for simplicity.

[9] The SIP specification specifies all times in multiples of a default timer T1. In order to simplify reading, we calculated all timer values using the T1 default value of 500 ms recommended in RFC 3261.

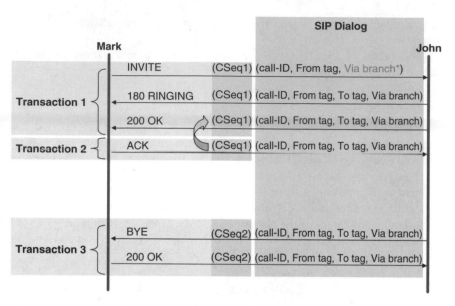

Figure 3.3 A dialog is identified by the Call-ID, From tag, To tag, and Via branch. * If a forking proxy is in the path, this branch parameter may be changed. For this reason the branch value found in the response identifies the dialog.

then the retransmission still occurs until a final response is received, but the retransmission interval is immediately set to 4 s. The client will wait up to 64∗500 ms, or 32 s, for a final response. After this delay it will consider the transaction has failed. In the case the transport layer indicates a failure (e.g., an ICMP-unreachable IP packet has been received), then the transaction fails immediately without trying any retransmission.

If the server response is lost, then the client will retransmit the request. For this reason the server side of the transaction simply retransmits its last response each time it receives a retransmitted request indicating the response has been lost. Note that because only the *last* response is retransmitted, some **provisional responses** may be lost. SIP does *not* guarantee delivery of provisional responses (this causes problems in certain cases, but RFC 3262 has introduced the new PRACK request as an extension to SIP to solve the issue).

Figure 3.4 is an example of retransmission of a BYE transaction. Even after it has sent a final response, the server will keep the response in memory for 32 s (64∗500 ms) in the case an incoming client request requires a retransmission after the server has sent the final response. This happens if the final response is lost.

It is important to realize that this basic two-way handshake works only if the final response to the request arrives quickly after the request has been sent. SIP expects and requires non-invite transactions to complete within a couple of seconds. If this two-way handshake was used for transactions that take a longer time to complete, then the request would be resent multiple times, which would obviously be very inefficient. Because of this, a different strategy is required for INVITE transactions.

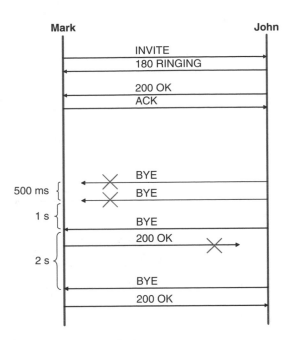

Figure 3.4 Non-invite request retransmission.

3.3.1.1.8.2 *Invite transactions*

(a) A three-way handshake

The handling of the INVITE transaction in SIP is completely different from the handling of other transactions. The handling of the INVITE is one of the most complex aspects of SIP.

The two-way transaction used for non-invite transactions presents a number of issues:

- The initial transaction is resent every 4 s until the final response arrives. In a telephony application, the 200 OK signaling that a user has picked up the phone can arrive up to 3 min after the INVITE has been sent (this is the amount of time most telephone networks let a phone ring before canceling the call). In some PSTN interworking applications (prepaid calling card, network prompts), the 200 OK may never arrive. In order to avoid unnecessary overhead the retransmission of the INVITE request stops as soon as a provisional or final response arrives.

- If the 200 OK response is lost, it will be retransmitted 4 s later in the previously described retransmission scheme. This is obviously unacceptable because no audio path can be established during this delay between the caller and the callee. Instead, if the caller has received a 180 RINGING provisional response, the status of his call line would continue to appear as 'ringing'. This situation is avoided in an INVITE transaction, because the server expects to receive an immediate ACK after it sends a 200 OK. If the ACK is not received, then the response is retransmitted after 500 ms.

Figure 3.5   INVITE request retransmission.

The three-way handshake works as follows for the client side:

- Over UDP the INVITE request is resent after 500 ms if no provisional or final response is received. The INVITE retransmission continues, but with a retransmission interval that doubles every time, until a provisional or final response arrives (Figure 3.5). The transaction attempt aborts after seven retransmissions. Note that the server side is required to send back a 100 TRYING provisional response within 100 ms, unless it knows that it is going to send a final response within 200 ms. So, the INVITE retransmission mechanism should normally trigger only if a message gets lost, not if the server side is slow.

- Over TCP or a reliable transport the INVITE request in not retransmitted.

The server side of the INVITE transaction over UDP will retransmit the last provisional response if it receives a retransmitted INVITE request (this means one or more provisional responses have been lost). The handling of the final response is different:

- In RFC 2543 it is retransmitted after 500 ms if an ACK has not been received, then continues to be retransmitted, with intervals that double every time until an ACK is received. The retransmission of the final response is aborted after seven unsuccessful retransmissions, or if a BYE request is received for this dialogue, or, in the case of 3xx, 4xx, and 5xx responses, if a CANCEL request is received. If the retransmission that was aborted was for a 200 OK response, the server should generate a BYE request, in case the 200 OK response did arrive to the client user agent.

- RFC 3261, with its cleaner specification of the transaction layer, modified RFC 2543. It does not require retransmissions over reliable transports for 3xx responses. The 200 OK response needs to be retransmitted even on reliable transports because it may be routed through several proxies, which may forward it on UDP. Since UDP is unreliable the 200 OK could get lost, but SIP does not allow proxies to participate in the 200 OK reliability mechanism, which is ensured end to end only (see the following subsection on the ACK request). Therefore if the user agent server did not retransmit the 200 OK response, it could get lost in the network. RFC 2543 extended this behaviour to all final responses, not just 200 OK, in order to avoid adding exceptions to the already complex specification. In RFC 3261, which has a more formal definition of transactions, only 200 OK responses are retransmitted over reliable transports.

(b) The ACK request

The ACK request which completes the three-way handshake is formed as follows:

- Most of the headers are identical to the original INVITE headers—in the case of route header fields they must be identical.
- The To header will probably contain a tag that was added by the server side in its response.
- The CSeq header serial number is identical to the CSeq serial number of the INVITE request, which is an exception to the general rule to increase sequentially the CSeq header for new requests. This is so we can correlate the ACK to the INVITE transaction. The CSeq method portion is 'ACK'.

This is an example of an INVITE request and its ACK:

```
INVITE sip:john@192.190.132.31 SIP/2.0
    Via: SIP/2.0/UDP 10.11.12.13;branch=z9hG4bK776asdhds
    Max-Forwards: 70
    To: ''John'' <sip:john@192.190.132.31>
    From: :''Mark'' <sip:mark@10.11.12.13>;tag=1928301774
    Call-ID: a84b4c76e66710@10.11.12.13
    CSeq: 314159 INVITE

ACK sip:john@192.190.132.31 SIP/2.0
    Via: SIP/2.0/UDP 10.11.12.13;branch=z9hG4bK776asdhds
    Max-Forwards: 70
    To: ''John'' <sip:john@192.190.132.31>;tag=12344235
    From: :''Mark'' <sip:mark@10.11.12.13>;tag=1928301774
    Call-ID: a84b4c76e66710@10.11.12.13
    CSeq: 314159 ACK
```

The rules applying to the ACK request in a SIP network are also a bit complex. The handling of 200 OK final responses and 3xx, 4xx, and 5xx final responses is different. If an INVITE transaction is routed through several proxies, then each proxy can acknowledge 3xx, 4xx, 5xx, and 6xx responses by sending their own ACK. This applies only to proxies capable of acknowledging such requests locally (e.g., stateless proxies do not have this intelligence and remain passive, acting as simple message routers). But 200 OK responses are *always* acknowledged by the client user agent. This difference was made because the SIP specification wants SIP clients to know all the servers that have accepted an INVITE transaction. Because of forking proxies, more than one server may accept an INVITE, if this INVITE has been duplicated in the networks and a single INVITE request can generate more than one dialog.

In Figure 3.6, one INVITE generates two dialogs. The Figure also shows that, while provisional responses are not transmitted reliably, the final 200 OK response is transmitted reliably. Because multiple 200 OK responses can be received for a single INVITE, in theory the client could continue to wait for 200 OK responses for ever after sending an INVITE request. The SIP specification limits the waiting time to 32 s after the first 200 OK was received.

In Figure 3.7, a forking proxy generates an ACK for a 486 final response, while it forwards the 200 OK response to the client user agent.

The ACK is a very special type of request:

- It is never acknowledged. If an ACK gets lost, the server will retransmit the final response, and the client will retransmit the ACK when receiving the duplicate final response. This mechanism is very similar to the two-way handshake of non-invite

Figure 3.6   Forking an INVITE request.

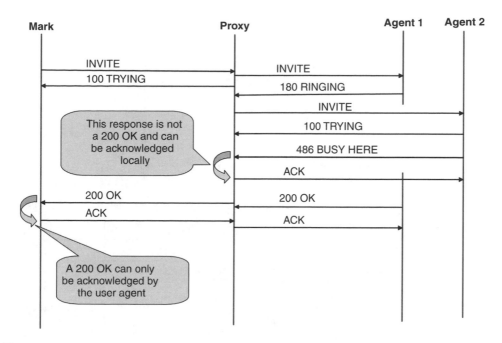

Figure 3.7 200 OK responses must be acknowledged end to end.

transactions, with the final response playing the role of the request, and ACK playing the role of the response.

- Because the ACK is never acknowledged, proxies cannot signal any failure back to the client. This requires a specific procedure for authentication: since the ACK cannot be challenged, it must contain the same credentials as the INVITE (see Section 3.6.2 for further details).

- If the response contains a Contact header with a URI, the ACK can be forwarded directly to this URI.

3.3.1.1.8.3 Provisional responses

(a) Reliability

Provisional responses are not delivered reliably in SIP. Figure 3.8 gives two examples of this. In the first example, a provisional response arrives soon enough to prevent the INVITE retransmit timer from firing. The two previous provisional responses (RINGING and QUEUED-1) are lost and never received by Mark user agent.

In the second example a forking proxy duplicates the INVITE of Mark and sends it to two user agents. The 180 RINGING provisional response of Agent 2 is lost, and Mark does not know that Agent 2 was contacted when the 200 OK from Agent 1 arrives.

(b) PRACK method

The new PRACK method was introduced in RFC 3262 (June 2002) to fix the issue of reliability of provisional responses. This issue was in fact a showstopper for any real

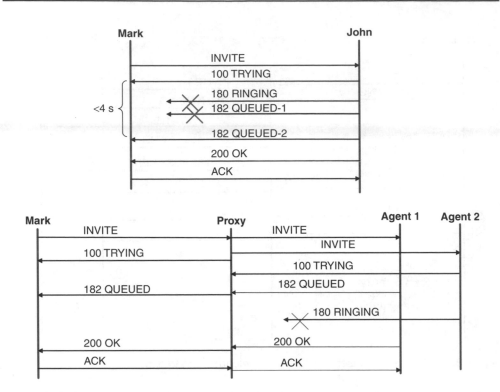

Figure 3.8   Lack of reliability for provisional responses in baseline SIP.

deployment of SIP for telephony, as many interworking scenarios with the PSTN were not supported. Mobile networks, for instance, make extensive use of pre-call announcements, which could not be delivered reliably in SIP prior to RFC 3262.

The optional PRACK extension mimics the end-to-end reliability algorithm of the 200 OK response. Provisional responses are retransmitted periodically until the acknowledgment (PRACK) arrives. The major difference is that PRACK, unlike ACK, is itself a normal SIP method, acknowledged hop by hop by each stateful proxy, and requiring its own response from the server. PRACK transmission reliability is ensured as for a normal request by the expected 200 OK; this means that if the client receives a retransmitted provisional response, it should not retransmit the PRACK, but rely on the PRACK response to decide whether it should retransmit it or not.

Each provisional response contains a serial number in a RSeq header, mirrored in the corresponding PRACK method RAck header (formatted as RAck: <response number> <Cseq number> INVITE). There should be a PRACK for each provisional response, unlike reliability mechanisms such as TCP, also based on serial numbers, which are cumulative (the acknowledgment serial number validates all received messages with lower values).

The sender of the INVITE should indicate that it supports the PRACK mechanism by including a 100rel option tag in the Require header field. This option can be rejected by the receiver with a 420 BAD EXTENSION response (unsupported header field with 100rel option tag). If it accepts the PRACK option, the receiver may send all 1xx responses (except 100 TRYING) reliably. In reliable provisional responses, it needs to include a Require header field with option tag 100rel, and a Rseq header (unique within the transaction).

Unacknowledged provisional responses are simply retransmitted with an exponentially increasing delay, with the same serial number. New provisional responses are sent with the next higher serial number (see Figure 3.9).

If a media offer (see Section 3.3.2.3.2.3) appears in a reliable provisional response (e.g., 183 SESSION PROGRESS), the PRACK should contain an answer to that offer. A PRACK can also contain an offer, in which case the corresponding 200 OK must contain an answer.

3.3.1.1.8.4 *Managing the complexity*

The exact procedure of the SIP reliable transmission mechanism therefore depends on the transport protocol used (reliable or not), the method used (special handling of invite requests), the type of response in the case of an invite request, and the existence of a special optional mechanism for provisional responses (PRACK). This complexity is typical of a monolithic design, and unfortunately tends to lead to spaghetti code and subtle bugs. RFC 3261 attempts to reintroduce functional layers more formally in the specification, and isolates a functional block responsible for retransmissions: which handles **transactions**.

If the request is an INVITE, the transaction includes not only the provisional responses and the final response, but also, *if* the final response was not a 2xx response, the ACK

Figure 3.9 Reliability of provisional responses using the PRACK request. Example of a gateway to a cellular network transmitting in-band announcements with a 183 SESSION PROGRESS (ISDN Progress); one PRACK is lost.

acknowledging the final response. The transaction layer of user agent clients (and some types of proxies) handles ACKs of non-200 OK responses for invite requests. If the final response is a 2xx, then the ACK is an independent transaction. This is because the SIP specification requires the 200 OK response to be handled by the application, and not by the transaction functional block. The transaction functional block is not allowed to retransmit 200 OK responses; this task must be handled by the user agent application (called the 'user agent core' in the specification).

For all other requests a new transaction is created, which includes all the responses to that request, up to the final response (and its retransmissions if any). There is no ACK, and retransmissions are handled completely by the transaction functional block.

This formalism improves the robustness of the specification, but there are still an unusual number of exceptions to the protocol-layering principles in the current SIP specification. As SIP evolves, this will be an obstacle to managing specification complexity. A possible way to simplify the SIP specification would be to drop direct support for UDP, and use instead a reliable transport layer on top of UDP, but this would cause backward compatibility problems to SIP implementations. Note that restricting SIP to TCP only would be simpler, but TCP does not always have the right deterministic latency properties to support high-quality telephony applications. Dropping support for forking proxies would also simplify SIP significantly;[10] there is no significant application using forking proxies yet, and most forking functions can be handled by 'back to back user agents'.

3.3.1.2   *Terminating a call*

The above example is a simple and successful call set-up. Figure 3.10 is a more complete example (in which Mary calls John) that includes the call termination by John. If Mary had terminated the call, she would have sent the BYE request, and the From and To fields would be reversed. The media flows are not shown, but the signaling messages include all mandatory headers.

Some SIP headers have abbreviated forms that can help in keeping the total size of a message below the MTU. In this example John's terminal is using the abbreviated form.

3.3.1.3   *Rejecting a call*

There are occasions when John may be unable to receive a call from Mary. He may not be at home, may not be willing to answer, or he may be already in another conversation. Some of these situations can be expressed in the reply message. SIP provides codes for the usual causes, but also defines more sophisticated replies, such as GONE, PAYMENT REQUIRED, or FORBIDDEN.

Figure 3.11 is an example of a simple BUSY HERE reply. This reply tells Mary that John cannot be reached at this location (but she might try to reach another location, such as John's mobile phone through a gateway, or via voicemail). Another reply, 600 BUSY

---

[10] This restriction was adopted by 3GPP.

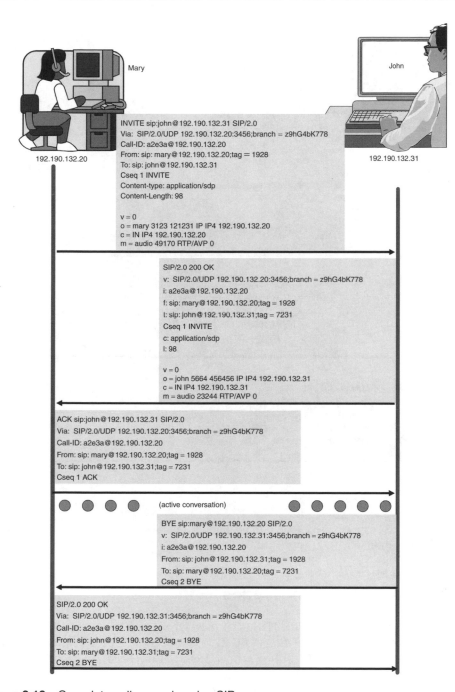

INVITE sip:john@192.190.132.31 SIP/2.0
Via: SIP/2.0/UDP 192.190.132.20:3456;branch = z9hG4bK778
Call-ID: a2e3a@192.190.132.20
From: sip: mary@192.190.132.20;tag = 1928
To: sip: john@192.190.132.31
Cseq 1 INVITE
Content-type: application/sdp
Content-Length: 98

v = 0
o = mary 3123 121231 IP IP4 192.190.132.20
c = IN IP4 192.190.132.20
m = audio 49170 RTP/AVP 0

SIP/2.0 200 OK
v: SIP/2.0/UDP 192.190.132.20:3456;branch = z9hG4bK778
i: a2e3a@192.190.132.20
f: sip: mary@192.190.132.20;tag = 1928
t: sip: john@192.190.132.31;tag = 7231
Cseq 1 INVITE
c: application/sdp
l: 98

v = 0
o = john 5664 456456 IP IP4 192.190.132.31
c = IN IP4 192.190.132.31
m = audio 23244 RTP/AVP 0

ACK sip:john@192.190.132.31 SIP/2.0
Via: SIP/2.0/UDP 192.190.132.20:3456;branch = z9hG4bK778
Call-ID: a2e3a@192.190.132.20
From: sip: mary@192.190.132.20;tag = 1928
To: sip: john@192.190.132.31;tag = 7231
Cseq 1 ACK

(active conversation)

BYE sip:mary@192.190.132.20 SIP/2.0
v: SIP/2.0/UDP 192.190.132.31:3456;branch = z9hG4bK778
i: a2e3a@192.190.132.20
From: sip: john@192.190.132.31;tag = 1928
To: sip: mary@192.190.132.20;tag = 7231
Cseq 2 BYE

SIP/2.0 200 OK
Via: SIP/2.0/UDP 192.190.132.31:3456;branch = z9hG4bK778
Call-ID: a2e3a@192.190.132.20
From: sip: john@192.190.132.20;tag = 1928
To: sip: mary@192.190.132.31;tag = 7231
Cseq 2 BYE

Figure 3.10 Complete call scenario using SIP.

Mary

192.190.132.20

John

192.190.132.31

INVITE sip:john@192.190.132.31 SIP/2.0
Via: SIP/2.0/UDP 192.190.132.20:3456;branch = z9hG4bK778
Call-ID: a2e3a@192.190.132.20
From: sip: mary@192.190.132.20;tag = 1928
To: sip: john@192.190.132.31
Cseq 1 INVITE
Content-type: application/sdp
Content-Length: 98

v = 0
o = mary 3123 121231 IP IP4 192.190.132.20
c = IN IP4 192.190.132.20
m = audio 49170 RTP/AVP 0

SIP/2.0 486 BUSY HERE
Via:SIP/2.0/UDP 192.190.132.20:3456;branch = z9hG4bK778
Call-ID: a2e3a@192.190.132.20
From: sip: mary@192.190.132.20;tag = 1928
To: sip: john@192.190.132.31;tag = 7231
Cseq 1 INVITE

ACK sip:john@192.190.132.31 SIP/2.0
Via:SIP/2.0/UDP 192.190.132.20:3456;branch = z9hG4bK778
Call-ID: a2e3a@192.190.132.20
From: sip: mary@192.190.132.20;tag = 1928
To: sip: john@192.190.132.31;tag = 7231
Cseq 1 ACK

Figure 3.11   Case of a busy call.

EVERYWHERE, can be used to let it be known that John cannot be reached at any location at this moment.

3.3.2   Syntax of SIP messages

SIP messages are encoded using the HTTP/1.1 message syntax (RFC 2068). The character set is ISO 10646 with UTF-8 encoding (RFC 2279, see Section 3.5.1.1.1 for more details).

Lines are terminated with CR+LF (Carriage Return, Line Feed), but receivers should be able to handle CR or LF as well.

There are two types of SIP messages: REQUESTS and RESPONSES. They share a common format as indicated in Figure 3.12.

Some header fields are present in both requests and answers; they are part of the 'General Header':

Figure 3.12 SIP message format: requests and responses (SP is an abbreviation for 'single space').

- **Call-ID** (example: Call-ID: f81d4fae-7dec-11d0-a765-00a0c91e6bf6@foo.bar.com). The Call-ID parameter serves many purposes. In REGISTER and OPTIONS requests it serves to match requests with the corresponding responses. For INVITE and REGISTRATION requests it also helps to detect duplicates (duplicate invite requests can occur when there is a forking proxy in the path). Successive INVITE requests with the same Call-ID but a different payload can be used to dynamically change parameters in a conference. The first part of the Call-ID is meant to be unique within each host, and the last part, a domain name or host IP address, makes it globally unique. In case an IP address is used, it must be routable (i.e., private addresses such as 10.x.x.x cannot be used). A new Call-ID must be generated for each call.

- **Cseq** (example: Cseq: 1234 INVITE). Every request has to have a Cseq header field, which is composed of an unsigned sequence number and the method name. Within a call, the sequence number is incremented at each new request (unless the request is a retransmission of a strictly identical previous request, as shown in Figure 3.13), and starts at a random value. The only exceptions are the ACK and CANCEL requests which keep the Cseq number of the acknowledged reply (for ACK) or the cancelled request (for CANCEL). The server must copy the Cseq value of the request in the corresponding replies.

- **From** (example: From: "MyDisplayName"<sip:myaccount@company.com> ;tag=221411414). This field must be present in all requests and responses. It contains an optional display name and the address of the originator of the request. Optional tags can be appended. Note that the From field contained in SIP replies is simply copied from the request and therefore does not designate the originator of the reply (the tag became mandatory in RFC 3261).

Figure 3.13   Usage of CSeq header.

- **To**   (example: To: Helpdesk <sip:helpdesk@company.com>;tag = 287447). This field must be present in all requests and responses and indicates the intended destination of a request. It is simply copied in responses. The tag is used mainly when a single SIP URI designates several possible endpoints (as in the case of a helpdesk). In this case a random tag is appended in the replies to allow a client to distinguish replies from individual endpoints (this tag became mandatory in RFC 3261).

- **Via**   (example: Via: SIP/2.0/UDP PXY1.provider.com; received 10.0.0.3). The Via field is used to record the route of a request, in order to allow intermediary SIP servers to forward the replies along the same path.

In order to achieve this, each proxy adds a new Via field with its own address to the list of existing Via fields. The request receiver can add optional parameters of the Via field (e.g., to indicate that it received the request from an address that is not the address contained in the previous hop's Via field it adds a 'received' parameter). Using this information, a proxy can forward the replies to the original sender, even if there is a **Network Address Translation (NAT)** device in the path.[11] In Figure 3.14, the caller only uses private

---

[11] This always works if the NAT device uses a permanent one-to-one mapping between internal and external IP addresses. In the more frequent case of N-to-one mapping using the external source port as an index to the internal address (Network Address and Port Translation NAPT), this works only when using a fixed port forwarding, or when the NAT function opens a reverse forwarding path for UDP packets arriving at the external port which lasts long enough for the SIP response to arrive. In general, NAPT traversal is a very complex topic, see Chapter 5.

Figure 3.14   Usage of Via header for messages exchanged though NAT.

addresses (10.x.x.x), and is connected to the Internet through a router doing NAT. When PXY1 receives the INVITE request, the request's source IP address is no longer 10.0.0.6, but has been changed by the NAT device to 192.9.10.10. So, the destination proxy records this information in a 'received' parameter. When it receives the reply for this request, it will discard the topmost Via header, and parse the next one: from the value of the 'received' parameter, it knows that it must forward the request to 192.9.10.10.

In fact the mechanism just described only works for NAT functions which translate IP addresses, not for NAPT functions which also translate the source port of the packet (see Chapter 5 for more details). The default SIP behaviour as described in RFC 3261 is to respond not to the apparent source IP address, but to the port contained in the Via header (1020 in our example). Unfortunately, this port is changed by NAPT functions. In order to fix this problem, RFC 3581 (August 2003) now specifies that when a client adds a 'report' parameter in the request, the proxy should store the apparent source port in this parameter, then respond to this port. Figure 3.14 uses RFC 3581.

The Via header also takes into account multicast transmission of signaling messages (not shown in the Figure): when a maddr parameter is present in a Via field, the reply is forwarded in multicast using the maddr address (and the ttl value stored in the ttl parameter).

Since RFC 3261, it is also mandatory to add the parameter **'branch=z9hG4bK...'**, which must begin with these letters, in order to identify an RFC 3261-compliant endpoint:

- **Encryption** (example: Encryption: PGP version = 2.6.2,encoding = ascii). This header field specifies that the message body, and possibly some message headers, have been encrypted. For more details see Section 3.6 on security.

Some header fields apply directly to the message body; these are part of the Entity header:

- **Content-Type** (example: Content-Type: application/sdp). This describes the media type of the content of the message body. In the example, of Figure 3.10 the message body contains a session description using the IETF SDP protocol. Another example is text/html.
- **Content-Length:** the number of octets of the message body. The CR+LF separating the header part and the payload part is not counted. If the payload is SDP, the CR+LF at the end of each SDP line is counted. Some vendors omit the last CR+LF in SDP.

3.3.2.1 SIP Requests

SIP requests are sent from the client terminal to the server terminal. The following methods exist:

- ACK: an ACK request is sent by a client to confirm that it has received a final response from a server, such as 200 OK, to an INVITE request.
- BYE: a BYE request is sent either by the calling agent or by the caller agent to drop a call.
- CANCEL: a CANCEL request can be sent to abort a request that was sent previously as long as the server has not yet sent a final response.
- INFO: defined in RFC 2976, the INFO request is used to carry information which does not change the call state. Some vendors use it for DTMF out-of-band transport (see Section 3.3.2.3.1).
- INVITE: the INVITE request is used to initiate a call.
- MESSAGE: this request is defined in RFC 3428 and is used for instant messaging (see Section 3.5.3).
- NOTIFY: this request is defined in RFC 3265 and is used to send event notifications (see Section 3.5.2).
- OPTIONS: a client sends an OPTION request to a server to learn its capabilities. The server will send back a list of the methods it supports. It may also in some cases reply with the capability set of the user mentioned in the URL, and how it would have responded to an invitation.
- PRACK: defined in RFC 3262, the PRACK request is used to implement reliable provisional responses.
- REFER: this request is defined in RFC 3515 and can be used to redirect sessions (see Chapter 5).

- REGISTER: c lients can register their current location (one or more addresses) with the REGISTER request. A SIP server that can accept a REGISTER message is called a registrar.
- SUBSCRIBE: this request is defined in RFC 3265 and is used to request specific event notifications (see Section 3.5.2).
- UPDATE: defined in RFC 3311, the UPDATE request is used to change media session parameters in early dialogs (i.e., before the final response to the initial INVITE).

Figure 3.15 illustrates the SIP request message format.

In addition to the mandatory fields of the General header, requests can carry additional fields in the Request header:

- Accept (example: Accept: application/sdp, text/html). This optional header indicates what media types are acceptable in the response. The syntax is specified in RFC 1288.
- Accept-Language (example: Accept-Language: fr, en-gb;q = 0.8, en;q = 0.7). This indicates the preferred languages of the caller. The syntax is specified in RFC 1288.
- Expires (INVITE and REGISTER) (example: Expires: Thu, 01 Dec 2000 16:00:00 GMT or Expires: 5 (in seconds)) For a REGISTER message, this header field indicates for how long the registration will be valid. The **registrar** can shorten the desired value in its reply. For an INVITE message, this can be used to limit the duration of searches.

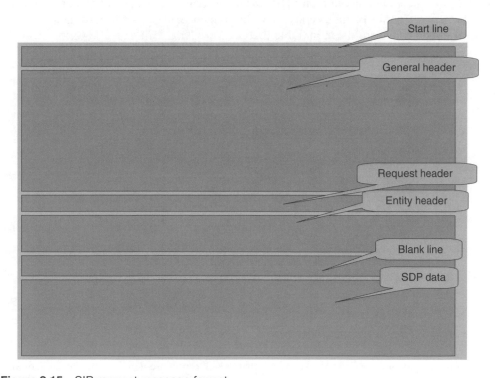

Figure 3.15 SIP request message format.

- Priority   (example: Priority: emergency). The values are those of RFC 2076, plus 'emergency'.

- Record-Route (also a response header field)   (example: Record-Route: sip:acd.support. com;maddr = 192.190.123.234,sip:billing.netcentrex.net;maddr = 192.194.126.23). Some proxies (see Section 3.4.2) may add/update this header field if they want to be on the path of all signaling messages. Some entities need to monitor the state of calls in order to work properly (e.g., a billing server might control a firewall to enforce the billing policy).

- Subject   (example: Subject: 'Conference call on the SIP chapter'). This is free text[12] that should give some information on the nature of the call.

3.3.2.2   SIP responses

A SIP server responds to a SIP request with one or more SIP responses. Most responses (2xx, 3xx, 4xx, 5xx and 6xx) are 'final responses' and terminate the SIP transaction. The 1xx responses are 'provisional' and do not terminate the SIP transaction. The SIP response format is illustrated in Figure 3.16.

The first line of a SIP response always contains a status code and a human-readable reason phrase. Most of the header section is copied from the original REQUEST message. Depending on the status code, there may also be additional header fields, and the response data part may be empty, contain an SDP session description, or contain explanatory text.

So far, six categories of status codes have been defined, classified according to the first digit. Common codes are listed in Table 3.1.

Figure 3.16   SIP response message format.

---

[12] This text should not contain any character that could cause problems to the parsing of the SIP request.

Table 3.1 Common codes

1xx	Informational		Request received, continuing to process the request.
		100	TRYING
		180	RINGING
		181	CALL IS BEING FORWARDED
		182	QUEUED
		183	SESSION PROGRESS (used to provide in-band network announcements, equivalent to ISDN progress message)
2xx	Success		The action was successfully received, understood, and accepted
		200	OK
3xx	Redirection		Further action must be taken in order to complete the request
		300	Multiple choices: several possible locations in contact headers
		301	MOVED PERMANENTLY: user can no longer be found at the address specified. New address is in Contact header field
		302	MOVED TEMPORARILY: alternative address in Contact header, which may also specify duration of validity
		305	USE PROXY: the specified destination must be reached though a proxy
		380	For future use: alternative service, described in the message body
4xx	Client error		The request contains bad syntax or cannot be fulfilled at this server
		400	BAD REQUEST
		401	UNAUTHORIZED
		402	PAYMENT REQUIRED
		403	FORBIDDEN
		404	NOT FOUND
		405	METHOD NOT ALLOWED
		406	NOT ACCEPTABLE
		407	PROXY AUTHENTICATION REQUIRED
		408	REQUEST TIMEOUT
		409	CONFLICT
		410	GONE
		411	LENGTH REQUIRED
		413	REQUEST MESSAGE BODY TOO LARGE
		414	REQUEST-URI TOO LARGE
		415	UNSUPPORTED MEDIA TYPE
		420	BAD EXTENSION
		480	TEMPORARILY NOT AVAILABLE
		481	CALL LEG/TRANSACTION DOES NOT EXIST
		482	LOOP DETECTED
		483	TOO MANY HOPS
		484	ADDRESS INCOMPLETE
		485	AMBIGUOUS
		486	BUSY HERE
		487	REQUEST TERMINATED (by a CANCEL request)
		488	NOT ACCEPTABLE HERE: used if an unacceptable offer is received in an UPDATE

(continued overleaf)

Table 3.1   (continued)

		491	Offer received in an UPDATE while an offer was already pending
5xx	Server error		The request contains bad syntax or cannot be fulfilled at this server
		500	INTERNAL SERVER ERROR: also used when an UPDATE is received before an answer has been generated to a previous UPDATE
		501	NOT IMPLEMENTED
		502	BAD GATEWAY
		503	SERVICE UNAVAILABLE
		504	GATEWAY TIMEOUT: also used if an UPDATE request is not acceptable because the user must be prompted; therefore, no immediate response can be generated.
		505	SIP VERSION NOT SUPPORTED
6xx	Global failure		The request is invalid at any server
		600	BUSY EVERYWHERE
		603	DECLINE
		604	DOES NOT EXIST ANYWHERE
		606	NOT ACCEPTABLE

The classification in Table 3.1 makes it easier to add new status codes: in case an old terminal does not understand a new Cxx code, it should treat it as a C00 code. Therefore, even old terminals will be able to react 'intelligently' when facing unknown status codes. These terminals can also give some additional information to the user if a reason phrase is present. RFC 3326 added a new header 'Reason' to facilitate interoperability with the PSTN by encapsulating the cause codes defined by Q.850:

Reason: Q.850 ;cause $= 16$;text $=$ "Terminated"

3.3.2.3   Mid-dialog requests

Once a dialog is established, many situations will require some control information to be transmitted in the middle of the call. For a real-time communication application, the most frequent uses of mid-dialog requests are:

- Transmission of DTMF information.
- Renegotiation of media streams.
- Redirection of media streams.

3.3.2.3.1   Transmission of DTMF and flash-hook information

Dual-Tone Multi-Frequency (DTMF) signals are those generated by modern analogue phones when you press one of the keys. Older rotary phones generate a series of small interruptions in the current loop through the phone, corresponding to the digit dialing. Such small interruptions are called **flash-hook**. They are also frequently used to control some class 5 features of the phone line, such as three-way calling.

The original SIP specification was focused on PC-based IP telephony and had over-simplified its specification for the transmission of DTMF and flash-hook signals for complex real-world telephony applications. Of course, the problem emerged quickly: without DTMF, you cannot call your answering machine, a prepaid telephony service, and most call centers, as these information systems frequently use DTMF tones to get information from you.

3.3.2.3.1.1 *The issues*

(a)   Telephony signals and low-bitrate voice coders

Low-bitrate voice coders (in practice anything lower than 32 kbit/s) usually cannot reliably transport DTMF tones. The reason is that these tones are composed of a mix of two pure frequencies that are almost impossible to find in the human voice. Many low-bitrate voice coders work by modeling a set of basic human speech components and transmitting only the model parameters on the other side, making it impossible to reproduce exactly pure frequencies. For this reason most of these coders also degrade music significantly, as they are designed for voice only.

DTMF signals that have been encoded and decoded using such low-bitrate coders will not be accurately recognized by DTMF-driven automatic systems (e.g., it will be almost impossible to enter a credit card number using DTMF, since at least one of the 16 or more digits will be misinterpreted).

Obviously, flash-hook, which is not a sound, is not transported using traditional voice-coding systems.

(b)   DTMF-driven call control services

The key advantage of VoIP over all other telephony techniques is the ability to control a phone call without ever being in the voice path, allowing the building of softswitches, as opposed to the traditional telephony systems which require a dedicated hardware-based switching matrix to route the media stream.

As an example, a traditional prepaid card system would connect the call from caller A, establish the media connection with A to get the PIN code and desired destination B for the call, then it would call B and continue to relay the media stream for the entire duration of the call (Figure 3.17). Some systems may optimize this slightly by using intelligent network commands to instruct an intelligent network telephony switch in the path of the call, the Service Switching Function (SSF), to make the call to B, but this SSF device will also route the media streams between A and B for the entire duration of the call. In such a traditional telephony system, two media streams of 64 kbit/s are established though the call control function for the entire duration of the call.

A properly designed VoIP network-based prepaid telephony server would establish a media stream with caller A only during the initial phase of the service, in order to get the PIN code, and destination of the call. The server would then call B but instruct A and B to exchange media streams directly over the IP network (see 'Redirection of media streams' in Section 3.3.2.3.2.5). It is no longer in the media path of the call.

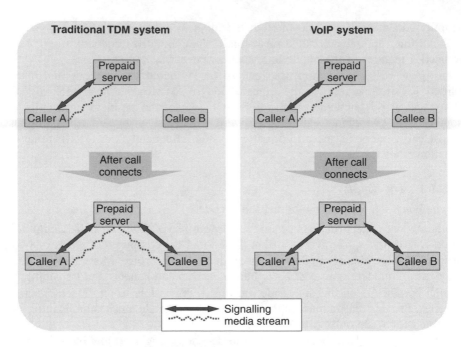

Figure 3.17   VoIP avoids media tromboning: example of a prepaid application.

The issue is that many such DTMF-driven call control services will still need to receive DTMF information, even when the media stream is passing directly between caller A and callee B. In most prepaid telephony services it is possible to stop the current call by pressing the '#' key, and get the opportunity to make another call to a new destination C without having to re-enter the PIN code. This requires DTMF information to be available on the call control link.

3.3.2.3.1.2   RFC 2833

A quick fix to the first issue (i.e., the transmission of DTMF and other events for communications using low-bitrate coders) was presented in RFC 2833 (RTP Payload for DTMF Digits, Telephony Tones and Telephony Signals), published in May 2000. RFC 2833 requires edge media devices to implement DTMF detection algorithms for all the media streams they generate. It is trivial for an IP phone because it obviously receives information about which keypad key is pressed, but for VoIP gateways connected to the PSTN this requires the implementation of DTMF detection algorithms in the G.711 stream received from the PSTN.

The idea is to send the DTMF information in the RTP stream as a named event, not as an audio-encoded signal. If the resulting RTP stream is received at another PSTN gateway, the PSTN gateway has enough information to regenerate the DTMF information as a waveform. The transmission of DTMF events in the RTP stream, with the same sequence number and timestamp reference as the rest of the RTP stream, allows perfect synchronization of the DTMF and media information, and avoids the possible duplication of DTMF signals at a VoIP gateway (one received directly from the media stream, the other received later in RFC 2833 encoded form).

Figure 3.18 shows the format of a telephony event encoded in an RTP packet. Such events should be generated as soon as a tone of more than 50 ms is detected. Each tone packet is sent three times for redundancy purposes, with the RTP sequence number incremented, while all other fields remain identical. Very short tones can be encoded in a single packet (by setting the 'end' bit). Longer tones can be sent either by continuously sending tone packets with a shorter duration until the tone stops, or by forming two packets: one signaling the beginning of the tone, one signaling the end of the tone. This prevents the sender from having to wait until the end of the tone to send the tone packet, which would obviously involve an unacceptable delay.

The volume is a value in negative dBm0 (e.g., the value 20 denotes a volume of -20 dBm0). The possible range is between 0 dBm0 and -63 dBm0, but values lower than -55 dBm0 should be rejected. The counter can encode durations up to 8 s if the timestamp unit is 1/8000 s, which is more than enough for most uses. A DTMF tone should always be longer than 40 ms in order to be properly recognized by in-band detectors.

As SIP uses the Session Description Protocol (SDP) to declare which type of media encoding, it was necessary to add a SDP payload format to declare which types of events a receiver can understand.

The following 'm' line can be used for receiving telephone events:

```
m=audio 44143 RTP/AVP 110
a=rtpmap:110 telephone-events/8000
a=recvonly
```

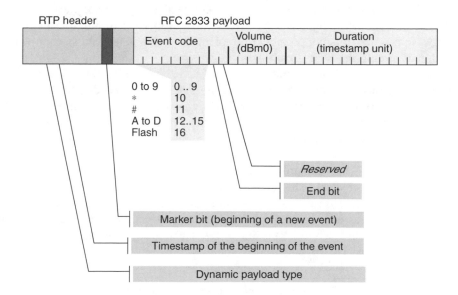

Figure 3.18 RFC 2833 RTP packet format for DTMF transport.

In addition the fmtp specifier can be used to detail which events can be received. The format is:

```
a=fmtp:<format> <list of values>
```

For instance, a receiver understanding all the events in the Figure except A, B, C, D, with dynamic payload type 100, would declare it using:

```
a=fmtp:100 0-11,16
```

In fact, all implementations are required to handle event 0 to 15, so the fmtp line is optional.

RFC 2833 also describes another format where tones are sent as a series of frequency, amplitude modulation, volume, and duration parameters.

One of the advantages of signaling DTMF information as an event is that all waveform analysis would then be performed by edge devices, making IP-based interactive voice response servers much easier to implement.

RFC 2833 is an interesting discussion of telephony signals in a VoIP network, which does solve the problem of transmitting DTMF and other signals in simple class 4 VoIP networks (these networks only route phone calls, without performing any complex service). It is a comprehensive reference to all types of tones and signals found on current networks, including DTMF, modem and fax tones, special information tones, etc.

RFC 2833 can also in principle solve the more complex problem of DTMF and call control, because any intermediary proxy can add its own 'm =' line requesting receipt of telephony events at a specific IP address, in which case a 'c =' line must be present in the media section of the SDP right after the 'm =' line,[13] while all other media streams are directed to the target user agent. However, many SIP user agent implementations have overlooked this requirement, and are unable to send media and telephony events to different destinations, let alone duplicate the events for transmission to multiple destinations. Right now in practice, it is not possible to implement a reliable prepaid system with RFC 2833 without routing the RTP stream.

RFC 2833 also caused some confusion as it leaves the implementer free to transmit DTMF simultaneously through the media stream and in event-encoded form, or to mute the media stream for the duration of the DTMF signal sent in event form. Only the latter is safe, as in many complex call flows, where the synchronization information may be lost, the simultaneous transmission of RTP in the regular media stream and using an event may cause duplicates.

Note that in H.323 it is mandatory to transmit all DTMF information out-of-band using the H.245 signaling channel. The audio signal is muted for the duration of the event transmitted out of band. However, some vendors can disable this mechanism and use RFC 2833 instead. This was introduced to allow some interworking between H.323 networks and SIP networks implementing RFC 2833 and lacking any signaling channel DTMF transmission function. This is not a good solution, however, and the use of one of the

---

[13] For servers which can handle multiple simultaneous calls, this also requires the allocation of a specific destination port for each call, which is very cumbersome.

methods described in Section 3.3.2.3.1.3 for signaling channel DTMF tone transmission in SIP networks is recommended.

The current weaknesses of RFC 2833 could be addressed by a more formal specification of how SIP should handle telephony events in complex call control applications, how events should be duplicated and sent to multiple destinations, and when the DTMF tones should be removed from the audio stream. Another issue is feature overlap of SIP application servers, as RFC 2833 possibly enables several call control devices on the signaling path to request telephony events, which may cause several of them to take incompatible actions simultaneously. This clarification work will probably be done in future revisions of SIP, but for now most vendors who face these issues have decided to solve the problem by using new SIP messages, as described below.

3.3.2.3.1.3   Alternatives to RFC 2833

VoIP is a technology breakthrough for the design of value-added services for telephone networks. The possibility of controlling calls without routing the media stream greatly enhances the density and scalability of application servers. It also decreases the cost of such servers, as many functions now do not require specialized telephony hardware and can run on standard computer platforms. Last but not least, the services are cheaper to operate, because the application servers no longer need to be located close to end-users, and therefore most services can be implemented using a single point of presence.

The implications of this paradigm shift are just beginning to be fully understood, and as expected many VoIP devices have been designed with the old TDM model in mind, assuming all servers that control the call are also controlling the media stream.

A properly designed VoIP edge device (gateway, IP phone) should be able to send all information that is possibly of interest to an application server over the signaling link, because only this link is guaranteed to reach all application servers. This is mandatory in H.323, using the H.245 channel.

Unfortunately, at the time of writing, there was no agreed standard way of doing this in SIP. Most network VoIP gateway vendors faced the problem and solved it using their own methods, some of which are described below. Interestingly, most SIP phones seem to be using only RFC 2833, most of the time without the ability to create separate UDP connections for telephony events, or sometimes even send the DTMF tones in-band without any form of coding. Unfortunately, this is a showstopper to any attempt to implement large scale class 5 services using such SIP phones, as many services (interactive voice response, prepaid, call centers, etc.) need access to the DTMF information and would require the application servers to route the media streams.

The methods used by VoIP gateway vendors to send events on the signaling link in SIP roughly fall in two categories:

- The use of the new INFO mid-dialog request defined in RFC 2976 to carry the telephone event. Some vendors use one of the MIME types defined by RFC 2833 (audio/telephone event and audio/tone MIME types), other vendors use encodings derived from H.323 or MGCP.

- The use of the new SUBSCRIBE/NOTIFY mechanism to carry the telephone event.

Unfortunately, since there is no common agreement of the exact encodings, there is no interoperability between the various SIP implementations, and most existing SIP networks use a single gateway vendor to overcome this problem. Some proxies are capable of understanding several formats and convert between them. The following sections describe the encoding used by some popular SIP gateway vendors.

(a) Cisco

Cisco uses a combination of the general INFO message defined in RFC 2976 (other methods are available as well, e.g., RFC 2833), and the SUBSCRIBE/NOTIFY method. Cisco implemented DTMF transport according to an Internet draft (draft-mahy-sip-signaled-digits-00, 'Signaled Digits in SIP'.

A SIP device can instruct a Cisco gateway to send a DTMF tone by sending it an INFO message formatted as follows:

```
INFO sip:1978551212@192.168.20.10 SIP/2.0
Via: SIP/2.0/UDP 192.168.0.1
From: 19785551234@192.168.0.1
To: 19785551212@192.168.20.10
Call-ID: 662606876@192.168.0.1
CSeq: 20 INFO
Content-Type: application/dtmf-relay
Content-Length: 22

Signal=9
Duration=250
```

The duration is in milliseconds. The gateway will confirm the receipt of this indication by responding to the SIP INFO message with a 200 OK response:

```
SIP/2.0 200 OK
Via: SIP/2.0/UDP 192.168.0.1
From: 19785551234@192.168.0.1
To: 19785551212@192.168.20.10
Call-ID: 662606876@192.168.0.1
CSeq: 20 INFO
```

In order to be notified of DTMF events from a Cisco gateway, a SIP application must first request to receive the DTMF events using the SUBSCRIBE mechanism. The advantage of using the SUBSCRIBE mechanism is that any SIP application server, in the case the call is processed by a chain of proxies, can request DTMF notification at any time during a call. The problem is that many application servers can react simultaneously and take incompatible actions:

```
SUBSCRIBE sip:1010@192.168.110.239 SIP/2.0
Via: SIP/2.0/UDP 213.56.166.173:5060
```

```
From: <sip:5500@213.56.166.173;user=phone>
To: <sip:1010@192.168.110.239>
Call-ID: 6CD8C67B-C0A011D3-806DB047-
B37E77AF@192.168.110.239
CSeq: 1 SUBSCRIBE
Contact: <sip:5500@213.56.166.173>
Expires: 3600
Events: telephone-event;duration=2000
User-Agent: NetCentrex IN Stack
Content-Length: 0

SIP/2.0 200 OK
Via: SIP/2.0/UDP 213.56.166.173:5060
From: <sip:5500@213.56.166.173;user=phone>
To: <sip:1010@192.168.110.239>;tag=A1E07C4-694
Date: Sun, 02 Jan 2000 23:08:57 GMT
Call-ID: 6CD8C67B-C0A011D3-806DB047-
B37E77AF@192.168.110.239
Server: Cisco-SIPGateway/IOS-12.x
Content-Length: 0
CSeq: 1 SUBSCRIBE
Expires: 3600
Contact:
<sip:1010@192.168.110.239:5060;user=phone>
```

If one of the requested events is received from the gateway, a SIP NOTIFY message with a representation of the signaled digits is sent to the requesting application server. In the following sample, the '9' key is pressed:[14]

```
NOTIFY sip:5500@213.56.166.173:5060 SIP/2.0
Via: SIP/2.0/UDP 192.168.110.239:5060
From: <sip:1010@192.168.110.239>;tag=A1E07C4-694
To: <sip:5500@213.56.166.173;user=phone>
Date: Sun, 02 Jan 2000 23:08:57 GMT
Call-ID: 6CD8C67B-C0A011D3-806DB047-
B37E77AF@192.168.110.239
User-Agent: Cisco-SIPGateway/IOS-12.x
Max-Forwards: 6
Timestamp: 946854581
CSeq: 102 NOTIFY
Event: telephone-event;rate=1000
Contact:
<sip:1010@192.168.110.239:5060;user=phone>
Content-Length: 10
```

---

[14] Expressed using the format defined by RFC 2833 (cf. Figure 3.18).

```
Content-Type: audio/telephone-event
0x0980010E

SIP/2.0 200 OK
Via: SIP/2.0/UDP 192.168.110.239:5060
From: <sip:1010@192.168.110.239>;tag=A1E07C4-694
To: <sip:5500@213.56.166.173;user=phone>
Call-ID: 6CD8C67B-C0A011D3-806DB047-
B37E77AF@192.168.110.239
CSeq: 102 NOTIFY
Server: NetCentrex IN Stack
Content-Length: 0
```

(b)   Nuera

Nuera also uses SIP INFO messages, encapsulating an MGCP-like syntax. A message body containing MGCP event information will be formatted as follows:

```
Content-Type: application/mgcp-event
Content-Length: <length of payload>
<MGCP event information>
```

An application requiring DTMF out-of-band information must request it using an MGCP notification request, embedded in an INFO message:

```
INFO sip:10.0.0.157 SIP/2.0
Via: SIP/2.0/UDP 10.0.0.168:5060
Route: NUERA-ID<sip:216.188.94.117>
From: 1003<sip:1003@10.0.0.157>
To:
NUERA-
ID<sip:216.188.94.117;user=phone>;tag=216.188.94.
117-eg101483118153
Call-ID:
tac1232002022709330152520S-
54444D0000000DD8BC5E753C7D184D10@216.188.94.117
CSeq: 1 INFO
User-Agent: NetCentrex IN Stack
Content-Type: application/mgcp-event
Content-Length: 15

R: [0-9*#] (N)

SIP/2.0 200 OK
Via: SIP/2.0/UDP 10.0.0.168:5060
```

```
Record-Route: <sip:10.0.0.157;maddr=10.0.0.157>
From: 1003<sip:1003@10.0.0.157>
To:
NUERA-
ID<sip:216.188.94.117;user=phone>;tag=216.188.94.
117-eg101483118153
Call-ID:
tac12320020227093301525205-
54444D0000000DD8BC5E753C7D184D10@216.188.94.117
CSeq: 1 INFO
Content-Length: 0
```

If one of the requested events is received from the gateway, a SIP INFO message with an MGCP event message body containing this observed event is sent to the application server. In the following sample, the '*' key is pressed:

```
INFO sip:1003@10.0.0.157;maddr=10.0.0.157 SIP/2.0
Route: <sip:1003@10.0.0.168>
To: "1003" <sip:1003@10.0.0.157>
From: "NUERA-ID"
<sip:216.188.94.117;user=phone>;tag=216.188.94.
117-eg101483118153
Via: SIP/2.0/UDP 216.188.94.117:5060
Via: SIP/2.0/UDP 216.188.94.117:5061
Call-ID:
tac12320020227093301525205-
54444D0000000DD8BC5E753C7D184D10@216.188.94.117
CSeq: 2 INFO
Content-Type: application/mgcp-event; version=1.0
Content-Transfer-Encoding: text
Content-Length: 5

O:*


SIP/2.0 200 OK
Via: SIP/2.0/UDP 216.188.94.117:5060
Via: SIP/2.0/UDP 216.188.94.117:5061
From:
NUERA-ID<sip:216.188.94.117;user=phone>;tag=216.188.94.
117-eg101483118153
To: 1003<sip:1003@10.0.0.157>
Call-ID:
tac12320020227093301525205-
54444D0000000DD8BC5E753C7D184D10@216.188.94.117
CSeq: 2 INFO
Server: NetCentrex IN Stack
Content-Length: 0
```

(c) Sonus

Sonus offers two mechanisms: DTMF relay and DTMF trigger.

   In DTMF relay, a mechanism similar to the signal and signal-update methods from H.245 is used for precise control of DTMF detection and generation. The signal parameter indicates the detected DTMF tone, the duration parameter indicates the total duration of the tone if known or an initial estimate of the tone duration, and signal-update subsequently updates the estimate of the total duration. The Content-Type header is set to 'application/dtmf-relay'. In the following example a DTMF is pressed for 250 ms:

```
INFO sip:1978551212@192.168.20.10 SIP/2.0
Via: SIP/2.0/UDP 192.168.0.1
From: 19785551234@192.168.0.1
To: 19785551212@192.168.20.10
Call-ID: 662606876@192.168.0.1
CSeq: 20 INFO
Content-Type: application/dtmf-relay
Content-Length: 22

Signal=9
Duration=250
```

The server is expected to confirm the receipt of this indication by responding to the SIP INFO message with a 200 OK response:

```
SIP/2.0 200 OK
Via: SIP/2.0/UDP 192.168.0.1
From: 19785551234@192.168.0.1
To: 19785551212@192.168.20.10
Call-ID: 662606876@192.168.0.1
CSeq: 20 INFO
```

The DTMF trigger mechanism is used for well-defined DTMF events not requiring timing information. The Content-Type header is set to 'application/dtmf'. In the following example the DTMF event '#' is pressed:

```
INFO sip:1978551212@192.168.20.10 SIP/2.0
Via: SIP/2.0/UDP 192.168.0.1
From: 19785551234@192.168.0.1
To: 19785551212@192.168.20.10
Call-ID: 662606876@192.168.0.1
CSeq: 20 INFO
Content-Type: application/dtmf
Content-Length: 1

#
```

The server confirms the receipt of this indication by responding to the SIP INFO message
with a 200 OK response:

```
SIP/2.0 200 OK
Via: SIP/2.0/UDP 192.168.0.1
From: 19785551234@192.168.0.1
To: 19785551212@192.168.20.10
Call-ID: 662606876@192.168.0.1
CSeq: 20 INFO
```

3.3.2.3.2   Negotiation of media streams

3.3.2.3.2.1   Session description syntax, SDP

SIP uses the Session Description Protocol (SDP) specified in RFC 2237. SDP is also
a product of the MMUSIC working group, and is mainly used today in the context of
the mBone, the multicast-enabled overlay network of the Internet. In order to be able to
receive an mBone session, a receiver needs to know:

- Which multicast address is going to be used by the session.
- What the UDP destination port will be.
- The audio and/or video coders that will be used (GSM, H.261, ...).
- Some information on the session (name, short description).
- Contact information.
- Activity schedule.

The primary purpose of SDP is to define a standard syntax for this type of information.
The SDP session description can be conveyed with various transport methods, depending
on the context: the Session Announcement Protocol (SAP) on the mBone, the Real-Time
Streaming Protocol (RTSP) for streaming applications, SIP to set up point-to-point and
multipoint interactive communications.

SDP is a human-readable protocol, consisting of several <type> = <value> lines
terminated by CR + LF. The field names and attributes use US-ASCII characters, but free
text fields can be localized since SDP uses the complete ISO 10646 character set. This
philosophy, as opposed to a binary encoding like ASN-1 PER used in H.323, facilitates
manual programming and analysis of network traces at the expense of greater bandwidth
usage. However, bandwidth usage of a signaling protocol is negligible compared with the
actual media flows, so the trade-off is very good. The only major drawback of this method
is that the generation of the serialization and parsing code needs to be manual, which
is less reliable (and also slower) than the automatic generation allowed by more formal
specifications like ASN.1. Manual programming also leads to more interoperability issues.

The session description is structured in one section which applies to the whole session
(starting with 'v = ...'), and several media description sections (each starting with 'm =
...'). Parameters in the media sections can override the default parameters of the session-
level section.

Table 3.2 describes the various field types described by RFC 2237 for each section.

Table 3.2 SDP session description parameters

Session-level field type	Sub-section-level field type	Usage	Format and example	
v =		Protocol version	v = 0	M
o =		Owner/creator and session identifier	O=<username> <session id> <version> <network type> <address type> <address> O=mhandley 2890844526 2890842807 IN IP4 126.16.64.4	M
s =		Session name	s = <session name> s = SDP Seminar	M
i =		Session information	i = <free text session description> i = A seminar on the session description protocol	O
u =		URI of description	u = <Universal Resource Identifier> u =http://www.cs.ucl.ac.uk/ staff/M.Handley/sdp.03.ps	O
e =		Email address	e=<email address> (Optional free Text) or e=<Optional free Text> ''<''email address''>'' e =mjh@isi.edu (Mark Handley) e = Mark Handley <mjh@isi.edu>	O
p =		Phone number	p =< phone number > (Optional free Text) or p = <Optional free Text> "<"phone number">" p = +44-171-380-7777	O
c =		Connection information—not required if included in all media	c = <network type> <address type> <connection address> TTL must be included for multicast sessions. c = IN IP4 224.2.17.12/127 c = IN IP4 224.2.1.1/127	O
b =		Bandwidth information	b = <modifier (CT Conference Total| AS Application-Specific Maximum>:<bandwidth-value in kilobits/s>	O

Table 3.2 (*continued*)

Session-level field type	Sub-section-level field type	Usage	Format and example	
One or more time descrip-tion sections	t =	Time the session is active	b = CT:120 t = \<start time> \<stop time>, using decimal NTP in seconds	M
	r =	Zero or more repeat times	t = 2873397496 2873404696 r =\< repeat interval > \<active duration> \<list of offsets from start-time>, by default in seconds r = 604800 3600 0 90000 means that the repeat interval is 1 week (604,800 s), active for one hour (3,600 s) after each offset from the start time T. Offsets are here 0 seconds and 90,000 s (25 h). That is, if *** represents active periods and—idle periods: T *** $T + 1$ h—$T + 25$ h *** $T + 26$ h—$T + 1$ week *** $T + 1$ week $+ 1$ h——$T +$ 1 week $+ 25$ h *** … The repetition is valid until the stop time. Unit modifiers can be used for compactness, and the previous record can also be written as follows: r = 7d 1h 0 25h	O
z =		Time zone adjustments		O
k =		Encryption key	k =\< method >:\< encryption key > or k =\< method >	O
a =		Zero or more session attribute lines	a=\<attribute> or a=\<attribute>:\<value> a = recvonly	O

(*continued overleaf*)

Table 3.2   (*continued*)

Session-level field type	Sub-section-level field type	Usage	Format and example	
Zero or more media descriptions	m =	Media name and transport address	m =< media > <port> <transport> <format list>	M
			m = audio 49170 RTP/AVP 0 3 means that the media is audio, can be received on port 49170 (RTP only uses even ports, the next odd port being used by RTCP). The transport is protocol RTP/AVT (IETF's Realtime Transport Protocol using the audio/video profile), and the format is media payload types 0 or 1 of the AVT profile (0 is μ-law PCM-coded single-channel audio sampled at 8 KHz, 3 is GSM) Other RTP profiles would be coded after the slash, (e.g., a hypothetical profile XXX would appear as RTP/XXX).	
	i =	Media title		O
	c =	Connection information— optional if included at session level		O
	b =	Bandwidth information		O
	k =	Encryption key		O
	a =	Zero or more media attribute lines		O

(a)   Dynamic and static payload types

Under a particular profile, some RTP payload types are static; in other words, their meaning is fully defined in the profile (e.g., RTP/AVP 0 is a 64-kbit/s μ-law PCM). Other RTP payload types only have a meaning in association with a particular session described in SDP. These are dynamic payload types. The SDP RFC gives an example of a 16-bit linear-encoded stereo audio sampled at 16 kHz. There is no static payload type defined that would exactly correspond to this. Instead, we will be using an arbitrary unused number for the payload type, say 98 (m = audio 49232 RTP/AVP 98), and describe the format of the transported data in SDP:

$$a=rtpmap:98\ L16/16000/2$$

The format is a = rtpmap : <payload type> <encoding name>/<clock rate>[/<encoding parameters>], which in our case translates to 16 linear, 16,000-Hz sampling, 2 channels.

By extension, the term 'dynamic payload type' applies to any RTP format whose media-encoding characteristics are by external signalling (e.g., through an H.245 Open-LogicalChannel message).

3.3.2.3.2.2   SDP in the context of interactive communications

SDP was initially designed in the context of multicast media transmissions over the mBone, the multicast overlay of the Internet. The problem it solved was simple: SDP simply needed to convey to all potential listeners the multicast IP address and port of the media transmission. This media session description using the SDP syntax was encapsulated in the higher level SAP (Session Announcement Protocol), and sent in a multicast packet to all potential listeners.

Simply replacing SAP by SIP does not make SDP a suitable protocol for interactive communications:

- The multicast sessions were essentially one-way, with IP address and port selected by the sender. An interactive communication uses two-way media streams, with IP address and port selected by the receiver.

- Multicast sessions are fairly static, they are advertised in advance, and all session changes do not have to occur in real time. In an interactive session, audio, video, and other media streams can be established, stopped, or changed at any time.

- Multicast sessions, because of the number of participants, do not even attempt to negotiate a common set of media encodings. The sender chooses one coder, and listeners simply have to adapt or fail to join the session. In an interactive communication, such as a phone call, the parties expect to be able to communicate, and this may require the negotiation of a common set of media coders.

- Interactive communications may require intermediary servers to charge the communications, and therefore to be aware of the types of media used by each party, and when media streams are established or stopped.

The initial version of SIP overlooked many of these issues. The sample SIP call flow consisted of a very simple case where the interactive communication:

- Was established between terminals which were assumed to have the same coders available. The issue of which coder to use first, how to maximize the chance of using symmetrical coders (the same coder for each direction of media stream), or how to negotiate a new coder were not really specified.

- Established the stream immediately after the stream has been proposed. In reality, an endpoint may want to use a proposed media (e.g., video), only some time after it has been proposed.
- Was not redirected, or re-established to a different endpoint in the middle of a communication. This situation may happen in real communications (e.g., communications to call centers, transferred calls, etc.).

The result was a situation where, despite marketing announcements, SIP interoperability was limited to basic calls only. A clarification on the exact procedures to use to open a media stream, close it, or renegotiate it was required. This was resolved in RFC 3264 under the umbrella of the new SIP RFC in June 2002.

3.3.2.3.2.3   The SDP offer/answer model for unicast streams

RFC 3264, "An Offer/Answer Model with the Session Description Protocol (SDP)", presents all the procedures that were in the previous SIP RFC, as well as new call flow discussions and examples. This complex portion of the interactive communication protocol is the equivalent of H.245 in the H.323 series.

RFC 3264 first defines the values of some mandatory SDP parameters that become useless or redundant in the context of SIP, where a lot of information is exchanged in the SIP message itself: the subject should be empty (s =< single space > or s = -), the time of the session should be set to 0 (t = 0 0), 'e' and 'p' parameters are not used, etc. RFC 3264 also restricts the number of sessions to only one per SDP message.

The media control model is based on 'offers' and 'answers'.

An 'offer' may contain zero, one or more media stream descriptions, each in a line beginning with 'm =', followed by optional attributes. The order in which the media formats appear in the 'm' line is the order of preference. The answerer should choose the first one that is acceptable to it.

For instance, the following offer proposes the G.711 A-law and μ-law for the audio media, with a preference for μ-law, and proposes H.261 and MPEG for video, with no preference:

```
v = 0
o = john 4898446519 4898446519 IN IP4 johnendpoint.anywhere.com
s =
c = IN IP4 johnendpoint.anywhere.com
t = 0 0
m = audio 41732 RTP/AVP 0 1
a = rtpmap:0 PCMU/8000
a = rtpmap:1 PCMA/8000
m = video 43221 RTP/AVP 31
a = rtpmap:31 H261/90000
m = video 49222 RTP/AVP 32
a = rtpmap:32 MPV/90000
```

An offer can contain multiple 'm =' lines for the various media types (e.g., audio, video). This means that the sender of the offer is willing to send or receive these media streams simultaneously. This is valid even if there are multiple 'm =' lines for the same media type. For instance, this can be used in conjunction with RFC 2833 to transmit telephony events to an application server, and regular audio to the destination user agent. (see Section 3.3.2.3.1.2 for more details). The expression power of this structure is equivalent to a single H.245 capabilityDescriptor, where each 'm =' line is an AlternativeCapabilitySet, and the union of all 'm =' lines is a SimultaneousCapabilitySet.

In this sample, all proposed media are bidirectional. A unicast stream offer will have the attribute 'a = sendonly' if the endpoint only wishes to send media to its peer. If it only wants to receive the indicated media, the attribute will be 'a = recvonly'. If it only wants to 'warn' the peer that it may establish a media stream at a later time, and provide a hint on the parameters that will be used, it will use 'a = inactive'. The default is 'a = sendrecv', if the endpoint wishes to establish a media stream to and from its peer.

The usefulness of 'sendrecv' and 'recvonly' is obvious. The 'inactive' or 'sendonly' attributes can be useful in more complex situations: for instance, in a SIP-based call center, when the call center application server wishes to ring an agent phone through a gateway, but does not want the agent to stream back audio immediately, because the call center application server is currently sending waiting music to the caller. Another common use of the 'inactive' parameter is to allow an appliance which need to initialize a DSP with some coder before using it to select the coder using an 'inactive' offer/answer, and then activate it through a new offer (see Section 3.3.2.3.2.5). These parameters can also be used to work around bugs in IP phone or media gateway implementations:

- Some endpoints do not support receiving SIP INVITE messages without SDP. Sending an inactive SDP media stream offer may work around the bug with the same effect.

- Some high-density VoIP gateways use multiple LAN cards for RTP streams, and do not support 'looped back' RTP streams sent and received on the same card. A way to work around this bug is to give a hint to the VoIP gateway of which source IP address may be used in the future with an inactive session, so it can avoid selecting the wrong LAN card.

The port number indicated in the 'm' line is the UDP port the endpoint wishes the peer to send media to (a = recvonly or a = sendrecv). For 'a = sendonly' and 'a = inactive' streams, the RTCP stream should be sent to the indicated port number plus one, unless a more specific indication is present in the media description.

Depending on the value of the 'a' attribute, the list of media formats indicates the media formats that can be sent (sendonly), are expected to be received (recvonly), or both (sendrecv). The sendrecv offer therefore contains enough information to allow the answerer to use symmetrical coders.

When RTP is used, the offer indicates the dynamic payload type that is expected to be used for the media. In sendonly and sendrecv offers, however, the payload type may be changed in the answer, in which case the answer value should be used instead of the one originally proposed. This may be required if the answerer is only able to receive certain payload types.

There may be fmtp parameters to include further information on media formats, such as supported events for RFC 2833.

In the case of RTP streams, all media descriptions *should* contain 'a = rtpmap' mappings from RTP payload types to encodings. If there is no 'a = rtpmap', the default payload-type mapping, as defined by the current profile in use (e.g., RFC 1890), is to be used.

The answer will usually be based on the offer, but change some elements: it may include a reception IP address and port, remove some media formats, etc. In this case the origin line ('o =') must be changed, and must contain a new version number. The time ('t =') line must be identical to the one of the offer.

There must be an exact one-to-one mapping between the number of 'm' lines in the offer and in the answer, which are matched based on the ordering. Rejected streams must contain a port number of 0. Although SDP requires at least one media type to be present in the 'm' line, it will be ignored. A stream must be rejected if there are no acceptable media formats.

This is the answer of Mark to the offer of John, where he selected PCMA and H.261:

```
v = 0
o = mark 4898446720 4898446720 IN IP4 markendpoint.anywhere.com
s =
c = IN IP4 markendpoint.anywhere.com
t = 0 0
m = audio 51762 RTP/AVP 1
a = rtpmap:1 PCMA/8000
m = video 53221 RTP/AVP 31
a = rtpmap:31 H261/90000
m = video 0 RTP/AVP 32
```

If a media offer is accepted, the answer should contain a unicast address if the offer was unicast, and should not change the media type. The answer should mark the direction of the stream from its point of view (e.g., 'recvonly' if the offer was 'sendonly', and vice versa). If the offer was 'sendrecv', the answer may choose to select only one of the send or receive modes (e.g., 'recvonly' if the answerer is not going to generate any media). In all cases, the answerer may want to mark the media as 'inactive' if it is not willing to use it right away. This is the only option if the offer was 'inactive'.

If a type of media has been accepted for a 'recvonly' or 'sendrecv' offer, it is still necessary to select from the offered media formats the ones that are acceptable. A 'recvonly' response should contain at least one media format selected from those present in the offer. When selecting a media format, it is important to remember that the ordering of the media formats in the offer represents the preference of the offerer, and therefore the selected format should be the first that is acceptable to the answerer. Note that once the answer has been sent the answerer must be prepared to receive media in any of the formats listed.

Similarly, for 'sendonly' or 'sendrecv' answers, it is necessary to select from the proposed media formats those that can be sent by the answerer ('sendonly'), or sent and received ('sendrecv'). At least one of the proposed media formats must be in the offer. In 'sendrecv' answers, changing the payload type should only affect the payload type of the stream received by the answerer and the answerer should still use the offer payload type to send media to the offerer.

The order of media formats in the answer represents the preference of the answerer. However, in order to maximize the chances to use the same coders in both directions, it should in general be the same relative order as the order of the offer. In 'sendonly' or 'sendrecv' answers, the media stream sent to the offerer should be the preferred format according to the offer from the formats listed in the answer.

Similarly, the offerer should use the first acceptable coder in the 'recvonly' or 'sendrecv' response. It may temporarily switch to another media type for special conditions (e.g., when it switches back to G711 for modem transmission, when it uses RFC 2833 for DTMF transmission, or if based on RTCP receiver reports it decides to switch to a lower bitrate coder).

In all cases, the IP address and port in the answer indicates where the answerer is expecting to receive media. In the case of a 'recvonly' or 'sendrecv' answer, the IP address and port will be the sink for the RTP stream. For all types of answers except 'inactive', the IP address and the port + 1 will be the sink for RTCP packets (i.e., RTCP sender reports for 'recvonly' answers, RTCP receiver reports for 'sendonly' answers, or both for 'sendrecv' answers).

3.3.2.3.2.4   *The case of multicast streams*

In the case of multicast stream offers, the meaning of the 'sendonly' or 'recvonly' attributes is no longer the direction of the media stream from the perspective of the offerer. If the multicast stream is marked 'recvonly', it means that all participants are only allowed to receive media on this multicast address and port, they cannot send to it. If the multicast stream is marked 'sendonly', it means all participants can send to this multicast address, but should not attempt to receive media from it.

The answer should be identical to the offer, except that some media formats may be removed to indicate that the answerer does not support them.

3.3.2.3.2.5   *Redirection or renegotiation of media streams*

There are many circumstances in which it is necessary to redirect media streams dynamically. A good example is a prepaid service: the caller initially needs to receive and send media to the prepaid service server, but, as soon as the prepaid server has established the communication with the desired called party, the server must redirect the media streams to flow directly between the caller and the callee. Note that this cannot be done by using call-level redirection, because the prepaid server must remain in the call-signalling path in order to monitor the duration of the call and eventually dynamically cut the call if the caller has exhausted his credit.

The reader may think that the server could simply serve as a relay for the media stream, so that the caller would never need to dynamically redirect the media streams. This is true ... in fact, this is the way traditional TDM prepaid services work using 'service nodes'. But the single most important technological advantage of VoIP over TDM voice is precisely this ability to do better than simply relaying media streams. A VoIP-optimized prepaid server will be able to handle many more conversations than its TDM counterpart, and it will also require a lot less bandwidth between the service platform and the rest of the network. Of course one can choose not to take advantage of this, but in this case

VoIP presents no advantage. In fact, TDM will work better, because media relaying adds to voice path delay, and this delay will be far more noticeable in VoIP applications than in TDM.

This ability of all VoIP endpoints to dynamically redirect media streams is a very important requirement. Devices not supporting this feature should be dismissed. It is also important to check that VoIP feature servers, such as prepaid servers or contact center servers, do take advantage of this feature. Many engineers are still 'thinking TDM' when they implement VoIP systems, and media relaying is one of the most frequent bugs found on feature servers. Unfortunately, it does not affect the external functionality of a system, which is only detectable in a lab by taking network traces. But, it is very important not to deploy such systems, as in real networks the increased delay is likely to cause echo perception problems, let alone scalability issues.

In H.323, the renegotiation of media streams is done by using the mandatory H.245 Null TerminalCapabilitySet procedure, or 'TCS = 0', (i.e., a procedure that has no impact on the call control state). In SIP, the renegotiation of media streams, whether required to change the target of a stream or to use an alternative media format, is done using the offer/answer model. The new offer should be encapsulated in a new INVITE message on the same Call-ID, often called a RE-INVITE. Soon after the release of RFC 3261, a bug was found in this procedure: the INVITE message impacts the call state; therefore, the use of a RE-INVITE does not allow media session changes before the first INVITE completes with a final answer. To solve this, a new UPDATE method was introduced in September 2002 in RFC 3311. The UPDATE method does not impact the state of an existing dialog and therefore can be used in an early dialog before the first INVITE completes (Figure 3.19). It should be used only in this context. Support for the UPDATE method must be specified in the Allow header field.

Both the initial offerer or the answerer can initiate a new offer if they wish to change anything in the existing situation: modify, add, or remove media streams. The 'o =' line must be identical to the initial offer, except the version number which must be incremented if anything in the SDP has changed. An 'm' line must be present for each 'm' line in the previous offer, but new 'm' lines can be added. Media streams are removed by setting the SDP port to zero.

In the following example, Mark resends an offer to change the port for the audio stream to 51780 and indicates that he has stopped sending video but can still continue to receive it (Mark has clicked the 'stop video' button):

```
v = 0
o = mark 4898446720 4898446721 IN IP4 markendpoint.anywhere.com
s =
c = IN IP4 markendpoint.anywhere.com
t = 0 0
m = audio 51780 RTP/AVP 1
a = rtpmap:1 PCMA/8000
m = video 53221 RTP/AVP 31
a = recvonly
a = rtpmap:31 H261/90000
m = video 0 RTP/AVP 32
```

Figure 3.19 Usage of the UPDATE method.

John accepts the change, and responds with:

v = 0
o = john 4898446519 4898446520 IN IP4 johnendpoint.anywhere.com
s =
c = IN IP4 johnendpoint.anywhere.com
t = 0 0
m = audio 51780 RTP/AVP 1
a = rtpmap:1 PCMA/8000
m = video 53221 RTP/AVP 31
a = sendvonly
a = rtpmap:31 H261/90000
m = video 0 RTP/AVP 32

This procedure is now to be used in all cases, even to put a stream on hold (e.g., by changing a 'sendrecv' stream to 'sendonly'). This is a change to the recommendation of RFC 2543 which uses the '0.0.0.0' IP address to hold a stream. This prevented the continued receipt of RTCP receiver reports when the remote party was on hold, but allowed the receipt of media to continue.

3.3.2.3.2.6 Fax

Just like H.323, SIP either transports the fax G.711 signal transparently (pass-through), or uses the ITU T.38 fax relay protocol. While the use of T.38 has been thoroughly documented in H.323, it has been left outside the main specification track in SIP. A draft was published in October 2000 (draft-mule-sip-t38callflows-02.txt) to document the way the Clarent Corporation was using SIP to establish T.38 sessions. The call flow was then added to a document listing sample call flows (draft-ietf-sipping-call-flows-00.txt) and was further documented in draft-ietf-sipping-realtimefax-00.txt. T.38 annex D also discusses SIP call establishment procedures for T.38.

(a) T.38

When a SIP gateway decides, after detecting the fax V.21 preamble flags, that it needs to encode a fax signal using T.38, it should use the offer/answer model to include all the T.38 parameters in SDP format. T.38 transmission replaces the normal audio transmission over RTP (only the CNG signal may be sent in-band if the originating gateway did not detect it). Once the fax transmission is complete, the normal RTP audio transmission should resume. During fax transmission, the normal audio transmission can either be stopped completely, or be put on hold.

The call flow presented in Figure 3.20 is a simple case of a gateway with a dedicated fax port; therefore, it knows that the media session is going to be T.38 in the first INVITE.

The response shown in Figure 3.21 also contains T.38 parameters, as well as the reception port for IFT packets. Note that the selected mode is 'UDP redundancy', not 'FEC', because the parameter a = T38FaxUdpEC:t38UDPFEC has been removed in the answer.

If the fax communication is detected in the middle of a voice call, a SIP RE-INVITE should be used with the T.38 parameters in the new SDP session. Once the fax communication terminates, another RE-INVITE is used to re-establish the normal RTP audio session.

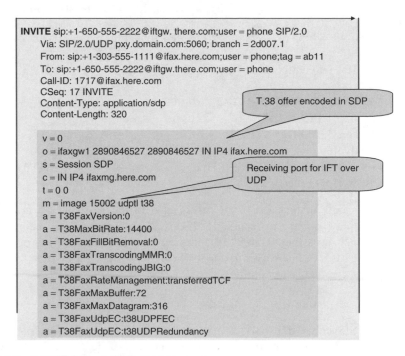

Figure 3.20   INVITE from a dedicated fax port.

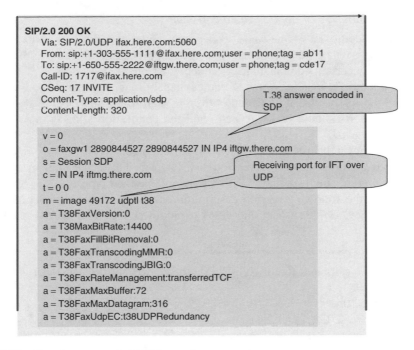

```
SIP/2.0 200 OK
   Via: SIP/2.0/UDP ifax.here.com:5060
   From: sip:+1-303-555-1111@ifax.here.com;user = phone;tag = ab11
   To: sip:+1-650-555-2222@iftgw.there.com;user = phone;tag = cde17
   Call-ID: 1717@ifax.here.com
   CSeq: 17 INVITE
   Content-Type: application/sdp                    T.38 answer encoded in
   Content-Length: 320                              SDP


   v = 0
   o = faxgw1 2890844527 2890844527 IN IP4 iftgw.there.com
   s = Session SDP                                  Receiving port for IFT over
   c = IN IP4 iftmg.there.com                       UDP
   t = 0 0
   m = image 49172 udptl t38
   a = T38FaxVersion:0
   a = T38MaxBitRate:14400
   a = T38FaxFillBitRemoval:0
   a = T38FaxTranscodingMMR:0
   a = T38FaxTranscodingJBIG:0
   a = T38FaxRateManagement:transferredTCF
   a = T38FaxMaxBuffer:72
   a = T38FaxMaxDatagram:316
   a = T38FaxUdpEC:t38UDPRedundancy
```

Figure 3.21   Receiver accepts T.38 fax call.

(b)   Fax pass-through

Fax pass-through is simple, and is the fallback mode if T.38 is not supported by the two gateways (488 NOT ACCEPTABLE HERE or 606 NOT ACCEPTABLE response). If a gateway attempted to initiate a T.38 media session in a RE-INVITE and this was rejected, it should initiate a new RE-INVITE with the pass-through session parameters. Fax pass-through only requires to dynamically or change the RTP payload type to G.711 μ-law or A-law, to disable any silence suppression, and to keep echo cancelation active. Specific SDP parameters have been defined in RFC 3108. For G.711 μ-law, the following SDP description can be used in the RE-INVITE:

> Content-Type: application/sdp
> Content-Length: 181
>
> v $= 0$
> o $=$ faxgw1 2890844527 171091 IN IP4 iftgw.there.com
> s $=$ Session SDP
> c $=$ IN IP4 iftmg.there.com
> t $= 0\ 0$
> m $=$ audio 12322 RTP/AVP 0
> a $=$ rtpmap:0 PCMU/8000
> a $=$ ecan:fb on -
> a $=$ silenceSupp:off - - - -

3.3.2.3.2.7   Exchange of capabilities using SDP

H.323 uses a specific set of messages to negotiate the capabilities of endpoints. These capabilities are very flexible and can take into account processing or bandwidth constraints which make only certain combinations of coders possible.

SIP does not have a specific syntax to express capabilities, and therefore also uses SDP for this purpose, except that it *may* omit both 'e =' and 'p =' lines. The session ID must be unique, the ports should be set to zero, and the connection address must be present despite being ignored.

An 'm' line must be present for each supported media type (audio, video, image, etc.), followed by the supported media formats. Each media format must also be further specified with an 'a' line associating it to a dynamic payload type.

For instance, the following session description can be used for an endpoint that supports G.711 μ-law, G.711 A-law, and GSM as audio codecs, and H.261 and H.263 as video codecs:

```
v = 0
o = sampleendpoint 465878951 465878951 IN IP4 192.0.0.1
s = −
t = 0 0
c = IN IP4 192.1.2.3
m = audio 0 RTP/AVP 0 1 3
a = rtpmap:0 PCMU/8000
a = rtpmap:1 PCMA/8000
a = rtpmap:3 GSM/8000
m = video 0 RTP/AVP 31 34
a = rtpmap:31 H261/90000
a = rtpmap:34 H263/90000
```

Unlike H.323, there is no way to specify that some coders of different media types cannot be used simultaneously (e.g., that a processor-intensive coder such as G.729 cannot be used with H.263). In H.323, it is possible to specify that the H.261 coder can be used with either G.711 or G.729, but H.263 can only be used with G.711.

Similarly, it may be impossible, for bandwidth reasons, to use video coders and data-sharing simultaneously, but this time a bandwidth constraint may be expressed in SIP by using the SDP session level 'b =' parameter.

3.4   CALL-HANDLING SERVICES WITH SIP

SIP defines many functional names for call-handling features, such as:

- Proxy (stateless, stateful, forking ...) server.
- Registrar server.
- Redirect server.

- Location server.
- Back-to-back user agent.

In addition to these names, the industry frequently also uses terms like 'application server' or 'feature server'.

Despite the fact that most of these functions are called 'servers', they really describe functions, and do not necessarily refer to separate servers. In fact, most of the time, a given SIP server will implement the features of many, if not all, of the entities listed above. For instance, a server could:

- Register SIP user agents in a certain area (registrar behavior).
- Reply to other SIP server location requests (location server behavior).
- Handle outgoing calls of locally registered SIP devices (stateful proxy).
- Propagate *simple* instant-messaging messages without modification (stateless proxy).
- Implement certain complex applications such as contact center call distribution for calls to certain numbers, using back-to-back user agent behavior.

The descriptions in the following sections refer to each of these behaviors and use SIP 'server' terminology for them, but the reader should keep in mind that all these functions can coexist in a single box. Typically, it is only in very large deployments that call control and registration features may be separated, requiring separate registrars and stateful proxy boxes.

3.4.1 Location and registration

3.4.1.1 The registrar function

A registrar is a server that accepts REGISTER requests. The same server may also implement other SIP functions (e.g., serve as a proxy). Registrars are needed to keep track of the current location of a user agent. The IP address of a user agent may change under a number of situations: connection via an ISP providing dynamic addresses, connection on a LAN that provides addresses via DHCP, or a roaming user. In order to be able to reach this user from its SIP address, an entity in the SIP network needs to maintain the mapping between SIP addresses and IP addresses: this is the purpose of the registrar.

In order to facilitate user mobility and avoid manual configuration as much as possible, SIP defines a well-known 'all SIP servers' address (sip.mcast.net: 224.0.1.75). A client can therefore in theory register his current IP address with a multicast register message (Figure 3.22). For some unclear reason SIP restricts the TTL (time to live) of this message to one, limiting the discovery method to the local subnet. This feature is roughly equivalent to the gatekeeper discovery method described in H.323. However, in H.323 the gatekeepers that are willing to handle the request can reply, allowing the client to

Figure 3.22   Proxy learns endpoint IP address through REGISTER message, and can route incoming requests to it.

select the appropriate gatekeeper and contact it directly later on. Currently, SIP servers cannot reply to a multicast REGISTER message; therefore, the client doesn't have the chance of learning the address of an appropriate SIP server, or even of knowing whether there was a SIP server to accept the registration.

The registrar can also be contacted by unicast if the address of the registrar is known. In this case the procedure is the same as for any other SIP request.

The registered state is not permanent. If not refreshed, it will time out after 1 h by default (this default value can change as specified in the Expires header field). In order to maintain its registration, a terminal needs to refresh it periodically.

If the terminal (or the user) moves and wants to modify the parameters of the registration, then it can cancel an existing registration by sending a contact value of '*', and send a new registration, as shown in Figure 3.23.

3.4.1.2   Locating users from SIP addresses

SIP addresses are called URIs (Uniform Resource Identifiers). URIs are really names (except those SIP addresses that use an IP host address, such as the address used in our simple call example), they do not refer directly to the transport address to be called but to an abstract entity that can reach the user directly or indirectly.

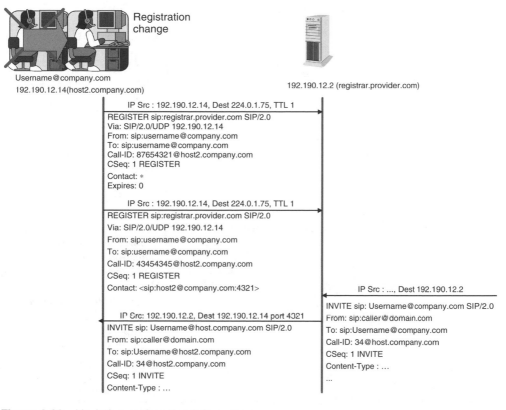

Figure 3.23 Updating registration information.

SIP URIs have two major forms, an email-like form, and a telephone number form:

- The general format of email-form SIP URIs is user@host, where host is usually a fully qualified domain name that can be resolved to an IP address using the DNS system. In many cases the SIP address of a user will be the same as his email address.

- The general format of a telephone number-form SIP is phone-number@host; user=phone. Because SIP is still mainly used for telephone calls, this is one of the most widely used formats in SIP networks. The host part is optional and may indicate a server that can reach this phone number, which can be used to specify a preferred service provider. Most telephony systems, however, can decide where to route the phone call based on the phone number only (through prefix analysis or a local number portability query); so, the domain name part is not present in most telephony applications.

There is also another type of URI which serves a different purpose; this is a sort of command line requiring some action from the user agent, described by a method attribute.

Table 3.3 Common SIP URL formats

John@netcentrex.net:1234	Vanilla SIP URI . . .
Userdomain.com	No user part, default port will be 5060
support@company.fr:2345;transport=UDP	Wants to be contacted using UDP
192.190.234.3:8001	Contact the server at this IP address
support@netcentrex.net;maddr=239.255.255.1;ttl=32	Override normal host name to transport address mechanism: use multicast to 239.255.255.1 with a TTL of 32 instead
+33-231759329@cybercall.com;user=phone	Global phone number
0231759329;isub=10;postd=w11p11@cybercall.com;user=phone	Local phone number with ISDN subaddress, wait for dial tone, then dial 11 (pause) 11 using DTMF
ACD@netcentrex.net?priority=hign&customercode=1234	Using proprietary extension headers to control priority in an ACD system . . .
Newcomer@reg.usergroup.com;METHOD=REGISTER	Previous URIs would trigger a SIP INVITE request, this one initiates a registration to the registrar of usergroup: reg.usergroup.com

Optionally, a SIP URI may also specify a port number and a transport mode if the SIP default transport (UDP) and port (5060) is not to be used. A list of example SIP URIs is given in Table 3.3.

Most of the extensions (headers, maddr, etc.) are not allowed in the To, From parameters of SIP requests and responses, but can be used in the contact parameters.

3.4.1.2.1 The original RFC 2543

RFC 2543 describes how to locate the physical endpoint using its SIP URI; this is done in two stages:

- First, the SIP URI allows the calling endpoint to locate a SIP server. This SIP server will be the destination of the initial INVITE message. The SIP server can be the final destination of the call, and even if it is not it is supposed to know how to reach the called endpoint.

- If the SIP server is not the final destination of the call, it will redirect the INVITE request to the called endpoint. This can be done in two ways: either by instructing the calling endpoint to send a new INVITE request to another location using the 302 MOVED reply (redirect server behavior), or by transparently relaying the INVITE message to the appropriate transport address (proxy behavior). The first model is similar to the H.323 direct call model, and the second similar to the H.323 gatekeeper routed call model.

In order to locate the SIP server, a SIP terminal will use DNS. A SIP URI domain name must have an SRV record, an MX record, a CNAME or an A record. The resolution algorithm is represented in Figure 3.24.

First, the terminal will retrieve the SRV resource records for the considered domain name. Then, it will only keep records of type `sip.udp` or `sip.tcp` (RFC 3263 seems to consider that it should be_sip._udp or_sip._tcp). If there is a `sip.udp` record, the terminal will contact the SIP server using UDP at the specified transport address. It will use the port specified in the SIP URL or default to the port specified in the `sip.udp` record. If there is a `sip.tcp` record, the same method will be used, but over TCP. If no SRV record is found, the terminal will try to retrieve the IP address of a SIP server by looking at the MX records first (normally used to point to a mail server), then CNAME records (pointing to an alias name), and finally an A record (pointing to an IP address).

Pointing to a SIP server instead of the called endpoint directly allows the called endpoint to move (the transport address changes), while enabling the use of DNS caching. If the address of the called endpoint was stored directly in DNS, there could be a lot of trouble with DNS caching. Normally, all DNS records can be cached by the DNS resolver. The cached record expires after a certain period once it has been first retrieved by a DNS

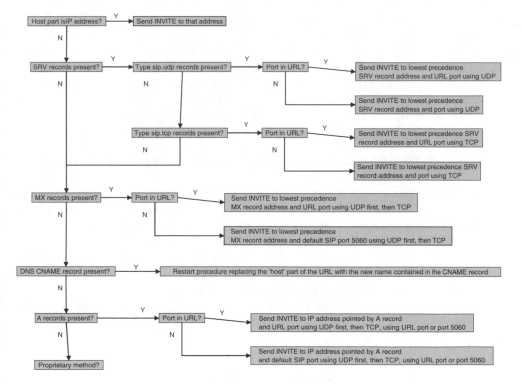

Figure 3.24 Location of a SIP server using DNS.

query; this period is called the **Time To Live** (**TTL**, in seconds). The value of the TTL is stored in the DNS record. Therefore, when the terminal moves, the caller could still have a wrong address in the DNS resolver cache, and the call would fail. The only solution is to set the TTL to zero and update the primary DNS record as the terminal moves; this is neither very easy nor cache-friendly, and therefore not scalable.

On the other hand, a SIP proxy server is not likely to move very often, and storing its address in a DNS SRV, MX, or A record does not cause any trouble. This SIP proxy server needs to know the current location of the called terminal (e.g., by implementing SIP registrar functionality), and can redirect the INVITE request to the appropriate location.

3.4.1.2.2   RFC 3263 and the use of NAPTR records

The newer RFC 3263 ('Locating SIP Servers') brings substantial changes to RFC 2543. It first states (for 'backward compatibility') that if the URI contains a numeric IP address (and optional port), but without a protocol specifier, the UDP should be used to reach this IP address. Similarly, if the target is not a numeric IP address, but a port is provided instead, then UDP should be preferred. This is because UDP was the preferred transport in RFC 2543.

In all other cases (i.e., if the URI is not a numeric IP address and contains no protocol specifier or explicit port), then the **Naming Authority Pointer Record** (**NAPTR**) DNS mechanism defined in RFC 2915 should be used to resolve the URI into a next hop address. NAPTR records are also used by ENUM (see Chapter 2).

When a SIP server needs to locate the appropriate resource to reach user@subdomain. domain.org, it will query the DNS for the new NAPTR of the DNS (DNS type code 35, defined in RFC 2168 and RFC 2915) for subdomain.domain.org.

The NAPTR record is used to attach a rewrite rule, based on a regular expression, to the DNS domain name. Once rewritten, the resulting string can be interpreted as a new domain name for further queries, or a URI (Uniform Resource Identifier) which can be used to delegate the name lookup. The syntax of the NAPTR RR is as follows:

Domain TTL Class Type Order Preference Flags Service RegExp Replacement

Domain, TTL, and class are standard DNS fields. The type field is set to 35 in the case of the NAPTR. The order and preference field specifies the order in which records *must* be processed when multiple NAPTR records are returned in response to a single query. The ordering is lexicographic, order is used first, then preference.

The 'S', 'A', and 'U' flags indicate that this NAPTR record is the last one and that the next query should be made using SRV records (flag 'S'), an A record (flag 'A'), a protocol-specific algorithm (flag 'P'). In all these cases the 'replacement' field will be used as the new name to fetch the corresponding resource record. If the flag is 'U', the regular expression[15] specified in the RegExp field should be applied to the domain name in order to get a new URI.

---

[15] An expression composed of a series of symbols each defining a specific modification to a string, and defined in POSIX.

The service field defines the protocol that should be used after this step of the resolution (H323, LDAP, SIP, TEL, SIPS), and the type of service that will be provided: D2U (UDP transport), D2T (TCP transport), or D2S (SCTP transport). For memory, ENUM uses E2U as the type of service.

A SIP server looking for a next hop protocol for a SIP call will therefore look for NAPTR resource records with the service field set to:

- SIP+D2U for a list of next hops that must be reached via SIP/UDP.
- SIP+D2T for a list of next hops that must be reached via SIP/TCP.
- SIP+D2S for a list of next hops that must be reached via SIP/SCTP.
- SIPS+D2T for a list of next hops that must be reached via SIPS/TCP.

All other service fields are discarded, as well as the options that are not supported by the requesting server (e.g., SCTP), the responses are sorted according to the preference value (lower values have a higher priority), and the server must try them out in the allotted order.

Resource records with the SIP+D2U, SIP+D2T, SIP+D2S, or SIPS+D2T service codes also have the 'S' flag: full resolution requires a request for SRV resource records for the resource indicated in the 'replacement'. The regular expression field will be empty. For instance, if we have the following NAPTR records for subdomain.domain.org:

```
; order pref flags service      regexp replacement
  IN NAPTR 50 50 "s" "SIPS+D2T"   ""  _sips._tcp.subdomain.domain.org.
  IN NAPTR 90 50 "s" "SIP+D2T"    ""  _sip._tcp.subdomain.domain.org.
  IN NAPTR 100 50 "s" "SIP+D2U"   ""  _sip._udp.subdomain.domain.org.
```

This indicates that the server supports TLS over TCP, TCP, and UDP, in that order of preference. If the client supports TCP and UDP, TCP will be used, targeted to a host determined by an SRV lookup of _sip._tcp..subdomain.domain.org. The lookup will, for instance, return this list of SRV records:

```
;;            Priority   Weight    Port    Target
    IN SRV    0          1         5060    proxy1.domain.com
    IN SRV    0          2         5060    proxy2.domain.com
```

In theory, the NAPTR record could change subdomain.domain.org into anything. RFC 3263 says that if this is the case then SRV records with the original name must also be present, for backwards compatibility with RFC 2543.

The whole procedure requires at least three DNS queries for each transaction (one to resolve the URI into NAPTR records, one to resolve the new resource name specified by the NAPTR resource field into a set of SRV resource records, and at least one to resolve the selected SRV record target into an IP address). In fact, most of the time, more requests will be required if the URI contains many subdomain components, because in the more general case, these domains will not all be in cache. For subdomain.domain.org, this is five DNS requests, each request lasting about 100 ms!

The procedure introduced by RFC 3263 requires one more DNS lookup than the previous RFC 2543 mechanism which looked for SRV records directly. However, it does not bring anything new because it does not make use of any of the possibilities of NAPTR records, which go beyond anything that can be achieved with SRV records. Overall, this an overly complicated and verbose RFC (with many exceptions to the general rule) for a very simple address resolution problem. It could evolve into something more interesting if the full power of NAPTR records, with the use of regular expressions, was used. It would then become very similar to ENUM. As it stands now, it seems to be chasing the same concepts as ENUM, keeping the complexity and leaving important features out.

3.4.1.3   Redirect server

A redirect server responds to an INVITE request with a 3xx reply (or rejects the call with a client error or server error):

- The 300 MULTIPLE CHOICES reply can be used when the SIP URL of the request can be contacted at several alternative addresses. The choices are listed as Contact fields. This can be used as a simple form of load balancing or, more interestingly, to let a caller know all the available means or media that can be used to communicate with the destination user. For instance, the returned Contact field could be:

    Contact: sip:John_gsm@company.com,sip:John_home@family.org

- The 301 MOVED PERMANENTLY reply indicates that the SIP URL of the request can no longer be contacted at this location. The client should try to contact the new location given by the Contact header field of the reply. This change is permanent and can be memorized by the client. The Contact header can also indicate several possible destinations.
- The 302 MOVED TEMPORARILY reply redirects the client to a new location, as above, but for a limited duration, as indicated by the Expires field.
- The 305 USE PROXY indicates that the specified location should be reached via the indicated proxy.
- The 380 ALTERNATIVE SERVICE is really for future use; it is not fully defined in the current SIP RFC. This reply is more complex, and may seem a bit redundant is light of the previous replies: in addition to providing a new destination in the Contact field, the reply can also contain a session description in the message body that represents the sending capacities of the new destination. The caller is expected to send an INVITE request to this new destination, and offer in its SDP session description the appropriate capabilities (which can be a copy of the SDP parameters of the 380 reply, except for receiving RTP ports).

Other replies (e.g., 303) were defined in early SIP drafts, but have become obsolete.

A redirect server can be used in conjunction with a registrar to redirect calls to the current location(s) of the caller. It can also act as a basic form of call distribution system, as shown in Figure 3.25.

Figure 3.25   SIP redirect server distributes calls, in a very basic contact center.

Redirect servers can be useful tools to improve the scalability of complex call management systems. Inserted as a front end, it can distribute calls among a pool of secondary servers, achieving load balancing. This is permitted by the maddr parameter of the Contact field:

<sip:originaladdress@callcenter.com:9999;maddr = sophisticatedACD3.callcenter.com>

By returning this, the redirect server indicates that the caller should send an INVITE with the same destination URI (originaladdress@callcenter.com), but send it to the third ACD server of the pool (ACD3.callcenter.com). The maddr parameter instructs the caller to bypass the normal procedure to find the appropriate SIP server from the domain part of the URL, and to use the domain name provided instead.

One of the most interesting uses of the redirect server in conjunction with a registrar is for the deployment of large-scale residential networks. A network serving hundreds of thousands of SIP endpoints cannot be realistically realized with a single server. The reason is that if N endpoints are sending a registration message every S seconds, the number of messages per second that would need to be processed by the central server would be on average N/S, and much worse when the network starts or restarts. With $N = 300,000$ and $S = 60$, the central server should process over 5,000 registration messages per second. Obviously, in this case it is necessary to use a number of separate registrar servers (e.g., one server per block of 6,000 users). The call control function can still be centralized because there are a lot less calls than registration messages (if each user makes 1 call every 5 hours, this is $300,000/(5*3,600) = 16$ calls per second). But, in order to terminate incoming calls to the right user agent, the central call control function will need to query the registrars, which can be done using an INVITE/REDIRECT transaction. This type of strategy is currently used with success in residential networks with over 200,000 users.

Redirect server functionality is very similar to the role played by the H.323 gatekeeper when using the direct call model.

3.4.2 The proxy function

3.4.2.1 Definition

A proxy server acts as a server on one side (receiving requests) and as a client on the other side (possibly sending requests).

Strictly speaking, a proxy should be mostly transparent to user agent messages, simply passing messages and changing them in very limited ways. A proxy can forward a request without any change to its final destination, it can decide to validate requests, authenticate users, fork requests, resolve addresses, and cancel pending calls, etc.

Depending on the level of control the proxy has over the SIP messages it processes it can be a stateless proxy, a stateful proxy, or even a back-to-back user agent:

- A **stateless proxy** simply chooses the next hop destination for an incoming SIP message using To header information; it keeps no state for the call or even the transaction (it will not handle retransmissions, but simply pass them on transparently). For instance, a stateless proxy will not carry out any local processing for a CANCEL request other than forward it, and will not even acknowledge locally any response, but simply pass it transparently to the original sender of the request. This behavior is made possible because SIP allows a proxy to store some state in the messages (e.g., in the Via header). This state is copied in the response, and therefore the stateless proxy server does not need to keep in memory any call parameter to be able to forward a response, it simply finds the information it needs in the response itself (e.g., the next hop is at the bottom of the pile of Via headers, after discarding the Via header corresponding to the proxy itself). Stateless proxies have often been presented as a technology breakthrough that would make SIP networks considerably more scalable than any other network. The reality is that a stateless proxy can serve only very limited purposes (e.g., it cannot do billing), and therefore can be used only in simple infrastructure call flows, such as performing load balancing or basic message routing in core networks. Even in this role it cannot do very much, as even a simple load-balancing function usually needs to keep in memory the number of calls it has sent to each destination and discard any destination that appears to fail frequently.

- **Stateful proxies** are much more useful, as they can keep any state relative to the call and all transactions involved in the call. Stateful proxies also manage locally some aspects of the transactions (e.g., they will handle retransmissions locally and acknowledge the final responses, except 200 OK and CANCEL requests). Stateful proxies can serve most call control purposes required in a SIP network, such as choosing an egress route for a phone call among multiple gateways by offering the call in sequence to multiple gateways and analysing the responses to eventually try another gateway if the current attempt fails due to congestion or any other reason. Since a stateful proxy memorizes when a call has begun, it can generate call detail records about the duration of the call when the call ends.

- Some applications need so much control on the call that they cannot be implemented within the restrictions set on proxies.[16] For instance the requirement to transparently forward any 200 OK response received from a destination may not be compatible with applications which need to filter the responses for security or any other purpose. Many of the most sophisticated applications, such as business telephony applications, contact centers, need the complete range of possibilities of a user agent. Therefore they act as a full user agent receiving a call and re-initiate a call as a user agent. Strictly speaking, such servers are no longer proxies, but should be called **back-to-back user agents**.

The names **feature server** or **application server** that emerged in marketing presentations are now widely used by the industry, but they have no precise meaning. A feature server is any server that implements an application! It could be a stateful proxy, a back-to-back user agent, or even an interactive voice response server that can receive, generate, and bridge calls. The closest to a feature or application server in traditional telephony terms would be a **service node**.

There is frequent confusion leading to the idea that a SIP application server can be used to replace the old intelligent network model. In fact, the intelligent network model refers to the ability of a Service Control Point (SCP) to use an abstract protocol-independent model of a call to remote-control a Service Switching Function (SSF), which is the only one that is part of the telephony network and implements telephony protocol stacks. The function implemented by an SSF with SIP stacks and the SCP together can be described as an application server. But, a simple, monolithic SIP-only application server does not replace an IN architecture, which has been designed precisely to facilitate the programming of protocol-independent application.

In Figure 3.26, the proxy is at least a stateful proxy because it locally sends a 100 reply and generates ACKs (transaction awareness). Note that a stateful proxy is not allowed to send ACKs locally to 200 OK responses (Figure 3.27). This message and its reliability must be handled end to end, ensuring that the call is established and media can start flowing only when the end-to-end handshake is complete. Only a back-to-back user agent can send an ACK locally to a 200 OK response (and it must understand the consequences).

Most useful functions (e.g., the ability to drop a call from the proxy) go beyond the strict definition of a 'proxy': most commercial server implementations are back-to-back user agents, according to SIP terminology. In section 3.4.2.2 we do not restrict ourselves to the strict 'proxy' terminology and describe the various functions of a server that has control over SIP signaling during the call (i.e., these that encompass the proxy and back-to-back user agent features).

---

[16] These theoretical restrictions do not serve any useful purpose; in fact, they imply a very poor feature set, and virtually all commercial products behave as back-to-back user agents.

Figure 3.26 Simple call through a stateful proxy.

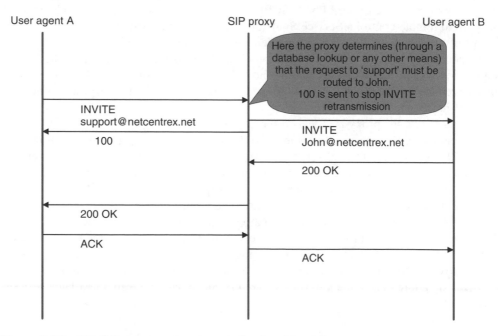

Figure 3.27 200 OK response is acknowledged end to end.

3.4.2.2 Examples of proxies

3.4.2.2.1 Call agent function

A call agent is a service that handles incoming and/or outgoing calls on behalf of a user. In traditional telephony this type of function is performed by the intelligent network infrastructure of the operator, or by the PBX of the company. The concept of 'call agent' was introduced in the IP telephony area in Scott Petrack's description of a Call Management Agent (CMA). A call agent can perform the following tasks:

- Try to find the user by redirecting the call setup messages (SIP INVITE or H.323 SETUP) to the proper location or several possible locations simultaneously.
- Implement call redirection rules, such as call forward on busy, call forward on no answer, call forward unconditional.
- Implement call filtering with origin/time-dependent rules.
- Record unsuccessful call attempts for future reference.

All these functions can be performed by the SIP proxy. Simple call redirection and filtering features (call forward unconditional, origin/time-dependent filtering) can also be implemented on a SIP redirect server. The SIP proxy server offers the most flexibility, because it can choose to relay all the call signaling, and therefore monitor and control all aspects of the call. In order to be able to use these services, the user must force all incoming call attempts to go through the appropriate SIP proxy. One way of doing this is to configure DNS records appropriately, as indicated in Section 3.4.1.2.

3.4.2.2.1.1 Sequential forking

Figure 3.28 shows a call forward on no answer (sequential forking). Note that if John is not logged on at desk 1 and the *proxy* also acts as a registrar, redirection can be immediate if John's registration has timed out. The example also shows an example of use for the CANCEL request, which is acknowledged with a 200 OK, and causes the initial INVITE request to be answered immediately with a 487 REQUEST TERMINATED answer.

The call agent can also be a functionality of end-user software, but this is usually less practical than using a separate centralized proxy server, because the end-user workstation can be switched off at any time and may have a dynamic IP address.

By accessing the database of a registrar, a SIP proxy can solve most user mobility/address change issues of the end-user terminal. For instance, each time a user connects to the Internet via an ISP, he gets a new IP address. But if his SIP software registers this new IP address, the proxy will be able to relay all calls to the new IP address.

3.4.2.2.1.2 Parallel forking

Forking proxies can duplicate a request and send copies of it to several hosts, each with a specific 'branch' parameter. This is called parallel-forking. Parallel-forking proxies are

Figure 3.28   Sequential forking example.

not necessarily transparent to responses and may filter out some replies before forwarding them back to the client. They may cause some hosts to receive duplicates of the same request with the same Call-ID (but different branch parameters); but, SIP clients must reply to each request.

Parallel-forking proxies can contact several endpoints belonging to the same person simultaneously. Some manufacturers call this as the 'simultaneous ringing' function. Although this call flow will work in a demo lab, it unfortunately cannot be implemented in a real telephony network, because in a real network each INVITE can cause the network to send back a one-way announcement (using a 183 SESSION PROGRESS response) which can be an error message ('network busy'), some information ('please type a PIN code', 'the cellphone you are calling is being located'), or sometimes even advertising ('welcome to the X network'). The forking proxy has no way of merging multiple audio sources to provide feedback to the user in the unlikely, but possible, case of multiple in-band messages, and therefore can only be used if such a possibility does not exist (e.g., if all called numbers are private extensions directly controlled by the proxy).

There can be other potential applications of parallel-forking proxies. A possible use is to handle NOTIFY messages when SIP is used for presence or alert applications (SUB-SCRIBE/NOTIFY methods). The forking proxy is then acting as a concentration point for notify messages. A forking proxy can also be used at the edge of a VoIP network to try to send a call simultaneously to multiple VoIP peer networks, which are expected to reject the call or redirect it using a 302 MOVED response (this is identical to the LRQ blast procedure used in H.323 networks).

3.4.2.3   *The Via and Record Route headers. Strict routing and loose routing*

A request from A to B can be routed through several proxies. It many cases it is desirable to force the response(s) to such a request to follow the exact same path as the request (e.g., a proxy might be billing the call or controlling a firewall and needs to have access to all the information regarding the call).

When a TCP connection is used for a SIP transaction, this is not generally an issue: the reply to a request automatically gets back to the other end of the TCP 'pipe', because TCP maintains a context throughout the connection.[17] On the other hand, when UDP is used some information must be present in the request datagram in order to allow the receiver to know where to send the reply.

Since SIP is Transport protocol-independent, all SIP requests and replies contain Via headers for exactly this purpose. This also helps avoiding routing loops (each proxy checks whether it is already in the Via list). Each time a SIP proxy forwards a request, it appends its name to the list of forwarding proxies recorded in the Via headers. When a proxy forwards a reply, it reverses the process and removes its name from the list. Additional details on the use of Via headers can be found in Sections 3.3.1.1.2 and 3.3.2.

If not only the requests and their associated replies (transactions), but also all requests within a dialog (e.g., ACK, NOTIFY), must be routed along the same path, the Via header is not sufficient and proxies must use the Record Route header. This is because SIP endpoints can add a Contact header field that enables other endpoints to send them requests (e.g., BYE requests) directly, and therefore proxies are not guaranteed to be on the path of all requests in a SIP dialog. When proxies update the Record Route header, they insert their SIP URL, with an optional maddr parameter, on the first line of the list. Requests can be routed on a predefined path by using the Route header. The routing model of RFC 3261 is called 'loose routing' because it allows proxies to route the message to additional hops not indicated in the Route list (Figure 3.29). The only constraint is that all proxies indicated in the Route list must be visited before the request is forwarded to the target indicated in the original Request-URI.

The old specification of routing (RFC 2543 before bis 05) specified that proxies should strictly route according to the Route list. In addition, the original Request-URI header was overwritten and could not be recovered. These problems were fixed by 'loose routing', and a work-around strategy was specified to enable loose routers to prevent loss of information when a message is routed through an old 'strict router' (Figure 3.30).

When SIP proxies are configured to route signaling messages, the call model is very similar to the H.323 gatekeeper routed call model.

---

[17] For this reason many SIP stacks do not support loose routing with TCP transport, which makes the 'UDP falls back to TCP for large messages' strategy impossible.

Figure 3.29   Loose routing using Route header information.

3.4.2.4 Loops and spirals

SIP implements the detection of loops via two mechanisms:

- Max-Forwards: this feature is mandatory in RFC 3261. The Max-Forwards header contained in every request is decremented at each hop (requests are initially sent with a default value of 70). If this counter reaches 0, the request should be rejected with a 483 TOO MANY HOPS response.

- Loop detection: proxies can detect that they have already processed a request by analysing the Via header list. If the Request-URI, From or To header fields have changed, this is a not a loop: it only indicates that the call has been processed by an application that changed the header elements that influence routing (e.g., the destination) and that the new destination is also routed by this proxy. This normal situation is called a **spiral**. On the other hand, if the headers listed above have not changed, this is a loop and the request should be rejected with a 482 LOOP DETECTED response.

Note that the loop detection mechanism is more complex but detects loops sooner than the Max-Forwards mechanism. Both mechanisms do not prevent loops involving a segment of the PSTN (i.e., call to a PSTN user who redirects the call book to the caller on a VoIP network).

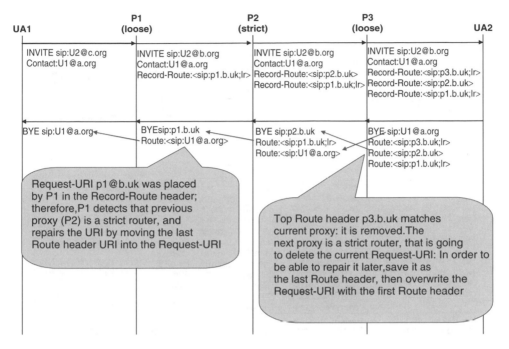

Figure 3.30 Loose routing work-around for older strict routers that erase original Request-URI headers.

3.4.2.5 Billing for SIP calls

By definition, all participants invited by a common source for a given session are in the same SIP 'call'. This call is identified by a globally unique Call-ID. Within a call, each leg can be identified by a unique combination of the To, From, and Call-ID fields. A proxy performing the call-accounting function should be able to distinguish different legs and create a CDR for each call leg. It should also be able to recognize RE-INVITE messages that only change the media description (see Section 3.3.2.3.2.5), not the participants; in this case it should not create a new call leg.

In the PSTN, a call is usually paid by the person who initiates it. A proxy relaying all signaling from the terminal of a user can create appropriate accounting records by logging the INVITE (RE-INVITE requests are ignored), CANCEL, and BYE requests, as well as the replies (Figure 3.31). The duration of each leg can be derived from the first accepted INVITE request (200 OK) up to the first BYE request.

In order to force the user to go through the proxy to make calls, one option is to control a firewall in the network from the proxy, as illustrated in Figure 3.31. This prevents the user from trying to bypass the call-accounting feature of the proxy. In the PacketCable® architecture for cable networks, the call management proxy dynamically sets 'gates' on the cable end **CMTS** (a sort of router with cable-specific features) for media channels using reserved quality of service.

Call-ID	From	To	Operation	Time Stamp	
4321@192.190.12.32	Mary@domain1	John@domain2	INVITE	11/11/1998	11:11:11
4321@192.190.12.32	Mary@domain1	John@domain2	200	11/11/1998	11:11:30
4321@192.190.12.32	Mary@domain1	John@domain2	BYE	11/11/1998	11:21:12
4321@192.190.12.32	Mary@domain1	Mark@domain3	INVITE	11/11/1998	11:14:13
4321@192.190.12.32	Mary@domain1	Mark@domain3	200	11/11/1998	11:14:40
4321@192:190:12:32	Mary@domain1	Mark@domain3	BYE	11/11/1998	11:21:13
4321@192.190.12.32	Mary@domain1	July@domain3	INVITE	11/11/1998	11:14:13
4321@192.190.12.32	Mary@domain1	July@domain3	486	11/11/1998	11:14:40
4444@192.190.12.32	Mary@domain1	Dilan@domain4	INVITE	11/11/1998	15:25:25
4444@192.190.12.32	Mary@domain1	Dilan@domain4	200	11/11/1998	15:25:40
4444@192:190:12:32	Mary@domain1	Dilan@domain4	BYE	11/11/1998	15:31:43

Figure 3.31   A proxy could control a firewall to ensure that all communications are logged.

In reality, many networks do not need this, as all VoIP devices in the network are configured to accept calls only if the INVITE comes from the service provider proxy–this can be done by simple **access control lists** (**ACL**s) restricting SIP signaling traffic on the routers connected to these resources. This way, if a user tries to bypass the network proxy, it will not be able to establish a call (e.g., to PSTN gateways). Without some form of dynamic firewall control, direct VoIP user-to-user calls on the IP network will be allowed on a best effort basis as long as user devices (e.g., softphones) are not under the control of the service provider. This is usually not a problem, as there are virtually countless ways of communicating without control in best effort mode.

3.4.3   Multiparty conferencing

SIP can be used to establish multipoint conferences, even in multicast mode (remember this protocol comes from the MMUSIC Group!). However, SIP does not currently provide any form of floor control.

3.4.3.1   Multicast conferencing

A multicast conference is a conference in which media streams are sent using multicast (for more details on multicast, see the companion book, *Beyond VoIP Protocols*, Chapter 6). The signaling related to this conference can be sent using multi-unicast or multicast (Figure 3.32).

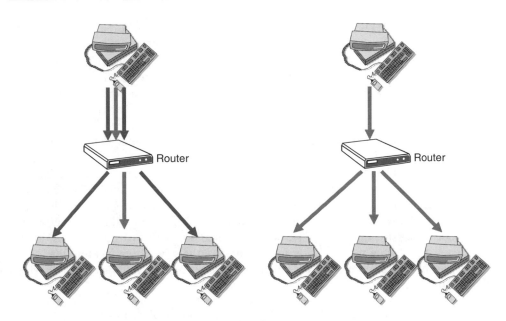

Figure 3.32   Multi-unicast and multicast signaling.

In the case of multi-unicast signaling, there is no significant difference from the point-to-point case, except that the SDP session descriptions indicate multicast addresses and the offer/answer model is also a bit modified compared with the unicast media case (see Section 3.3.2.3.2.4 for details).

When multicast signaling is used to establish multiparty conferences, SIP requests are carried using UDP, since this is the only transport protocol that can be multicast over IP. Multicast requests are expected to be used mostly to set up conference calls, and therefore the destination URL will generally be a conference name rather than an individual. However, the theory also allows usage of a multicast request with the URL of an individual (e.g., for multicast searches). The replies to a SIP request are then sent back to the sending UDP port on the same multicast address. In order to reduce network traffic and avoid a possible storm of synchronized replies, there are some modifications compared with the multi-unicast invitation procedure, including the following:

- 2xx replies are not sent.
- 6xx replies are sent only if the destination URL matches the name of a user on the host (i.e., the request is a multicast search rather than an invitation to a multi-party conference).
- Replies are sent after a 0–1-s random delay.

This form of multicast signaling was described in the first SIP RFC, but is not recommended in RFC 3261. It works in simple cases, but becomes very complex to manage if the full generality of SIP call flows is considered. Therefore, it seems SIP is headed more toward the use of multi-unicast to control multicast media sessions.

As long as all INVITE messages are sent from a central entity in unicast, RFC 2543 describes a basic form of floor control by sending new INVITE messages with the 'c' SDP parameter set by convention to null '0.0.0.0' to mute an endpoint, and re-invites the endpoint later (non-null 'c' parameter) when it is allowed to take part in the conference. Since the advent of RFC 3261 and its more formal description of media offers and answers, it is now prohibited to use this convention—the use of 'inactive' or 'recvonly' SDP attributes should be used instead.

SIP natively supports **layered encodings**. This class of coders encode the media information using several simultaneous data streams. One stream contains basic information (just enough to render a low-quality signal), and the other streams include additional information that can be used to reconstruct the signal with a higher quality (e.g., a video coder could send intra frames on one channel, and delta frames on another). Therefore, a receiver can choose the best bandwidth/quality trade-off by choosing to receive one, two, or more data streams. This is particularly suited for multicast conferences, allowing all receivers to tune the reception to their best settings, while preventing the sender from having to send customized data streams for each receiver. SDP describes a layered encoded stream as follows:

$$c = \text{<base multicast address>}/\text{<ttl>}/\text{<number of addresses>}$$

For instance:

$$c = \text{IN IP4 } 224.2.1.1/127/3$$

Multicast addresses used need to be contiguous (224.2.1.1, 224.2.1.2, 224.2.1.3). Unfortunately, there is no known commercial implementation yet using this facility.

3.4.3.2 Multi-unicast conferencing

The support of SIP for multi-unicast media conferences is limited. A central entity can be set up to act as an MCU to either mix or switch incoming media streams. The central bridge could implement very simple floor control by using RE-INVITES with the inactive, recvonly, or sendrecv SDP attributes. In practice, this is sufficient as most conferencing services use external, application-level user interfaces for floor control, and require the VoIP protocol only to implement basic mute/active/redirect functions, which can be readily provided by SIP and SDP.

However, SIP still lacks some messages for full support of video transmission control. An example is the request for full frames, present in H.245. Most video coders, (e.g., H.261 or H.263) send full frames only from time to time and deltas in-between. Most of the time, the instant at which a participant decides to speak will not coincide with the sending of a full frame. Therefore, if the MCU simply copies the incoming video stream to the output stream, the receivers will have to wait for the next full frame to get an image (Figure 3.33). So, the MCU needs to completely recode the stream in order to be able to send a full frame when the video switches. In a similar case, H.323 can mute the video stream of non-active speakers and request a full frame when it switches to an active speaker (VideoFastUpdate message). Such a message is also useful to quickly recover

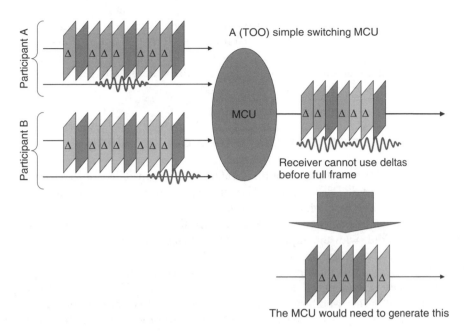

Figure 3.33 FullIntraRequest messages are helpful in all situations where the video source changes.

from video packet loss. Some RTCP messages (FullIntraRequest) or new SIP messages could be used for the same purpose, but video control is still not documented enough to allow for seamless high-quality interoperability across vendors.[18]

3.4.3.3 *Ad hoc conferencing*

SIP provides a simple and elegant way to switch from an existing point-to-point unicast call (A–B) to a multiparty multicast conference (A–B–C–...). The person (e.g., A) who wants to invite a new participant to the conference sends an INVITE message to the other party (B) and the new participant (C) with the parameters for the new session (i.e., a multicast address and eventually new coders instead of a unicast address), but keeps the old Call-ID. Keeping the same Call-ID tells B that this is not a new call, but new parameters for the existing call. This method can also be used to change session parameters in an existing call.

Ad hoc conferencing using unicast streams is also possible; in this case the new INVITE message redirects all streams to a media-mixing function. Many SIP phones implement such a mixing function locally for up to three media streams (three-way conferencing), in which case it is not necessary to redirect the media streams, the phone simply activates the mixing function for all the streams it receives. In fact, it is difficult in SIP to activate

---

[18] The most advanced initiative is the IETF work in progress draft: draft-levin-mmusic-xml-schema-media-control-03 'XML Schema for Media Control'.

a network-based conferencing function from an IP phone, because no standard conference activation message is defined to instruct the proxy to perform the required session changes. This is a significant obstacle for the deployment of SIP endpoints at the edge of public networks, because phone-embedded three-way functions are usually limited (G.711-only in most cases) and use twice the bandwidth of normal calls.

3.4.4   Configuring network-based call handling

Unconditional call forward call-handling features can be installed on a proxy/registrar simply by using REGISTER messages. For instance, a user who wishes to temporarily redirect his phone line to another extension just needs to send a REGISTER request with his name (or regular extension) in the To header field, and the new extension in the Contact header field, with the appropriate Expires value. This is roughly equivalent to a subset of the services offered by H.450.3 in H.323.

More sophisticated call-handling features (i.e., call agents) are outside the scope of SIP, and will probably be configured using other protocols, such as HTTP when SIP endpoints are multimedia PCs (the web browser is a perfect interface to customize the behavior of a sophisticated proxy). However, the XML **Call Processing Language** (**CPL**) has emerged as a very flexible way of expressing call-handling rules, and is supported by some SIP proxies. Most of the time, the CPL script is configured through a non-SIP interface; but, if the phone supports it, in certain cases it can be configured by using the REGISTER message to carry simple CPL scripts as a payload. Unfortunately, there is no standard way of doing this yet.

3.5   INSTANT MESSAGING AND PRESENCE

Instant messaging and presence (which has become a buzzword) are the most popular applications of a more general use of SIP for the subscription and exchange of stateless event messages. These capabilities have been added to SIP by RFC 3265 (SIP, 'Specific Event Notification') in a very general way that wasn't particularly targeted at instant messaging. In fact, it was first used by some VoIP vendors to implement out-of-band DTMF transmission during a call (as described in Section 3.3.2.3.1.3). With Windows XP, Microsoft introduced a new version of its Messenger® client, which included support not only for voice and video, but also instant messages. The client works primarily with proprietary protocols using a specific Microsoft central server as a message reflector, but can be configured to use SIP as well. Microsoft chose to use the mechanism defined in RFC 3265 for the subscription of presence information and notification of state information. This instantly made this extension of SIP a de facto standard. For users strongly in favor of a convergence of unified messaging protocols, the use of RFC 3265 was a step in the right direction, but still did not define the exact message content. The convergence of message contents is a more difficult step, as it may depend on the capabilities of each client, and has enormous implications for the large IM systems that currently 'own' their users. The convergence of the IM format is the main task of the IMPP Working Group of the IETF.

3.5.1   Common Profile for Instant Messaging (CPIM)

This specification of the IMPP (Instant Messaging and Presence Protocol) Working Group of the IETF defines a number of operations and features to be supported by instant-messaging systems. The profile aims at facilitating the interworking between various instant-messaging systems, by providing an intermediary canonical format which facilitates the design of transcoding gateways. Obviously, this format can also be used to format instant messages, not just as a conversion intermediary format. Today, two popular instant-messaging protocols follow the CPIM guidelines: XMPP (eXtensible Messaging and Presence Protocol) used by the Jabber IM client, and SIMPLE (SIP for Instant Messaging and Presence Leveraging Extensions).

3.5.1.1   *Common Presence and Instant Messaging message format*

The CPIM canonical message format specification is still a work in progress of the IMPP working group (draft-ietf-impp-cpim-msgfmt-07.txt). The draft defines a new Multipurpose Internet Mail Extension (MIME) format 'message/CPIM', intended to be a common format for CPIM-compliant messaging protocols. MIMEs are defined in RFC 2045, 2046, and 2048.

One of the key reasons to encourage instant message systems to support the CPIM message format natively is to allow a gateway between two instant-messaging systems to preserve the electronic signatures that can be added to a CPIM message. Signatures are lost if any transcoding has to occur.

Although the defined format complies with MIME, it does not allow for all the options of MIME. This simplification aims at suppressing or restricting all the options of MIME that can present an obstacle for interoperability or the verification of electronic signatures (e.g., suppression or addition of headers, extensibility of header formats, weak internationalization, etc.).

One of the key requirements for an instant-messaging system is to be able to support all character sets. CPIM uses UTF-8 encoding.

3.5.1.1.1   *The universal character set and the UTF-8 format*

Computer science discovered very late that the world was not just using the well-known US-ASCII characters, which are encoded using only 7 bits. In contrast, many operating systems were handling 8-bit character sets, which led to a system where each language required its own code page (Latin-1, Hebrew, Arabic, Greek) encoded on 8 bits, and where a given character could have multiple encodings depending on the character page. For instance, the Euro sign (€) was $0 \times A4$ in Latin-9 (ISO 8859-15), 0×80 in Latin-2 (CP1250), and 0×88 in Cyrillic (CP1251). Any program that needed to use characters from multiple pages simultaneously would have to be very cumbersome in design.

The Universal Character Set, defined by ISO/IEC 10646-1, contains in a single character set almost all the symbols used by all known writing systems on earth. This is a multi-octet character set: **UCS-2** contains the first 64,000 characters and is encoded on 2 octets (it is also called the Basic Multilingual Plane or BMP), **UCS-4** is encoded on 4 octets and can contain potentially many more characters beyond the first 64,000 (although there is currently no character defined beyond those already contained in the BMP). The UCS character set is identical to the Unicode character set defined by the Unicode Consortium, but Unicode defines more character properties, semantic conventions, and more character-rendering options. UCS and Unicode co-operate closely with each other and have so far used the same code points for each character.

Multi-byte character sets are not compatible with many current applications or systems that are byte-oriented. Many systems are also only able to handle correctly 7-bit US-ASCII characters. For instance, in any C program '\0' means 'end of the string', but this sequence can be found in the middle of a multi-byte character stream (e.g., resulting in commands like 'printf' not being used with UCS-4 character streams). Even recent systems that understand 2-octet characters cannot handle UCS-4 characters. In order to facilitate the use of UCS in such systems, **UCS Transformation Formats** (**UTF**s) have been defined:

- UTF-7 encodes all BMP characters using only octets with the first bit set to '0', and therefore is transparent even to older 7-bit mail systems.
- UTF-8, defined in RFC 2279, uses variable length (1–6 octets) encodings for UCS-2 or UCS-4 characters, but preserves all 7-bit US-ASCII characters, which are encoded on one single octet, with the usual 7-bit ASCII value. ASCII character values are encoded in UCS-4 as 0000 0000 to 0000 007F, and are encoded in UTF-8 as 00 to 7F. Because of this, UTF-8 is 'file system safe' (it was originally called UTF-FSS). For multi-octet sequences, the first octet indicates the number n octets in the sequence with n high-order bits set to '1'. All the following octets have the first two bits set to '10', and 6 variable bits. The remaining $(8-1-n)$ bits of the first octet and the $6*(n-1)$ bits of the following octets are used to encode the UCS character, as shown in Table 3.4.

Let's give some examples:

- The copyright sign © (Unicode character U + 00A9 = 1010 1001) is encoded in UTF-8 as 11000010 10101001 = 0 × C2 0 × A9.

Table 3.4 UTF-8 encoding of 4-octet characters preserves 7-bit ASCII values

UCS-4 range (hex.)	UTF-8 octet sequence (binary)
0000 0000 – 0000 007F	0xxxxxxx
0000 0080 – 0000 07FF	110xxxxx 10xxxxxx
0000 0800 – 0000 FFFF	1110xxxx 10xxxxxx 10xxxxxx
0001 0000 – 001F FFFF	11110xxx 10xxxxxx 10xxxxxx 10xxxxxx
0020 0000 – 03FF FFFF	111110xx 10xxxxxx 10xxxxxx 10xxxxxx 10xxxxxx
0400 0000 – 7FFF FFFF	1111110x 10xxxxxx ... 10xxxxxx

- The euro (€) sign (U+20AC) is encoded in UTF 8 as $0 \times E2\ 0 \times 82\ 0 \times AC$.
- A good list of Unicode fonts for the Microsoft Windows® operating system can be found at http://www.alanwood.net/unicode/fonts.html

3.5.1.1.2 Message format

The message/CPIM format is a multipart MIME format which encapsulates:

- Content and message-related metadata.
- The message itself in the form of any MIME content.
- Optionally, an electronic signature according to S/MIME, RFC 2633.

Figure 3.34 shows an example without an electronic signature. The end of the message body in Figure 3.34 is defined by the framing mechanism of the transport protocol used.

3.5.1.1.2.1 MIME header part

In Figure 3.34 the MIME header part is composed only of the mandatory message/CPIM Content-Type header, but other headers can be added before the blank line if necessary. Each line is ended with CR+LF characters.

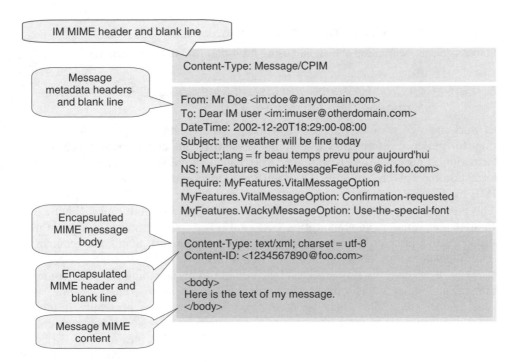

Figure 3.34 CPIM multipart MIME format.

3.5.1.1.2.2   Message header part

This part must remain intact end to end. The headers and their values must not be changed in any way, or even reordered. Each line has a 'key: value' form (with a single space after the ":"). The key must contain only US-ASCII characters (some control characters like or ' ' 'must be escaped), while any UTF-8 character (with the same escaped control characters) is allowed in the value portion. A header can be tagged to indicate that it contains a specific language by using the ';lang = tag' after the header name and colon, where 'tag' is a language identifying token (defined in RFC 3066).

From, To, Subject, DateTime (RFC 3339: date/UTC time/time offset) are headers defined in CPIM. Figure 3.34 also shows the extension mechanism for the CPIM format. A developer can define his own extension namespace (here MyFeatures), by using the NS (namespace) header. New header keys beginning with 'MyFeatures.' can then be used. They will be ignored if not understood. It is possible to indicate to the receiving system that it needs to support an extended header in order to understand the message by using the Require header followed by the header key that must be supported.

3.5.1.1.2.3   Encapsulated MIME object

This is the message itself, any MIME type can be encapsulated. Like any MIME-encoded object, it is composed of a header part and a content part, separated by a blank line. For simple, text-only IM systems, the text/XML MIME type using the UTF-8 encoding can convey any written symbol from any language.

3.5.1.1.2.4   MIME security multipart message wrapper

The message can be secured and signed using multipart MIME, as shown in of Figure 3.35.

3.5.1.1.3   Common Presence and Instant Messaging (CPIM) Presence Information Data Format (PIDF)

This work in progress of the IETF IMPP Working Group is attempting to define a standard format for presence information sent from a **presentity** (the entity about which presence information is generated) to a **watcher**. Note that multiple devices may send presence information for a given presentity. Draft-ietf-impp-cpim-pidf-06.txt defines the new XML MIME media type 'application/cpim-pidf+xml' and has an optional charset parameter.

The presence information of a *presentity* consists of one or more tuples (Figure 3.36) with status, an optional communication URI, and other optional presence markup information (relative priority, timestamp, human-readable comment). Status may contain one or multiple values, the 'open' and 'closed' values mean the entity is 'ready'/respectively, 'not ready' to receive instant messages, but does not imply anything for other communication means. Other status values may be defined in extensions (busy, off-line, away, on the phone, etc.). There may be more than one tuple for a presentity if multiple devices/applications can reach the presentity and each one creates a presence component in the form of a tuple. For instance, in a SIP REGISTER message, the To header field (address of record) would be considered the presentity, while each URI in the Contact header would be a point of communication for that presentity, each one identified

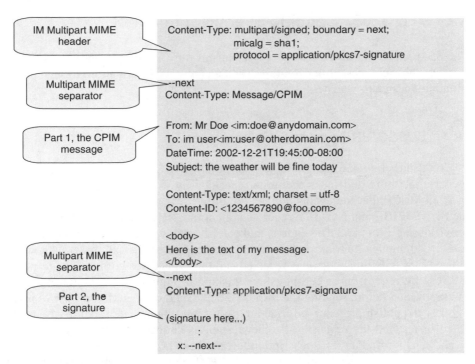

IM Multipart MIME header

Content-Type: multipart/signed; boundary = next;
         micalg = sha1;
         protocol = application/pkcs7-signature

Multipart MIME separator

--next
Content-Type: Message/CPIM

Part 1, the CPIM message

From: Mr Doe <im:doe@anydomain.com>
To: im user<im:user@otherdomain.com>
DateTime: 2002-12-21T19:45:00-08:00
Subject: the weather will be fine today

Content-Type: text/xml; charset = utf-8
Content-ID: <1234567890@foo.com>

<body>
Here is the text of my message.
</body>

Multipart MIME separator

--next
Content-Type: application/pkcs7-signature

Part 2, the signature

(signature here...)
         :
    x: --next--

Figure 3.35   CPIM message electronic signature using multipart MIME.

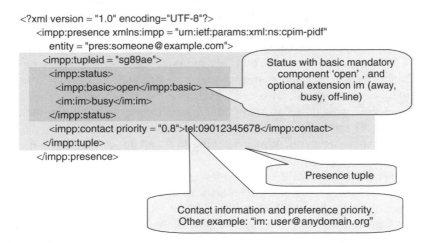

```
<?xml version = "1.0" encoding="UTF-8"?>
    <impp:presence xmlns:impp = "urn:ietf:params:xml:ns:cpim-pidf"
      entity = "pres:someone@example.com">
    <impp:tupleid = "sg89ae">
     <impp:status>
       <impp:basic>open</impp:basic>
       <im:im>busy</im:im>
     </impp:status>
     <impp:contact priority = "0.8">tel:09012345678</impp:contact>
    </impp:tuple>
   </impp:presence>
```

Status with basic mandatory component 'open' , and optional extension im (away, busy, off-line)

Presence tuple

Contact information and preference priority. Other example: "im: user@anydomain.org"

Figure 3.36   PIDF presence information format.

in a separate tuple. The 'q' values from the Contact header field could be translated into 'priority' values for the tuple.

The CPIM-PIDF format includes mechanisms for integrity, confidentiality, and authentication, independently of SIP.

3.5.2   RFC 3265, 'Specific Event Notification'

RFC 3265 specifies SIP-based transport mechanisms for watchers to subscribe to presence information, and for 'presentities' (entities sending presence information) to send presence information updates to watchers.

3.5.2.1   SUBSCRIBE and NOTIFY requests

RFC 3265 defines two new optional requests: SUBSCRIBE and NOTIFY. These requests are generic in nature and must be further specified by **event packages** (the name was taken from MGCP, but the events defined are different from MGCP events).

The SUBSCRIBE and NOTIFY requests are normal SIP requests, which can be routed by proxies using From and To headers, and can be acknowledged by a 200 OK or a 202 response. 200 implies that the request has been accepted, while 202 only acknowledges that the SUBSCRIBE message was received and the syntax was correct. The response to a SUBSCRIBE request must be immediate, making it impossible to ask for any form of user authorization before sending the response. 202 would be used, in the case of a buddy list request to publish presence information, to respond immediately before the user has accepted or rejected the request. All other responses defined for other SIP requests are also valid, and imply that the subscription has not been accepted as is.

The SUBSCRIBE request can be sent by a SIP client willing to receive certain events (the **subscriber**) to a SIP server generating these events or already receiving these events (the **notifier**). The SUBSCRIBE request contains an Expires header limiting the duration of the subscription, which can be shortened if the response contains a shorter period in its own Expires header. In order to improve scalability for heavy load notifiers (e.g., voicemail systems), longer periods can be requested by rejecting the SUBSCRIBE with a 423 INTERVAL TOO SMALL; however, by convention intervals above 1 h must be accepted. With the Expires header, the subscription becomes 'soft-state', which is a very common approach of Internet protocols, also used by RSVP. Soft-state subscriptions are more tolerant of protocol errors or network instability, avoiding any undesired accumulation of state in any network entity.

When a subscriber wishes to stop subscribing to a certain set of events, it can do so by setting the Expires header value to 0.

The Event header in the SUBSCRIBE request specifies the category or set of events that are requested. The exact syntax of the Event header is free and must be defined by specific Event-Packages. Optionally, the body of the SUBSCRIBE request can also be used to further specify the subscription, but again it must be specified by the Event-Package.

Both SUBSCRIBE and NOTIFY can create a SIP dialog, as defined above for INVITE requests. Therefore, these requests do not require any prior INVITE request and can be sent asynchronously at any time. Alternatively, they can be sent within an existing dialog: in this case the Event header must contain an 'id' parameter to distinguish between the various subscriptions. Sending multiple SUBSCRIBE requests with identical 'id' parameters within an existing dialog can be used to refresh subscriptions (if it does not correspond to an active subscription, it will be rejected with a 481 response).

Because the SUBSCRIBE/NOTIFY mechanism was primarily defined to handle state change notifications, the first SUBSCRIBE will trigger an immediate NOTIFY within the same dialog, to synchronize the initial state status of the subscriber. This is also true even if the Expires header has a value of 0. This allows simple state polling with SUBSCRIBE requests having an Expires value of 0 to be carried out. Examples of commonly used state information include voicemail box status, busy state of a user (for call completion on busy), and buddy lists with presence status.

The notifier can decide to terminate a subscription at any time by sending a NOTIFY message with a Subscription-State header with a value of 'terminated' and a reason parameter. One useful reason parameter is 'rejected', which can be used when a user has decided not to accept a subscription. This mechanism is often used, as shown in Figure 3.37, because the initial SUBSCRIBE request has been acknowledged by a 202 response.

3.5.2.2 *Use of RFC 3265 for presence*

The SIMPLE working group (SIP for Instant Messaging and Presence Leveraging Extensions) is attempting to get the various instant messaging and presence implementations to converge in an interoperable standard based on SIP. SIMPLE works in close co-operation with the IMPP Working Group. At the time of writing all SIMPLE documents were still in draft state.

Draft-ietf-simple-presence-09.txt specifies how to use RFC 3265 for presence. It defines **presence agents** as SIP devices able to receive presence subscription requests and to send presence information for a given presentity. Presence is handled by creating a specific 'presence' Event-Package. In the future other specific types of events may be created to

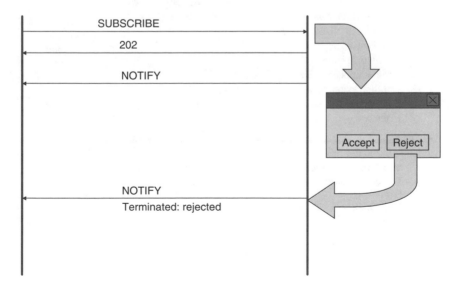

Figure 3.37 Case of a rejected subscription.

handle the requirements for a buddy list (a party is typing a message, message delivery confirmation, typical party states).

According to RFC 3265 the name of the Event-Package 'presence' must be in the Event header field of SUBSCRIBE and NOTIFY requests. No SUBSCRIBE body is yet defined and therefore should normally be empty (Figure 3.38).

In the example the subscription has been accepted immediately with a 200 OK. As soon as the presence user agent receives the subscription, it must, according to RFC 3265, immediately send back the current presence state in a NOTIFY message. The notification data should use the CPIM body type defined by IMPP: application/cpim-pidf+xml. In Figure 3.39 the presentity is *closed* and therefore not ready to receive instant messages. A non-standard extension explains the cause: the user is busy.

As soon as the presence information changes, the presence user agent sends a new NOTIFY message. In Figure 3.40 the presentity is now *open* and can receive instant messages. The rate at which presence notification updates can be sent is limited to at most one every 5 s.

3.5.2.2.1 Watcher information

Draft-ietf-simple-winfo-format-03.txt defines an XML format for the **watcher information**, and defines a new payload type for it: application/watcherinfo+xml. The watcher

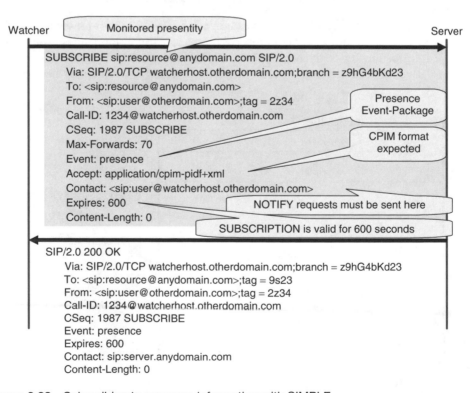

Figure 3.38 Subscribing to presence information with SIMPLE.

Figure 3.39 Notification of presence information using SIMPLE.

Figure 3.40 Updated presence information sent in a new NOTIFY request.

information is the list of all active and pending requests to receive event notifications (subscriptions) for a specific resource. The watcher information (Figure 3.41) includes the URIs of the watchers, an id, the current status of the subscription, the event that caused transition to that status, and optionally other parameters, such as the duration of the subscription.

In order to receive the watcher information, a normal SUBSCRIBE request can be sent to the presence server, as illustrated on Figure 3.42. If the subscription is accepted, the updates to the watcher information are reported in NOTIFY requests (Figure 3.43).

Figure 3.41   Watcher information format.

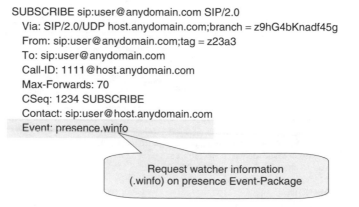

Figure 3.42   A SUBSCRIBE request for watcher information.

```
NOTIFY sip:user@host.anydomain.com SIP/2.0
   Via: SIP/2.0/UDP server.anydomain.com;branch = z9hG4bKnf44z
   From: sip:user@anydomain.com;tag = r34s3
   To: sip:user@anydomain.com;tag = z23a3
   Call-ID: 1111@host.anydomain.com
   Max-Forwards: 70
   CSeq: 1288 NOTIFY
   Contact: sip:user@server.anydomain.com
   Event: presence.winfo
   Content-Type: application/watcherinfo + xml
   Content-Length: ...
```

> Requested watcher information, update at each change

```
<?xml version = "1.0"?>
<watcherinfo xmlns = "urn:ietf:params:xml:ns:watcherinfo" version = "0" state = "full">
    <watcher-list resource = "sip:user@anydomain.com" package = "presence">
       <watcher id = "7234c7s" event = "subscribe"
            status = "pending">sip:watcherA@foo.com</watcher>
    </watcher-list>
</watcherinfo>
```

> This is a new candidate watcher

Figure 3.43 Watcher information update sent through a NOTIFY request.

3.5.2.2.2 *Procedure for new subscriptions, presence authorization*

Draft-ietf-simple-winfo-package-04.txt defines a framework for the authorization of presence subscriptions. When a request for presence information arrives at a presence server, the presence server will usually require the user to authorize the new subscription. Of course, this is only possible if the user is first made aware of the new subscription, which is relatively easy if the presence server is the user agent of the user, but becomes more complex if the presence server is a network-based device. The idea behind the draft is to always allow a user to subscribe to any modification of the watcher information relative to his own presence. In Figure 3.44 the presence server is the SIP proxy for domain "anydomain.com".

New subscriptions will update the watcher information, and therefore the user will receive a NOTIFY with the update for watcher XML information, as shown in Figure 3.45. Only the changes are included; therefore, the user needs to cumulate these changes to get a complete up-to-date view of watcher information.

Part of this authorization framework relates to new subscriptions. Note that some standard states have been defined for subscriptions. Most of the states are self-explanatory (see Figure 3.46). The 'waiting' state has been added to allow a user to learn about subscription requests even if they have expired; this enables the user to set a specific policy on the presence server to accept any retry for that subscription.

This framework enables a user to know which subscriptions need to be authorized; but, at this moment there is no standardized way to tell a presence server to authorize a subscription. It could be a web page; but, obviously, a set of SIP messages to do this

Figure 3.44   Subscribing to watcher information.

Figure 3.45   Watcher information update sent in a NOTIFY request.

would be a much better option. One idea is to use an XML policy syntax similar to the Call Processing Language (CPL), embedded in REGISTER requests, to do this.

The consequence is that only distributed presence models work today in a standard way. These models co-locate the user end-point and the presence agent so that authorizations remain local. This is the model used by Microsoft in Messenger®.

3.5.3   RFC 3428, 'SIP Extensions for Instant Messaging'

RFC 3428 defines a new MESSAGE request which carries MIME body parts representing the content of an instant message. The MESSAGE request is usually sent outside the

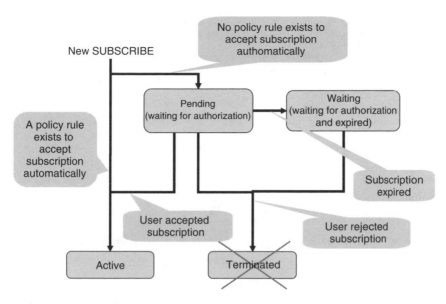

Figure 3.46 Subscription states.

context of an existing dialog, but does not create its own dialog. It can also be sent as part of an existing dialog in some circumstances (e.g., if an instant message is sent as part of an existing voice call). The response can be a provisional or a final response; usually, it will simply be 200 OK if the message has been received, or 202 if it has been stored for presentation to the target user as soon as possible. Neither the MESSAGE request nor the 200 OK reply are allowed to have a Contact header (they do not create a dialog). A MESSAGE request can have an Expires field, which is a simple indication of its validity for proxies that may try to store it if the target user is not immediately available.

RFC 3428 defines an instant message URI, im:user@domain, which is independent of the underlying instant message transport protocol. For all practical purposes, however, it is translated into a SIP URI immediately and placed in the Request-URI of the message request before sending. An instant message can be sent to an instant message URI (im:someone@domain.org), or to a SIP URI. If the IM URI is used the next hop server and SIP transport method can be found by performing a SRV DNS request for _im._sip.domain.org which should return a resource record of SIP proxy that can route the message (this is still an IETF draft: draft-ietf-impp-srv-01).

The size of the message payload is limited to 1,300 bytes or 200 bytes less than the path MTU if known (this is usually not the case), in order to avoid the message segmentation problems of SIP. A basic form of congestion control is ensured by requiring clients to never send a MESSAGE until the previous MESSAGE request has been acknowledged with a response. Figure 3.47 gives an example of the MESSAGE request. The text/plain content type only allows US-ASCII characters. By using the 'char/xml; charset = utf-8' payload type it is possible to send any type of character:

```
MESSAGE sip:user2@domain.com SIP/2.0
  Via:SIP/2.0/TCP user1pc.domain.com;branch = z9hG4bKdfg45
  Max-Forwards: 70
  From: sip:user1@anydomain.com;tag = 4sd83
  To: sip:user2@anydomain.com
  Call-ID: ef4234@10.10.10.10
  CSeq: 1 MESSAGE
  Content-Type: text/plain
  Content-Length: 41

  Hello. This is a sample instant message!
```

Figure 3.47   Sample MESSAGE request.

```
Content-Type: char/xml; charset=utf-8
<body>
This is a utf-8 message, it can contain all non US-ASCII characters
  like € or ©!
<body>
```

RFC 3428 also requires the 'message/CPIM' content type to be supported, and therefore instant messages can carry any type of MIME content.

The framework defined by RFC 3428 is still quite basic compared with GSM SMS specifications (e.g., it does not cover confirmation of message receipt). However, RFC 3428, together with the CPIM format, provides a good foundation for sending instant messages across SIP-based instant-messaging systems from various service providers. In the future it will do so across non-SIP-based instant-messaging systems, like the popular GSM short Message Service and Multimedia Message Service.

3.6   SIP SECURITY

3.6.1   Media security

Media encryption is specified by SDP. The 'k' parameter of SDP stores the security algorithm in use as well as the key. The following formats are defined in RFC 2327 (SDP):

- k = clear : <encryption key> This format refers to the encryption algorithms described in RFC 1890 ('RTP Profile for Audio and Video Conferences with Minimal Control', January 1996). RFC 1890 first describes how to extract a key from a pass phrase in a standard way. The pass phrase is put in canonical form (i.e., leading and trailing white spaces removed, characters made lower case, etc.), then hashed into 16 octets by the MD5 algorithm. Keys shorter than 128 bits are formed by truncating the MD5 digest. The name of the algorithm in use is concatenated before the key and separated from the key with a single slash. Standard identifiers for the most common algorithms can be found in RFC 1423 (DES-CBC, DES-ECB, ...), the default being DES-CBC. RFC

1423 also describes how to store additional parameters needed for particular algorithms, such as the 64-bit initialization vector of DES-CBC: for example the following line can be used to initiate a DES-CBC-encrypted session:

$$k = clear:DES-CBC/aZ25rYg7/12eR5t6y$$

- k = base64 :< encoded encryption key > The format is the same as above, but base64-encoded to hide characters not allowed by SDP.
- k = prompt This, prompts the user for a key. The default algorithm is DES-CBC.

3.6.2 Message exchange security

3.6.2.1 Authentication

Most vendors support the mechanisms defined by RFC 2617 ('HTTP Authentication: Basic and Digest Access Authentication') for basic authentication (clear password) or digest authentication (hash code derived from the message content and a challenge sent by the server or proxy).

User agents, registrars, or redirect servers should use response code 401 to indicate that they cannot accept the request without further authentication information. Proxies should use response code 407.

3.6.2.1.1 Basic authentication

The use of basic authentication has been deprecated in RFC 3261, as the password is sent in clear form over the network. An RFC 2543 server user agent willing to authenticate a client user agent using the 'basic' method will respond with the following header:

WWW-Authenticate: Basic realm = "realm_information_here"

The realm is simply context information that should be presented to the user in order to allow him to select the proper username and password. SIP requires that it be globally unique.

The client must re-issue the request, sending back the user ID and password, separated by a single colon and encoded as a base64 string, using the Authorization header:

```
Authorization: Basic QWxhZGRpbjpvcGVuIHNlc2FtZQ==
```

Note that since the ACK accepts no response, any authentication information that was accepted for an INVITE must be accepted also for the corresponding ACK (same Authorization and Proxy-Authorization headers). This is true of all authentication methods.

3.6.2.1.2 HTTP digest for user agent authentication, registrars, and redirect servers

In order to avoid sending the password in clear form, many user agents support the HTTP digest method. By default, the Authorization header field contains the MD5 digest of:

- The Request-URI.
- The username.
- The nonce value.
- Optionally, the message body.

Use of the digest method is also specified in the WWW-Authenticate header of the 401 Unauthorized response:

```
WWW-Authenticate = Digest realm="realm_information_here",
qop="auth,auth-int",
nonce="Od1128f1806872deac4e01029b7c96b3",
stale=FALSE, algorithm=MD5,
  opaque="5ccc069c403ebaf9f0171e9517f40e41"
```

The nonce should be generated randomly for each 401 response and must not contain any double quote. If an 'opaque' string is included by the server, it should be passed back by the client in the response. The stale flag, when set, indicates that the previous Authentication data were correct, but were rejected because the nonce information was stale. The 'quality of protection' parameter indicates that the server supports authentication only (auth) and authentication with integrity protection (auth-int), in which case the message body can be included in the hash value calculation.

The client then re-issues the request (keeping the same Call-ID), including the requested authentication information in the Authorization header. The MD5 hash value is in the response parameter:

```
Authorization = Digest username="81@realm information here",
   realm="realm_information_here",
      nonce="0d1128f1806872deac4e01029b7c96b3",
      uri="destination.test.org",
      qop=auth,
      response="2923fb70ddfdf57f7ffe5cc436ab4889"
      opaque="5ccc069c403ebaf9f0171e9517f40e41"
      algorithm=MD5
```

In response to the resubmitted request, the server can provide some feedback regarding successful authentication in the Authentication-Info header. In addition, it may provide a new nonce parameter (nextnonce), and may even include a hash value proving it also knows the client secret (rspauth parameter):

```
Authentication-Info: nextnonce="47364c23432d2e131a5fb210812c",
rspauth="29364c52832d2e131a545211212c"
```

Note that another authentication method based on PGP, now deprecated, was defined in RFC 2543. This semantic allowed some variable fields (such as the Via field) to be

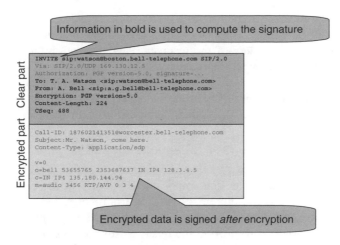

Figure 3.48   Scope of PGP signature in RFC 2543.

excluded from the signed data. Figure 3.48 shows that the PGP signature could protect both the clear part and the encrypted part of the SIP message.

3.6.2.1.3   HTTP digest for proxy servers

A proxy may decide to authenticate a request by using the 407 PROXY AUTHENTI-CATION REQUIRED response, which contains a Proxy-authenticate header that issues a challenge. The client must then resend the request with a Proxy-authorization header providing the credentials matching the challenge. The content of these headers is similar to that of WWW-Authenticate and Authorization.

In order to avoid this round trip, the client can of course provide the credentials in the first message, if the authentication replay protection mechanism allows it.

On subsequent responses, the server sends a Proxy-Authentication-Info header, with the same parameters as those of the Authentication-Info header field.

Proxies must be completely transparent to the WWW-Authenticate, Authentication-Info, and Authorization headers, and must forward them without any change.

3.6.2.2   Encryption of messages

If the media encryption key must be protected, then the SDP requests and replies must be encrypted. There are many other reasons for protecting SIP messages (e.g., to hide the origin or destination of calls and the related information fields such as Subject, etc.). In general, however, SIP messages only need to be authenticated, which is useful not only to prevent call spoofing, but also for accounting and billing.

SIP messages can be encrypted hop by hop (e.g., using IPsec). They can also be transported over a secure transport layer such as TLS (in this case "sips:" URIs are used). SIP also describes an end-to-end encryption strategy based either on a shared secret key between the sender and the receiver or on a public key mechanism. If a common secret key is used, then the receiver of the message is able to decrypt a message encrypted by the sender by using the shared password. If a public key scheme is used, the sender encrypts the message using the public key of the receiver. This encryption can be performed by the sender of the request or by an intermediary security proxy.

RFC 2543 also defined an encryption mechanism based on PGP, which has been deprecated. The request line and unencrypted headers were sent first, followed by an Encryption header field, which indicates the encryption method in use; for instance:

```
Encryption: PGP version=2.6.2,encoding=ascii
```

The encrypted part began after the first empty line (CR + LF of the previous line immediately followed by CR + LF). Figure 3.49 is taken from RFC 2543.

If just the message body has to be encrypted, an extra empty line had to be inserted in the body before encryption to prevent the receiver from mixing up message body data and encrypted headers. There are specific issues with the Via header, since it is used by proxies to route the request back to the source.

3.7 SIP AND H.323

The SIP versus H.323 debate has been a very heated one among VoIP engineers. Behind the technology facade of some arguments, the debate has been fueled and biased by the interests of many telecom manufacturer companies who roughly fall in the following categories:

- Early VoIP players, with a strategy based on standards, who have already captured the largest VoIP market share and defend H.323. In this camp there is also the vast majority of PBX manufacturers, who like the similarity between ISDN and H.323 and offer H.323 WAN interfaces.

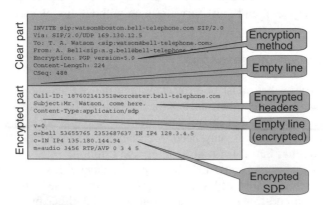

Figure 3.49   SIP message encryption in RFC 2543, using PGP.

- Early VoIP players, with proprietary products, who are trying to extend the life of their existing products, by criticizing H.323 and announcing a leapfrog to future standards in their roadmap. This is also the case of most traditional central office manufacturers, who must pay lip service to VoIP but also wish to extend the life of their existing systems and most of all the 'closed model' which ensures comfortable maintenance fees. These manufacturers usually build mostly proprietary VoIP systems, but very often label proprietary protocols with a reassuring 'pre-XXX', where XXX is a still-immature protocol.

- Start-ups arriving too late to catch the first H.323 wave, who engaged all their marketing resources to promote SIP and present H.323 as obsolete.

Let's now turn to discuss some of the features of SIP and H.323 that are most frequently used for or against them. The conclusion is that the 'protocol war' has been very beneficial to both protocols, stimulating many improvements, to the point that SIP and H.323 are now virtually identical protocols!

In 2004, H.323 still has the lion's share of VoIP deployments, in telephony carrier networks. It is also used by virtually all corporate IP-PBXs for the VoIP trunk interfaces. (In 2002, over 2 million such H.323 PBX trunk ports were sold in the USA alone.) All versions of Microsoft Windows® still include the NetMeeting® H.323 client, as well as a TAPI H.323 implementation. NetMeeting is also used as a VoIP and collaboration component in most other client-side and server-side programs of the company. However, SIP has become more prevalent on the desktop since the XP version, which hides NetMeeting and exposes Messenger®, a general purpose instant-messaging and VoIP client, based on proprietary protocols but where SIP can also be used.

SIP seems to be winning the battle for instant messaging, and this will probably drive the adoption of SIP for multimedia user interaction on a PC desktop. One can only guess what the future will be, but it is likely that SIP will become the de facto standard on PCs, while H.323 will remain dominant inside carrier networks and for the interconnection of IP-PBXs for some years, until SIP matures enough to include all the required features for PSTN interworking. In time, H.323 and SIP will probably continue to coexist in the market for carrier telephony over IP, but with a more balanced share.

The winner for IP phones and analog residential gateways is likely to be neither SIP nor H.323 (today most IP phones implement a proprietary PBX stimulus protocol, but among standard phones most support H.323, with a growing number offering SIP IP). IP phones controlled by a PBX, a hosted PBX, or a virtual PBX share common requirements with today's PBX business phones. These requirements led the market to implement dumb stimulus phones as opposed to smart phones, allowing the PBX to control any aspect of the phone user interface, including lamps, buttons, screens, etc. A stimulus phone makes it easier to implement any service without waiting for a standardized way of implementing this standard on a smart phone. It also facilitates centralized management. In VoIP, the leading stimulus phone protocol is MGCP, and it is likely that most IP phones will have to support MGCP is the coming years. MGCP is also required to interwork completely with analog gateways, as it supports services on- and off-hook. MGCP is the dominant protocol in the cable market.

3.7.1   Arguments in favor of SIP

3.7.1.1   *Speed and simplicity*

The simplicity of the basic SIP call is striking, compared with H.323v1 which required four or five messages. This is a strong incentive for developers to choose SIP for new products, as H.323 is a bit 'scary' to begin with and certainly requires a steeper learning curve due to the use of advanced techniques, such as ASN.1. However, when looking at the call flows more carefully, this initial conclusion becomes more balanced:

- H.323v2 implemented the Fast Connect method, which is nearly identical to the offer–answer model of SIP and gives the same performance.
- The simple SIP call flow overlooks many aspects which are more thoroughly addressed in H.323, such as codec negotiation. The new additions of SIP in the last RFC relative to codec negotiation and the management of media streams make SIP a lot closer to H.323 in terms of complexity. The offer/exchange model is very similar to the H.323 open logical channel procedure.
- Some of the simplicity of SIP came from real bugs in the initial standard. A good example is the inclusion of media management data in the INVITE call control message (instead of a separate message unrelated to the call state, as in H.245). SIP had to introduce the UPDATE message to carry media changes without implying a call state change (this is identical to the function of message FACILITY in H.323 when tunneling H.245 information).

The new set of SIP RFCs is at least as complex as H.323v4. A specific problem of H.323 is the number of options not only allowed by the standard (Fast Start, Tunneling, Early H.245, etc.) but introduced during the evolution of the standard. SIP was initially much more homogeneous. However, the situation changed with the latest SIP RFCs, which present many more options (four transport options, each with some impact on high-level protocol procedures), and a number of backwards compatibility problems. Backwards compatibility is much better documented and specified in H.323, partially due to the design method using structured SDL state machines.

The use of UDP is another attractive aspect of SIP, because UDP makes it easier to scale servers and also allows implementations to better control latency and facilitates the traversal of NAT functions (see Chapter 5). Although H.323 can also be carried over UDP, the vast majority of implementations use TCP. We would argue that UDP is still not perfect, however. It has become clear recently with the new requirements on SIP (e.g., tunneling of SS7 ISUP messages) that using UDP was problematic for long messages with the current SIP specification. RFC 3261 therefore recommends in some cases to use TCP instead of UDP, depending on the MTU; but, this is obviously not a clean way of solving the problem. The 'ideal' VoIP protocol should probably use neither TCP nor UDP, but a newer protocol like SCTP (one of the transport options introduced for SIP in RFC 3261), which offers a better compromise between reliability and flexibility for latency control. We bet that most VoIP implementations in 10 years will use such a protocol, with better latency control properties than TCP, and yet provide a strong, reliable transport layer.

3.7.1.2 Multicast

The IETF has gained a lot of experience on multicast. There are thousands of regular users of the mBone, and more and more multicast applications. SIP was originally designed to work on a multicast-enabled backbone, not only for media streams, like H.323, but also for signaling messages: for instance an INVITE message can be sent, in theory, to a multicast group. H.323 needs to use multi-unicast for the same purpose. In practice, however, multicast signaling is not very useful except for large conferences, and in such applications the SAP protocol[19] is an optimal choice, also recommended by H.332 for H.323 conferences with large audiences. In fact, the set of requirements for large conferences is not well covered simply by allowing SIP messages to become multicast. In addition, some of the call flows required by SIP multicast become unnecessarily complex.

The recent SIP RFC concludes that SIP signaling should be used in multicast only in very specific circumstances. Even the forking of messages is discouraged in 3GPP's current SIP profile. In fact, the best combination for large conferences is probably multi-unicast signaling, combined with the use of SAP, as described in H.332. For this type of application, SIP and H.323 offer the same level of support. Nevertheless, we still give a slight advantage to SIP because SDP is used both in SIP and SAP, and the MMUSIC Group has a lot of experience with multicast.

Multicast can also be used in media streams, but here SIP and H.323 have the same level of support, both allowing the opening of multicast media channels and the use of layered coders (useful for multicast transmissions where not all receivers have the same capabilities).

3.7.1.3 URL usage

The usage of URLs as identifiers is powerful. At first sight there may seem to be no big difference between an H.323 email alias (john@name.com) and a SIP URL (sip:john@name.com). In fact, there is: an H.323 email alias assumes the protocol used is H.323, whereas SIP actually specifies the protocol in the URL itself. Because of this, a SIP server can redirect a call to non-SIP servers in a very flexible way: for instance, a SIP terminal, when called by another SIP terminal, may redirect the call to a web page, or to a mailto URL. This facilitates the integration of audio and video applications with other multimedia applications.

This feature is now available in H.323 with the url-ID type of AliasAddress introduced in H.225v2, and the H.323 URL introduced in H.323v4. However, H.323 does not clearly explain how such URLs can be used to redirect a call to different media or protocols. We believe this is the most significant advantage of SIP over H.323, one that may lead SIP to eventually replace H.323 completely for desktop applications. Outside of this scope, however, this feature is of little interest, as it is hard to imagine how an analog gateway could be redirected to, say, an email address.

---

[19] An IETF protocol, see companion book, *Beyond VoIP Protocols* multicast chapter.

3.7.1.4   Loop control

Loop control was overlooked in early H.323 versions, while it has been thoroughly specified in SIP. Most H.323 vendors implement a form a loop control, but there is still not
good interoperability between these implementations. This will be improved over time,
but as of today remains one of the advantages of SIP.

3.7.1.5   Text encoding

Text encoding is a feature for some, and an issue for others. This is one of many,
seemingly endless 'religious' wars between programmers. Text encoding has a lot of
advantages: it is simple, can be debugged easily using simple network sniffers, and makes
interoperability problems detectable 'visually'. This is very attractive for students and
programmers discovering telecom applications.

More experienced telecom programmers with a background in ASN.1 may have a different view. The biggest difficulty of telecom environment programming is that, unlike
PC software programming, reliability and bug-free implementations are a must. At the
same time, this is very challenging because protocols evolve quickly, and open telecom
environments cause various implementations from different programmers to interoperate.
As not all implementations and not all call flows can be tested in advance during development, many such communications between different vendors will occur for the first time
in live networks.

After the telecom world put a lot of research into this problem, it decided that 'free-
text' specifications, manual encoding, and parsing were not optimal to maximize the
reliability of implementations. Instead, programs are specified using formal description
languages, such as SDL, and the **protocol data unit** (**PDU**) syntax can be expressed
formally using an unambiguous, abstract syntax notation,[20] which can be automatically
compiled into serializers and parsers. Such automation avoids most of the opportunities
for bugs and frequently leads to faster implementations. It also becomes much easier to
add new parameters or options, because the generation of updated parsers is automatic.

There is no final argument to this debate: PC application developers and IP phone
vendors will probably prefer text encodings, but carrier-class proxy developers with experience in ASN.1 will feel more comfortable to ensure the quality of their code with ASN.1
and automatically generated code. Protocols specified in SDL can also be thoroughly tested
automatically, something that is impossible with a verbal specification.

3.7.1.6   Presence and instant messaging

SIP now supports presence and instant messaging applications with its SIMPLE extensions. H.323v4 also has support for instant messaging by using the T.140 protocol and

---

[20] Such as ASN.1, which is the most widely used syntax notation in the telecom world, allowing
powerful expression of data structures, constraints on data values, extensions, etc. Unfortunately,
ASN.1 is also complex to learn and optimized commercial ASN.1 tools are expensive. But,
once a developer is familiar with ASN.1, the productivity of ASN.1 for telecom applications
is phenomenal.

RFC 2793, but offers fewer possibilities for presence (only registration is supported). Clearly, the market has not selected H.323 for implementing presence and instant messaging applications.

However, it is still unclear who the winner will ultimately be for presence and instant messaging. SIMPLE and the open source JABBER protocol still compete; JABBER even has a slightly larger marker share. In fact, SIMPLE and SIP are really loosely coupled protocols (SIMPLE is implemented using only new messages that were not present in initial specification of SIP). In principle, it wouldn't cause any problem to use one protocol for telephony (e.g., SIP or H.323), and a separate protocol for presence (e.g., JABBER or SIMPLE). Any combination will work fine.

3.7.2   Arguments in favor of H.323

3.7.2.1 Logical channels

H.323 makes a clear distinction between the media types that can be sent or received (capabilities), and the media types that are active and actually sent over the network (logical channels). The first version of SIP did not have such a distinction, as SIP endpoints only advertised the coders they could receive and there was no clear procedure to open a media connection apart from actually sending the media.

This was improved with the new offer–answer model, which is now as powerful as the H.245 logical channels. SIP and H.323 are now completely equivalent regarding management of media channels, but SIP is still lacking a proper framework for the announcement of capabilities, independently of the media channels that it immediately wishes to send or receive, which causes problems (e.g., for fax transmission, see Section 7.2.3). There is still some work in progress in this direction.

3.7.2.2 Conference control, handling of video signals

H.323, alone or in combination with H.332, has powerful conference control features. SIP was not designed to carry out conference control, and consequently many of the features required to do so were not developed. In reality, however, this problem is minor, since conference applications usually implement their own conference control tools, independently of the underlying VoIP protocols.

H.245 also allows much more control of video signals (e.g., Video Fast Update requests), because of the experience in H.320 ISDN videoconferencing. It is frequently heard that 'SIP is the protocol of choice' for video; this is clearly a misleading statement. In fact, SIP still lacks many of the features already present in H.323 for comprehensive management of multiparty interactive video communications (e.g., H.323v4 describes how to control a far-end camera). Another example is the more precise way of declaring capabilities in H.323: an H.323 device that supports continuous presence video with several images decoded simultaneously can advertise exactly how many images it can receive and display, along with the audio coder used, not just the type of coder supported.

3.7.2.3   Capabilities negotiation

H.245 capabilities negotiation is very sophisticated and can express constraints on the simultaneous use of codecs. SDP was never designed to be used for capabilities negotiation; in fact, work is in progress to define a new 'SDP-ng' for this purpose. The current use of SDP by SIP for capabilities negotiation simply allows endpoints to state that they can support multiple coders by listing them in the 'm =' line of the session description. But, by doing so, the endpoints also indicate that they can support these coders simultaneously (which may not be true) and that they are prepared to receive the media immediately. This can create ambiguity and problems (e.g., for fax), as a gateway can support T.38 and G.711, for instance, but only be prepared to receive G.711 media at this moment. Some limited extensions of SDP have been defined to help solve this problem (RFC 3407, 'SDP Simple Capability Declaration'), but so far most implementations of SIP are still weak regarding capabilities negotiation. A more complete capabilities-enabled SDP would probably look like the existing H.245 CapabilitySet message!

3.7.2.4   Backwards compatibility, installed base

Newer versions of H.323 are fully backwards compatible, and interoperability in the H.323 world is now excellent. The millions of lines that have been installed allow manufacturers to test their implementations and gradually agree on the procedures that were still unclear and ambiguous in H.323. This was made quicker by the references to ISDN (Q.931) and the dominance of a few H.323 players, who have been able to quickly impose their views on the H.323 ITU Study Group, without endless discussions. This is also due to the fact that many service providers are members of the ITU Study Group, which implies that backwards compatibility issues are taken very seriously, as they create nightmares on operational networks.

   H.323 is well ahead of SIP in this regard, as the SIP market is still very small, mostly limited to trials or single-vendor closed networks (i.e., class 4 telephony applications). In addition, the standardization process of the IETF makes it possible to quickly publish an RFC, without the lengthy consensus approach of ITU. This is good in most cases, but in the middle of the Internet bubble this led to the publication of documents of insufficient quality, which now causes many interoperability issues, ambiguity, and lack of backwards compatibility. The situation is quickly improving as most telecom manufacturers recognize that their priority is to deploy real networks with real customers on a massive scale, and not to compete for sexy trials of 'new applications'. The focus is now on fixing issues first and then introducing new features.

3.7.2.5   Binary encoding

The argument of binary encoding generated from formal syntax descriptors versus text formats has already been discussed in Section 3.7.1.5. We reuse this argument here in favor of H.323 because 50% of developers consider the ASN.1 syntax and binary encoding more suited for telecom applications.

3.7.2.6 *Use in telephony applications*

This is the key argument in favor of H.323. SIP still lacks many essential features in order to be deployed in a carrier-class telephony network (with the exception of pure class 4 applications requiring SIP only to tunnel SS7 ISUP messages). Many features, such as the handling of ISDN information elements (e.g., CLIR information, type of number, number verification, etc.), are not present in SIP and require proprietary extensions. These elements are required whenever PBXs need to be connected to a VoIP backbone. Other extensions of SIP required for telephony applications (e.g., progress messages with 183 responses, reliability of provisional responses, etc.) have been added very recently and, therefore, are still not implemented by many vendors.

DTMF transmission is one of the biggest issues in SIP. Many SIP VoIP networks started by implementing class 4 networks, where DTMF transmission with RFC 2833 is sufficient, but were stuck later as they tried to deploy more complex class 5 or contact center applications. SIP proposes many solutions to this problem; but, what the industry needs is one solution, not many. Until this happens, many applications will require single-vendor networks.

H.323 starts with a significant advantage due to its Q.931 (ISDN) heritage and deployment experience, but in principle most of the issues could be fixed easily in SIP. One of the possible fixes is to include all ISUP parameters in SIP (e.g., the Generic Transparency Descriptors proposal from Cisco Systems®) which would make SIP a superset of H.323. One of the issues, however, is the maximum size of the payload that can be transmitted over UDP.

It seems that it will be at least 2 years before SIP manufacturers agree on standard ways to properly handle all telephony applications. It will probably also require more market consolidation before some vendors have enough leverage to efficiently promote their views.

3.7.3 H.323 to SIP gateways

H.323 and SIP are now so similar that protocol converters can easily be built for audio, fax, and video calls. Even instant messages could easily be added to existing H.323 standards like H.450.7 ('Message Waiting Indication'). Several commercial implementations exist, which are mainly used to connect SIP-based PC softphones to existing H.323 telephony networks.

An H.323 slow-start call flow (which has become very rare) must be mapped to a SIP call flow where media streams can be established soon after the INVITE message. One possible solution is to send an INVITE without SDP, wait for the offer in a provisional or the final SIP response, then send the response in the ACK message. In the rare cases where the H.245 channel would not be connected in time for the ACK response, the ACK can put all media channels on hold, and a RE-INVITE would need to be sent as soon as the H.245 channel is connected and the logical channels have been established.

An H.323 fast-connect call flow is identical to a SIP call flow (mapping is direct as shown in Table 3.5).

Table 3.5 H.323 fast-connect to SIP protocol conversion

No.	H.323 side of proxy	SIP side of proxy	Comment
1	→ Setup with FastStart		Contains proposals for backwards-logical channels
2	← Call proceeding		The proxy acknowledges the receipt of the setup messages
3		INVITE →	Contains proposals for backwards-logical channels in SDP format
4		180 RINGING ←	
5	← Alerting		
6		200 OK ←	User picked up, contains forwards-logical channel info if not sent before
7	← Connect with FastStart		Forwards-logical channels copied in H.323 CONNECT message
8		ACK →	
	H.245 can be opened at any time		
N		BYE ←	On-hook
N + 1	← Close H.245 + release complete		Call released on H.323 side
N + 2		200 OK →	

The SIP offer–answer model has made it much easier to map the H.323 formal management of logical channels into SIP. If new logical channels are opened during the conversation by the H.323 endpoint, the protocol converter would send a new INVITE message to the SIP endpoint, as illustrated in Table 3.6.

If a new offer is received from the SIP side, it is best to reset the H.323 or H.245 negotiation using the Null Capability Set (TCS = 0) message, and then resend capabilities corresponding to the codecs in the new offer, especially since it may contain new coders that have not been negotiated before. The protocol converter can then send an OpenLogicalChannel message with the codec selected during the negotiation.

The only remaining difficulty for an H.323–SIP proxy is DTMF handling. Since many SIP endpoints still do not support out-of-band DTMF, but rather exchange DTMF information end to end using RFC 2833 over the RTP channel, in many cases the protocol converter is not aware of DTMF information from the SIP side. Most SIP gateway vendors, however, implemented the INFO or NOTIFY message as described in Section 3.3.2.3.1.3, and this message can be mapped to an H.323 UserInputIndication. In the case of RFC 2833, some H.323 gateways can be configured to also receive DTMF via direct RFC

Table 3.6   The new offer–answer model is identical to the logical channel model of H.323

OpenLogicalChannel		
	INVITE $\rightarrow$	Same Call-ID as the previous INVITE (but Cseq is incremented). The SDP payload describes the new offer by modifying the media channels list
	200 OK $\leftarrow$	Contains the SDP answer and confirms the media channels list selected in the offer
$\leftarrow$ OpenLogicalChannelAck + OpenLogicalChannel		The OpenLogicalChannelAck contains the forward channels that have been confirmed in the SDP answer. The OpenLogicalChannel contains the backwards channels that have been selected in the answer
OpenLogicalChannelAck		

2833 RTP packets, but it is impossible to add any DTMF-controlled service into such a network in a scalable way (the service node must relay RTP packets, unless RFC 2833 is thoroughly implemented by all devices and they can separate the DTMF RTP stream destination from the media destination).

3.7.4   Conclusion on the future of SIP, and its relation with H.323

SIP has covered a lot of ground since the first SIP 'bake-off' meeting in April 1999. The protocol, born as a result of the Internet bubble, grew with it and was constantly expanded to absorb in just 4 years many of the 'new applications' that emerged during this period. Besides videoconferencing, the most interesting application is obviously instant messaging, which is one of the most popular applications on the Internet today. Unfortunately, SIP, like other protocols that were born at the same time (e.g., WAP on cellular phones), also inherited many of the typical problems of the Internet bubble era: it was frequently 'oversold' and presented as a telecom revolution, while in reality it mostly provided capabilities that were already available with other protocols. SIP was also designed too quickly to ensure a sufficient level of quality in protocol specifications, and this lack of methodology later forced the specifications to introduce awkward 'patches' and non-backwards compatible extensions to the original protocol.

Another more fundamental issue of SIP is that, unlike other telecom protocols, its design is driven only by very few manufacturers, with much less control from actual users than ITU protocols (the ITU is mostly driven by telecom operators). This gave unique characteristics to the protocol's evolution, not all of which are necessarily good:

- SIP quickly absorbs new applications and extensions (IETF drafts are a good way of documenting most of these extensions before they actually get to the standard—if indeed they do).

- There is no stability to the protocol: little or no attention is paid to actual deployment and migration issues.

- Priority is given to the addition of features, as opposed to securing a robust design first and then add features. The ITU approach is almost the exact opposite: ITU authored many fundamental tools for telecom products designs, like the ASN.1 syntax or the SDL state machine description language, and only then designed protocols to build on this robust set of foundation tools. SIP has a very broad scope, but is still very fragile as a protocol.

- Many 'buzzword' applications (e.g., presence) are getting much of the attention and standardization effort, while more fundamental revenue-generating features, but more boring, still have issues. This is the "99% complete" syndrome. SIP still has many open issues for its application to basic telephony.

Controlled by start-up companies, SIP has been 'demonstration'-oriented, with a lot of marketing activities but few significant deployments. The end of the telecom bubble will probably give a lot more importance to actual operational issues, slowing the development of new features and focusing on stability, completeness, and interoperability. This will be of benefit to SIP. If SIP begins to leverage the traditional design methods (formal state machines, layering, etc.), and to integrate the latest advances of IP networking technology (such as SCTP), while managing a smooth transition, the protocol truly has the potential to become the dominant multimedia session control standard of IP network. But it is not quite there yet.

4

The Media Gateway to Media Controller Protocol (MGCP)

4.1 INTRODUCTION: WHY MGCP?

4.1.1 Stimulus protocols

SIP and H.323 are very similar session-based, stateful protocols. The similarity is hidden behind all the cosmetic differences due to different ways of serializing essentially the same information, but basically both protocols share the same characteristics:

- They are composed of a call control protocol (H.225.0, SIP) and a media control protocol (H.245, SDP offer–answer model), with the media control protocol encapsulated in the call control protocol.
- The call control protocol is a slightly simplified version of ISDN Q.931 (H.225.0), with a more basic way of closing connections (three messages in Q.931, only one message in H.225.0 and SIP—although this is likely to change since the single-message closing sequence causes some issues).
- Both the call control protocol and the media control protocol assume a stateful or 'intelligent' endpoint (i.e., an endpoint which implements its own call-state machine, and its own logic, such as for the handling of call waiting, providing a ring-back tone while off-hook, etc.)

From a marketing point of view, having an 'intelligent' client is always a good thing, since it seems so obvious that an intelligent client will be able to do 'more things' than a 'dumb' client.

IP Telephony   O. Hersent, J.P. Petit, D. Gurle
© 2005 John Wiley & Sons, Ltd   ISBN: 0-470-02359-7

The problem is that, when looking at the most sophisticated corporate phone installations today, *none* uses ISDN phones (the equivalent in traditional telephony of a smart phone)! In fact, most, if not all, of the corporate PBXs use another class of protocols called **stimulus protocols**, optimized for the control of dumb phones. It is easier to understand why with an analogy. If SIP or H.323 were programming languages, they would be very similar to the BASIC language. You can do a lot of things with BASIC as long as you do things for which BASIC has the proper instructions, but you can do many more things with the C language, or with an assembly language. If a stimulus protocol were a programming language, it would be a low-level assembly language: certain things take longer to code, but there is nothing you cannot do. For instance:

- When you pick up the handset of an H.323 or a SIP phone, you get ring-back. When you pick up the handset of a PBX phone, sometimes you get a message like 'you have voicemail'.

- On an H.323 or a SIP phone, you have feature buttons or lamps, hard-coded by the phone manufacturer, for hold, transfer, three-way calling, message-waiting indication, etc. On a PBX phone, you may want to assign any feature to any button, to control any lamp, exactly as you like.

- On an H.323 or a SIP phone (without proprietary extensions), you need to pick up the handset or press the loudspeaker button to get a call. On a stimulus phone, the loudspeaker can be remotely activated by the PBX.

A stimulus protocol carries lower level instructions than ISDN, H.323, and SIP. For an incoming call, all these protocols simply send a 'you have a new call' message, and the phone is expected to ring all by itself. It is also expected to send ring-back as soon as you pick up the handset. A stimulus protocol would send a 'ring with ring type X' command, then a 'notify me if someone picks up the handset' command (or it could send an 'activate loudspeaker' command directly). For an outgoing call, once notified that the handset is off-hook, the PBX would send a 'play dial-tone command', followed by a 'notify me of the digits that have been dialing' command (but it could also send a 'play this audio message command').

In general, stimulus-based protocols have the following attractive characteristics:

- They simplify the endpoint software design and therefore minimize the number of endpoint bugs that can affect a PBX application: for instance, a common bug in SIP or H.323 endpoints is the inability to alternate between normal ring-back and network-generated prompts, because the programmers did not think of this unusual, but nevertheless mandatory, transition. This is controlled by the PBX in a stimulus protocol, and therefore cannot be an endpoint bug.

- They facilitate management of large numbers of endpoints, by minimizing the problems caused by the diversity of software flavors deployed at endpoints.

- They facilitate the centralized deployment of new features or applications, even those interacting with the endpoint. Usually, such deployments do not require any change in endpoint capabilities. Once the possibilities of the hardware endpoint have been properly mapped to stimulus commands, all services can be designed without requiring any addition to the device firmware.

- They make it easier to program applications or advanced services which require the co-ordination of multiple endpoints, by centralizing the state of all endpoints at the PBX. A typical example is the manager–secretary feature, where the manager screen needs to show that a call is coming (but does not ring), while the secretary phone rings. Such a service would require additions to the standard with H.323 or SIP endpoints.

The downside of stimulus protocols is that they absolutely require centralized resources: two stimulus protocols cannot communicate without a PBX. In addition, since the granularity of communications with the call controller is at a very low level, services require significantly more control messages than with more intelligent endpoints.

With only H.323 and SIP, VoIP would be lacking a stimulus-based protocol—MGCP fills the gap.

4.1.2　Decomposed gateways

In the early days of VoIP, most VoIP gateways were based on PCs, with some hardware boards handling media processing. Such gateways were already 'decomposed' in the sense that call control processing and media control resources were running on different modules, with some proprietary APIs between the telephony boards and the main PC-based gateway software.

The early fully embedded gateways usually retained this architecture, with a central processor handling call control, while dedicated **Digital Signal Processor** (**DSP**) boards handled media processing. But when the size of gateways grew to handle hundreds, or thousands, of channels, this architecture began to be problematic:

- Once the maximum number of DSP 'daughter boards' in a chassis was reached, another chassis needed to be installed with not only DSP boards, but also a new instance of the gateway call control software. This made it impossible to have centralized control of all the channels, and forced the duplication of call control resources.

- In the PSTN, carrier interconnections with thousands of channels typically use dozens of media-only trunks, and a single signaling-only channel (mostly SS7 ISUP) carrying the call control information for all the media trunks. If on the VoIP side the required capacity was big enough to require multiple gateway chassis, then with each gateway having its own local call control software, there would be a need for one SS7 call control-signaling link and ISUP stack per chassis (Figure 4.1), which is far too expensive.

One of the early proprietary protocols used to solve the problem was called Q.931+. One master call control device took SS7 ISUP signaling and distributed call control to each media gateway, in a Q.931-like form, over IP tunnels: this solution still required an instance of call control in each media gateway.

Figure 4.1   Which box should receive SS7 signaling in an SS7 configuration?

Some vendors quickly found the best solution, which was to have one master call control module, and physically separate media-processing modules with only DSP resources, a TDM media-only interface, and an IP interface; this is a bit like the PC-based architecture with separate DSP boards and call control on the PC, except that since the modules are now physically separate, they do not communicate though an API as in the early PC days, but through a protocol (Figure 4.2). Since both the DSP modules and the call control module have an IP interface, logically the media resources remote control protocol had to be over IP as well.

In order to implement such a solution, there was a need for a standard protocol between a call control function and a media gateway with no call control. The de facto standard today is MGCP, logically named the 'Media Gateway Control Protocol'.

It seems surprising that the same protocol could be used to control stimulus phones and dense media gateways. In fact, a phone *is* a media gateway between a microphone + speaker and the VoIP media stream, plus some user interface components (a handset with a hook, a keypad, buttons, etc.). Therefore, an IP phone stimulus protocol should comprise a pure media gateway control portion, plus some user interface control optional commands. This is exactly what MGCP is: a core set of commands for pure media control (we will frequently use the term 'MGCP trunk', or MGCP/T, although this is not strictly correct), *plus* a set of optional commands (we will frequently refer to the MGCP protocol plus the extensions to control an IP phone as 'MGCP line', or MGCP/L).

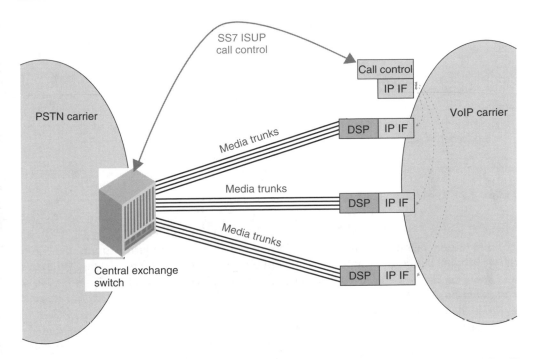

Figure 4.2 Centralized ISUP control box, controlling remote media gateways via an IP protocol.

4.1.3 Some history

The first proposal came from Bellcore (now Telcordia) and Cisco to address the needs of cable operators that wanted to become competitive local exchange carriers (CLECs) by using VoIP on top of their HFC infrastructure. The **Simple Gateway Control Protocol (SGCP)** was introduced early May 1998 by Cisco during a PacketCable™ meeting (and in other standards bodies, IETF, ITU-T SG 16 and ETSI TIPHON) as a cost-effective alternative and better suited protocol to implement and deploy than the then-current H.323 implementations in the context of cable operators' market.

The second proposal, the **Internet Protocol Device Control (IPDC)** was presented to ITU-T SG 16, ETSI TIPHON and IETF a month later. IPDC addresses more or less the same requirements as SGCP but with a different transport approach. While SGCP relied solely on UDP, enhanced with application-level reliability features, IPDC proposed the use of **DIAMETER** (an extension and replacement of **RADIUS**) to carry **protocol data units (PDUs)** between respective entities.

It was not long before the forces behind these two protocols realized that by unifying their efforts they could get bigger consensus and foster the adoption of their position. Bellcore and Level3 played a key role in merging these two proposals into one (Figure 4.3): MGCP. MGCP was proposed to all main standards groups: the IETF's

Figure 4.3   The Media Gateway Control Protocol family tree.

Media Gateway Controller (MEGACO) Working Group, ETSI TIPHON, and ITU-T SG 16. In addition, companies supporting this protocol created an industry forum, the Multi Service Switching Forum (http://www.msforum.org) to develop complementary protocols and services. In particular, MGCP was extended to also support ATM transport networks and voice on AAL2.

MGCP is was originally published as informational RFC 2705, 'Media Gateway Control Protocol (MGCP)' version 1.0; the specification was updated as RFC 3435 in January 2003. A variant of MGCP is also used by the PacketCable™ initiative under the name **Network Based Call Signaling Protocol (NCS)**, and the specification is available on the PacketCable™ website (currently PKT-SP-EC-MGCP-I06-021 127).

Later the ITU began to work on a new generation of stimulus protocols, called H.248. However, so far this protocol has not gained significant market acceptance. It is frequently mentioned in white papers or architecture documents, but most vendors still use MGCP. The reason is simple:

- The biggest user groups for stimulus-controlled media gateways, the Cable Television Laboratories (www.cablelabs.com) and their VoIP initiative PacketCable (www.packetcable.com) still use MGCP, and therefore virtually all media gateways and IP phones implement MGCP.

- H.248 doesn't add any significant capability to the MGCP protocol, and since MGCP satisfies the needs of stimulus phones and remote media gateways, the telecom manufacturer community doesn't feel there is a need to change something that isn't broken.

Like most IETF media protocols, MGCP uses the SDP syntax to express the format, source, and destination of media streams.

Overall, the quality of the MGCP specification was way better than the quality of the initial specifications of H.323 and SIP. The protocol is simple, focused on a clear scope, well-structured, with a cleanly separate transport layer, a well-defined connection model, and very few bugs in the standard. Without much marketing buzz, MGCP has made its way into the world of VoIP protocols and it is today one of the most widely implemented protocols in all of its original target markets: residential gateways, IP phones, and large-scale trunk gateways.

4.2 MGCP 1.0

The Media Gateway Control Protocol was first specified in draft-huitema-MGCP-v0r1-00.txt and was finally published as MGCP 1.0 in RFC 2705. In January 2003 an updated version, which corrected some ambiguities and inconsistencies in RFC 2705, was published as RFC 3435. RFC 3435 is also known as MGCP 1.0bis; however, it is still MGCP version 1.0 and is fully compatible with RFC 2705 (except for error fixes).

MGCP is designed to interface a media gateway controller and media gateway and supports a centralized call control model. The protocol is text-based, offering a set of simple primitives. The media gateway controller is called the **call agent** in MGCP terminology and the media gateways can be of different types:

- VoIP gateways. **Residential gateways** are designed to be customer premises equipment, usually connected to a couple of analog phone lines. These gateways, besides their pure media-processing capabilities, are also able, when connected to analog phones or PBXs, to generate ring voltages, to send specific signals required to set message-waiting indication lamps or to send caller ID information to a phone. **Trunking gateways** are high-density gateways interconnecting TDM media trunks and a VoIP network, with media-processing capabilities only.

- **Network Access Servers** (**NASs**). The MGCP protocol includes some extensions which allow a call agent to control modem banks. The protocol is also capable of driving **universal port** gateways, which can behave as a voice gateway if the detected signal is voice, and can also locally terminate modem connections if they detect a modem signal. NAS extensions are no longer part of the base protocol in RFC 3435; instead, they are provided as a separate package.

- Voice over ATM gateways.

Unlike H.323 or SIP, MGCP is a master–slave protocol. As shown in Figure 4.4, the call agent is a central controller, and the media gateways are slave devices that can only report events requested by the call agent and execute the commands of the call agent.

There is often some confusion between H.323/SIP and MGCP. 'I don't want to use MGCP in my network' is a frequently heard sentence. With SIP or H.323, which are call control protocols, it is natural to select only one of these protocols in the core network. Although SIP/H.323 gateways exist, very few service providers have both protocols running in their core network. Many network engineers assume MGCP is the same kind of protocol, and if you begin to implement MGCP somewhere in the network, you must have MGCP everywhere in your network. But MGCP is not a peer-to-peer protocol: two MGCP call agents cannot communicate using MGCP. MGCP call agents can only be at the edge of a network and must communicate between one another using a call control protocol (e.g., H.323 or SIP). Therefore, MGCP should be seen:

- At the customer access edge of the network, as the stimulus-mode option to drive IP phones and IP residential gateways. SIP and H.323 being the call-stateful, ISDN-like alternative.

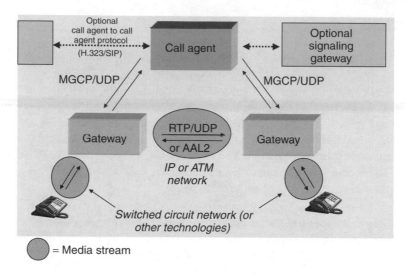

Figure 4.4   The MGCP ecosystem.

Figure 4.5   MGCP is an edge protocol, while SIP or H.323 must be used in the core network.

• At the PSTN interface edge of the network, as an internal protocol of large-scale decomposed trunk-side gateways.

MGCP is not an option for a core network call control protocol. It is an edge protocol as illustrated in Figure 4.5.

4.2.1   The MGCP connection model

The core component of a traditional telephony switch is a TDM bus. Each interface channel is connected to a time slot on the bus. If a channel A on an interface sends a media signal on time slot 231 of the TDM bus (let's call it channel 231), then any channel B on any interface listening to time slot 231 will receive a copy of this media signal. A full duplex connection for a phone call between channels 231 and 308 is established if channel 231 listens to time slot 308 and channel 308 listens to time slot 231. The traditional switch can perform all the media-switching functions it needs only by transmitting 'send' and 'listen' commands to interface channel 'objects' identified only by a time slot number.

A packet-based switch is more complex because, instead of using a TDM bus as a switching matrix, it uses a packet network. Destinations on a packet network are identified by an address and some parameters, not by a simple integer. Also, the media type for a TDM switch is always 64 K G.711 encoded speech or clear data, whereas it can be anything for a packet-based switch.

The MGCP connection model is built on two objects:

- Endpoints, which can originate (media-source) and/or terminate (media-sink) a media flow. A circuit which can receive RTP packets, decode them, and send the resulting G.711 data to a time slot on a TDM trunk is an endpoint. An entity which can receive RTP packets and relay them to another destination is an endpoint. An IP-based media mixer (MP in H.323 terminology) which can receive RTP streams, mix them, and send back the resulting streams to other RTP sinks is an endpoint. Similarly, on ATM networks any entity which can receive or originate AAL2 media streams is an endpoint.

- Connections (Figure 4.6). Each endpoint can have one or multiple connections, each of which can be inactive, send media, receive media or both. Of course, a connection is useful only if it collects media from an endpoint and sends it to another endpoint (in MGCP, connections are each attached to a named endpoint, but send media to an SDP-defined address, so it is possible to build connections from an MGCP endpoint to an IP address that is not necessarily a declared MGCP endpoint). Connections can be point-to-point or point-to-multipoint. A point-to-multipoint connection will typically send to a multicast IP address.

This model is generally much more powerful and scalable than the TDM time slot model. However, MGCP (like all other VoIP protocols) has one weakness compared with the TDM model: in TDM, point-to-multipoint connections are native, very easy to establish (each destination 'listens' to the source time slot), and invisible from the source. In VoIP, they require endpoints to support multicast or need to be emulated through a special-purpose endpoint that receives media from the source, then duplicates and sends media to all the destinations. Because of this, features like silent monitoring in contact centers, or lawful interception in residential telephony, are *much* more difficult to build than in the TDM world.

= Endpoint (media stream conversion/relay point)

Figure 4.6 MGCP endpoints and connections (the call agent controls connections on gateway endpoints).

MGCP uses a simple syntax for endpoint identifiers, with two components separated by the '@' character:

- A prefix which should be a unique identifier of the endpoint within the gateway. The prefix structure can be hierarchical (interface, channel on interface), with each component of the hierarchy separated by a slash ('/'). Local endpoint identifier components can be composed of any visible character except a space, '@', '/', '*', or '$'. '*' has the special meaning of '*all* defined values of this component', and '$' means '*any* of the values defined for this component'.
- The DNS domain name of the gateway. If the gateway is not registered in the DNS and the call agent expects non-DNS names, then the name should be any string composed of letters, numbers, and '.' or '-', as long as it is unique on the network. Some vendors also use the numeric IP address of the gateway, between square brackets (e.g., [10.10.10.10]).

A two-port analog gateway typically uses:

- aaln/1@analog-gateway.anydomain.org
- aaln/2@analog-gateway.anydomain.org

On a TDM trunk interface, where each interface handles multiple trunks and each trunk has multiple circuits, each with its circuit identifier (referenced by the ISUP CIC), vendor may use:

- IF3/2/1@large-trunk-gateway.anydomain.org for circuit 1 on trunk 2 of interface IF3.
- IF5/4/$@large-trunk-gateway.anydomain.org for *any* circuit on trunk 4 of interface IF5.

In practice, each vendor uses its own convention for local endpoint names; but, as long as sequences are identified by consecutive integers, it is very easy to configure the corresponding masks on a call agent.

4.2.2 The protocol

4.2.2.1 Overview

MGCP call agents and gateways exchange 'transactions', composed of one command, optional provisional responses, and a final response (Figure 4.7).

Commands and responses use a simple text format. MGCP 1.0 is composed of nine commands exchanged between the call agent and a media gateway or a NAS (generically called 'gateway' from now on). Each command is composed of a header, optionally followed by an empty line, and a session description. The header is composed of multiple lines (separated by a CR, an LF, or a CR + LF):

- A command line, composed of the command code, the transaction identifier, the target endpoint name (optionally with local endpoint identifier wildcards '*' or '$'), and the MGCP protocol version, separated by the ASCII space (0×20) or tab (0×09) character (e.g., the following is a valid command line 'RQNT 1207 endpoint/1@rgw-2567.anydomain.net MGCP 1.0'). Some MGCP gateways still use MGCP 0.1 (the MGCP version immediately after the merger of IPDC and SGCP), but the differences from MGCP 1.0 are minimal.
- A set of parameter lines, each using the 'name:value' format. Parameter names consist of one or two letters, followed by a colon (e.g., B:e:mu). The most common parameters are listed in Table 4.1.

Figure 4.7 MGCP commands and responses (each transaction refers to one or more gateway endpoints).

Table 4.1   Common MGCP parameters

K	ResponseAck
B	BearerInformation
C	CallId
I	ConnectionId
N	NotifiedEntity
X	RequestIdentifier
L	LocalConnectionOptions
M	ConnectionMode
R	RequestedEvents
S	SignalRequests
D	DigitMap
O	ObservedEvents
P	ConnectionParameters
E	ReasonCode
Z	SpecificEndpointID
Z2	SecondEndpointID
I2	SecondConnectionID
F	RequestedInfo
Q	QuarantineHandling
T	DetectEvents
RM	RestartMethod
RD	RestartDelay
A	Capabilities
ES	EventStates
MD	MaxMGCPDatagram

Command lines and parameter lines are case-insensitive. A sample command is shown in Figure 4.8. Each command targets one or more endpoints. Table 4.2 lists the verbs for each of the nine MGCP commands. Experimental verbs can also be added, whose names should start with an 'X'.

RFC 2705 also described the verb 'Move' for 'MoveConnection' in an appendix. This command disappeared in RFC 3435; it is now provided in a separate package (draft-andreasen-mgcp-moveconnection-00.txt). Also, RFC 3435 defines the 'Mesg' verb for the 'Message' command defined in RFC 3435, appendix B.

Each of these commands triggers a response, including a 3-digit response code. Commands and responses are associated by a TransactionID. The most common codes are listed in Table 4.3 for reference.

4.2.2.2   Events and signal Packages

4.2.2.2.1   Definitions and syntax

A call agent that only manipulates media connections on endpoints cannot easily interact with media information (e.g., in order to send back a dial tone to an analog gateway, the call agent would need to connect the gateway endpoint with an IP-based

```
RQNT 1207 endpoint/1@rgw-2567.anydomain.net MGCP 0.1
X: 0123456789
R: hu
S: v
```

The call agent sends a NotificationRequest to the gateway:

- The NotificationRequest verb is RQNT.

- The TransactionId is 1207.

- The target endpointName is endpoint/1.

- The gateway domainName is @rgw-2567.anydomain.net.

- MGCP protocol version is 0.1.

- The requestIdentifier is 0123456789.

- The gateway will need to notify the call agent when it detects hang-up 'hu'.

- The gateway must immediately generate an alerting tone 'v'.

Figure 4.8   A sample MGCP command.

Table 4.2   MGCP command verbs

Verb	Code	Direction: Call agent ($\rightarrow$) Gateway ($\leftarrow$)
EndpointConfiguration	EPCF	$\rightarrow$
NotificationRequest	RQNT	$\rightarrow$
Notify	NTFY	$\leftarrow$
CreateConnection	CRCX	$\rightarrow$
ModifyConnection	MDCX	$\rightarrow$
DeleteConnection	DLCX	$\rightarrow$ and $\leftarrow$
AuditEndpoint	AUEP	$\rightarrow$
AuditConnection	AUCX	$\rightarrow$
RestartInProgress	RSIP	$\leftarrow$

dial tone generator). MGCP uses **signals** to provide a simpler way to allow a call agent to give instructions to endpoints that have some local signal generation capabilities. A signal is simply an identifier known to the gateway which corresponds to some action on the media bearer (e.g., playing the dial tone). Many applications also require the call agent to be aware of certain events that are present in-band (e.g., DTMF signals), or through some means accessible only to the gateway (e.g., an off-hook transition). The gateway can report this type of information to the call agent using MGCP **events**.

Various types of gateways can report different events and generate different signals. Some signals and events, however, are very common and most gateways support them. In order to give some flexibility and extensibility for MGCP to support new types of gateways that require specific events or signals, without interfering with already-defined

Table 4.3 Common MGCP response codes

Code	Description
100	The transaction is currently being executed. An actual completion message will follow later (provisional response)
200	The requested transaction was executed normally
250	The connection was deleted
400	The transaction could not be executed, due to a transient error
401	The phone is already off-hook
402	The phone is already on-hook
403	The transaction could not be processed, because the endpoint does not have sufficient resources at this time
404	Insufficient bandwidth at this time
500	The transaction could not be executed, because the endpoint is unknown
501	The transaction could not be executed, because the endpoint is not ready
502	The transaction could not be executed, because the endpoint does not have sufficient resources
510	The transaction could not be executed, because a protocol error was detected
511	The transaction could not be executed, because the command contained an unrecognized extension
512	The transaction could not be executed, because the gateway is not equipped to detect one of the requested events
513	The transaction could not be executed, because the gateway is not equipped to generate one of the requested signals
514	The transaction could not be executed, because the gateway cannot send the specified announcement
515	The transaction refers to an incorrect connection-id
516	The transaction refers to an unknown call-id
517	Unsupported or invalid mode
518	Unsupported or unknown package
519	Endpoint does not have a digit map
520	The transaction could not be executed, because the endpoint is 'restarting'
521	Endpoint redirected to another call agent
522	No such event or signal
523	Unknown action or illegal combination of actions
524	Internal inconsistency in LocalConnectionOptions
525	Unknown extension in LocalConnectionOptions
526	Insufficient bandwidth
527	Missing RemoteConnectionDescriptor
528	Incompatible protocol version
529	Internal hardware failure
530	CAS signaling protocol error
531	Failure of a grouping of trunks (facility failure)

names, MGCP introduced the notion of 'packages'. A package is simply a namespace, identified by one or more letters. Events and signals can be defined within this namespace without risk of ambiguity with other namespaces. An event or signal name must be prefixed by its package name, separated by a slash (e.g., 'L/HU' refers to event 'HU' within package line 'L'). Package names and events are not case-sensitive (i.e., 'L/HU' is equivalent to 'L/hu' or 'l/hu').

There are some special cases in which the package prefix can be omitted:

- A call agent can omit the prefix for events and signals that are part of the default package, when sending commands to an endpoint that is known to support a default package (e.g., if the package line is the default package for an analog gateway, then dl and L/dl are equivalent signals).
- Digit events can be sent without a prefix, although it is recommended to send them with the appropriate prefix (e.g., L/8 for digit 8). Package names are not allowed to contain digits in order to prevent any ambiguity.

These exceptions are provided mainly to allow MGCP to remain tolerant of older implementations, but it is recommended to always include the package name prefix.

Events and signals are detected/applied by default on the bearer channels connected to the endpoints. It is also possible to request the event/signal to apply to a connection; in this case, the connection identifier is added after the event name, separated by an '@' sign (e.g., G/rt@234A2).

MGCP packages is an area where RFC 3435 expanded significantly on RFC 2705; many things (reason codes, actions, etc.) can be expanded in a package now, not just events and signals (see Section 6 in RFC 3435).

4.2.2.2.2 Categories of signals

There are three signal categories depending on the way they persist or not after being applied:

- **On–off (OO) signals** last until they are explicitly turned off by a NotificationRequest with an empty Signals line (or the endpoint restarts). These signals can be turned 'on' (resp. 'off') repeatedly, they simply remain 'on' (resp. 'off'). A message-waiting indication is a typical example.
- **Timeout (TO) signals** last until they are explicitly canceled or after a timer. An 'operation complete' event is generated when such a signal expires (e.g., the ring-back tone provided when a handset goes off-hook will typically expire after 3 min). Once applied, ring-back will stop (1) if canceled (new NotificationRequest without ring-back in the Signals line), (2) if an event requested by the NotificationRequest occurs (this is the default behavior, although it is possible to override it), or (3) if it times out. The timeout value must be defined by the package.

- **Brief (BR) signals**. These very short signals always complete once the endpoint has begun to execute them, regardless of subsequent events or Notification requests.

4.2.2.2.3 Common packages

Many packages have been defined. Here is a list of some of the reference documents:

- RFC 3064 (CAS)
- RFC 3149 (Business Phone)
- RFC 3441 (ATM)
- draft-foster-mgcp-basic-packages-10.txt
- draft-foster-mgcp-bulkaudits-08.txt
- draft-andreasen-mgcp-fax-01.txt
- draft-foster-mgcp-lockstep-00.txt
- draft-andreasen-mgcp-moveconnection-00.txt
- draft-aoun-mgcp-nat-package-02.txt
- draft-foster-mgcp-redirect-01.txt

There are still more on the way. The most common packages are listed in Table 4.4.

The following sections list the contents of some commonly used packages, defined in RFC 3660 (Basic MGCP packages). An X in the 'R' column denotes an event that can be requested by the call agent. The S column specifies the type of signal (on–off, Timeout, Brief).

4.2.2.2.3.1 Generic media package

As of RFC 3435, packages now have a version number, the original version is zero. The original version of the generic media package was version 0, the current version is version 1.

Table 4.4 Main MGCP packages

Generic media package	G
DTMF package	D
MF package	M
Trunk package	T
Line package	L
Handset package	H
RTP package	R
Network access server package	N
Announcement server package	A
Script package	S

Symbol	Definition	R	S
cf	Confirm tone or 'positive indication tone' of ITU E.182		BR
cg	Network congestion tone of ITU E.180/E.182		TO
ft	Fax tone detected (V.21 fax preamble and T.30 CNG tone)	X	
It	Intercept tone (ITU-T E.180 supplement 2)		TO
ld	Long-duration connection (>1 hour)	X	
mt	Modem detected (V.25 ANSwer tone, V.8 modified answer tone)	X	
oc	Operation complete	X	
of	Report failure	X	
pat(###)	Pattern ### detected (answering machine, tone, etc.). To be defined administratively on the gateway	X	OO
pt	Pre-emption tone (ITU-T E.180 supplement 2)		TO
rbk(###)	Ring-back on connection		TO (180 s)
rt	Ring-back tone or 'ringing tone' (ITU E.180 and E.182)		TO (180 s)

4.2.2.2.3.2   DTMF package

Symbol	Definition	R	S
#	DTMF #	X	BR
*	DTMF *	X	BR
. . .	. . .	X	BR
0	DTMF 0	X	BR
9	DTMF 9	X	BR
A	DTMF A	X	BR
B	DTMF B	X	BR
C	DTMF C	X	BR
D	DTMF D	X	BR
DD	DTMF tone duration exceeded, or generate DTMF in TO mode	X	TO
DO	DTMF signal generated in OO mode		OO
L	Long-duration indicator (over 2 s)	X	
oc	Operation complete		
of	Report operation failure	X	
T	Interdigit timer: 4 s (T critical) if it causes a final match of the digitmap rule; 16 s (T partial) if there is still ambiguity and multiple rules still apply		
X	Wildcard DTMF 0, 1, 2, 3, 4, 5, 6, 7, 8, or 9	X	

4.2.2.2.3.3 Trunk package

Symbol	Definition	R	S
as	Answer supervision	X	BR
bl	Blocking: bl(+) to block the circuit; bl(−) to unblock it		BR
bz	Busy as defined in ITU E.180		TO
co1	Continuity tone: 2,010 Hz (±30). When sending this tone during a continuity test, it is expected to receive this same frequency back	X	TO
co2	Continuity test: 1,780 Hz (±30). When sending this tone during a continuity test, it is expected to receive the co1 frequency back	X	TO
lb	Loop-back		OO
nm	New milliwatt tone (1,004 Hz)	X	TO
oc	Operation complete	X	
of	Report operation failure	X	
om	Old milliwatt tone (1,000 Hz)	X	TO
ro	Reorder tone (ITU E.182 congestion tone)	X	TO
tl	Test line 2,225 Hz (±25)	X	TO
zz	No circuit tri-tone	X	TO

4.2.2.2.3.4 Line package

The exact frequencies corresponding to some signals of the line package may vary in each country. Vendors usually provide localization of these signals through the gateway-provisioning interface.

Symbol	Definition	R	S
adsi(string)	adsi display		BR
aw	Answer tone	X	OO
bz	Busy tone (ITU E.180)		TO (30 s)
ci(ti, nu, na)	Caller ID (time, calling number, calling name). If quoted, the string can be UTF-8 encoded		BR
dl	Dial tone (ITU E.180)		TO (16 s)
e	Error tone	X	BR
hd	Off-hook transition	X	
hf	Flash-hook	X	
hu	On-hook transition	X	
mwi	Message-waiting indication tone		TO (16 s)
nbz	Network busy (fast cycle busy)	X	OO

Symbol	Definition	R	S
oc	Report on completion, may contain completed signal as a parameter	X	
of	Report failure	X	
osi	Network disconnect		TO
ot	Off-hook warning tone (when phone has been left off-hook for too long, and there is no active call)		OO (∞)
p	Prompt tone	X	BR
r0 ... r7	Distinctive ringing		TO (30 s)
rg	Ringing		TO (30 s)
ro	Reorder tone (congestion tone ITU E.182)	TO (30 s)	
rs	Ring-splash (reminder short ring for call-forwarded lines when a call is redirected)		BR
s(###)	Distinctive tone pattern, to be defined on the gateway	X	BR
sit	Special information tone (ITU E.180)		
sl	Stutter dial tone used to confirm an action and require additional input		TO (16 s)
v	Alerting tone		OO
vmwi	Visual message-waiting indicator		OO
wt	Call-waiting tone (ITU E.180)		TO (30 s)
wt1 ... wt4	Alternative call-waiting tones (ITU E.180)		TO (30 s)
y	Recorder warning tone (ITU E.180)		TO
z	Calling card service tone		BR

4.2.2.2.3.5 Handset emulation package

This handset emulation package is the same as line package, but some handset-related events like 'off-hook' and 'on-hook' can be signaled as well as detected. This is useful to provide the automatic off-hook feature (activation of a speaker phone), for phones controlled via CTI (Computer Telephony Integration). This also allows providing features like paging or remote baby monitoring.

4.2.2.2.3.6 RTP package

These events can be used by a call agent to get a more dynamic view of gateway media processing: for instance, some gateways can automatically change their coders from a compressed low-bitrate coder to G.711 for fax; the UC event can be used to learn that this occurred.

Symbol	Definition	R	S
UC	Used codec changed (e.g., UC(15) indicates the codec has changed to μ-law). Codec numbers are according to RFC 1890.	X	
SR(###)	Sampling rate changed (e.g., SR(20) for 20 ms)	X	
JI(###)	Jitter buffer changed (e.g., JI(20) for 20 ms)	X	
PL(###)	Packet loss exceeded (e.g., PL(20) for 20 lost in 100,000)	X	
qa	Quality alert	X	
co1	Continuity tone	X	TO
co2	Continuity tone	X	TO
of	Report failure	X	

4.2.2.3 The MGCP transport layer over UDP

MGCP commands and responses are sent over UDP. The call agent MGCP default receive port is 2727 and the gateway default receive port is 2427.

UDP has many advantages, because it allows the call agent and gateways to control the retransmission of lost packets themselves, and therefore avoid the uncontrolled latency and head-of-line blocking issues associated with TCP. A call agent can be implemented using a single socket for multiple gateways and is not subject to operating system limitations on the number of sockets, which are unavoidable with TCP.

An MGCP stack must handle packet loss detection and retransmission, and must also detect the loss of a connection. The mechanism of MGCP is very sophisticated. It is based on retransmissions, but ensures that a given command cannot be executed twice ('at most once').

Each command may receive one or more provisional responses (1xx), and at most one final response. The command, provisional responses, and final response constitute a **transaction**. Each command contains a transaction identifier between 1 and 999,999,999 (both included), which is copied in all responses related to that command. Each MGCP entity maintains a list of recent commands in process of execution locally, and of recently sent responses. The processing of commands and responses is global for all the endpoints managed by the MGCP device (i.e., there is no separate transaction space or retransmission buffer for each endpoint: typically transactions are managed per MGCP gateway—identified by its gateway name).

New received commands, which are not in the 'command in process' list or which do not have any response in the pile of recent responses, are sent to the MGCP execution engine, which generates a response. The transaction identifier remains in the pile of commands in process until a final response has been generated. Once a final response has been sent, it remains in the 'recent responses' pile until it expires (see p 285 for details on the expiration). This command processing cycle is illustrated in Figure 4.9.

If a command is repeated by a call agent after it has been fully executed, the corresponding response is already in the pile of recent responses. The command is not executed

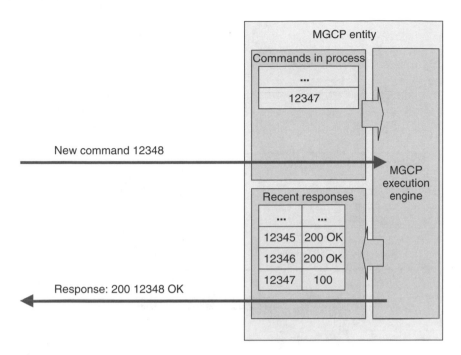

Figure 4.9   Handling of a new command by MGCP.

again, instead the stored response is sent as shown in Figure 4.10. If, on the other hand, the execution engine has not generated a response to the first command yet, the command is still in the commands in process pile, and a 100 PENDING provisional response is generated automatically (or 101 in the case of overload). As shown in Figure 4.11, the duplicate command is not sent to the execution engine.

A command in the commands in process pile expires as soon as the MGCP execution engine has generated a final response. An entry in the recent responses pile should not expire if there is still any chance that a duplicate command will be received. The expiration timer (T-HIST) should be greater than the maximum duration of a transaction which takes into account the maximum number of retransmissions, the delay between each retransmission, and the maximum propagation delay of a packet in the network. A typical value used is typically about 30 s; however, other values can be used as well, as long as the sender and receiver agree on the actual value.

MGCP also provides a way for the sender of a command to acknowledge previous responses: the response acknowledgment attribute contains a range of confirmed transactionIDs. In this case the corresponding response strings can be deleted immediately from the recent responses pile (there will never be a need to resend them), but the transactionID should still stay in the 'long-timer' pile in the unlikely case that some network element duplicates the UDP packet of the original command. This allows the MGCP entity to ignore the duplicate. No response at all is sent.

MGCP entities are required to evaluate dynamically the network round trip time from the time elapsed between the sending of a command and reception of a response: for

Figure 4.10   Handling of a duplicate command already executed by MGCP.

Figure 4.11   Handling of a duplicate command already in execution.

instance, they can evaluate the average acknowledgement delay (AAD) and the average delay deviation (ADEV). The first command retransmission timer can then be set to $AAD + N * ADEV$. Subsequent kth retransmission timers for the same command should be set to $AAD * 2^k + N * ADEV + $ random component (between 0 and ADEV), ensuring exponential back-off in case of network congestion, with an upper bound B set to 4 s typically. Once the upper bound for the retransmission timer has been reached, the implementation should also limit the number R of retransmissions. The recommended practice is to limit the total cumulated time during which the implementation attempts to resend. A complete retransmission scenario is shown in Figure 4.12.

In some special cases (e.g., transmission over satellite), the algorithm can also be modified to force a retransmission timer smaller than the round trip delay in order to ensure that the time to recover from packet loss is very small despite the link delay (but this uses more bandwidth). The sophisticated MGCP transport layer can therefore be tuned to show very network-friendly behavior, like TCP, but in a more application-controlled fashion, or to have a more aggressive behavior (very similar to the behavior of the SS7 transport layer MTP behavior over satellite links, called pre-emptive cyclic retransmission).

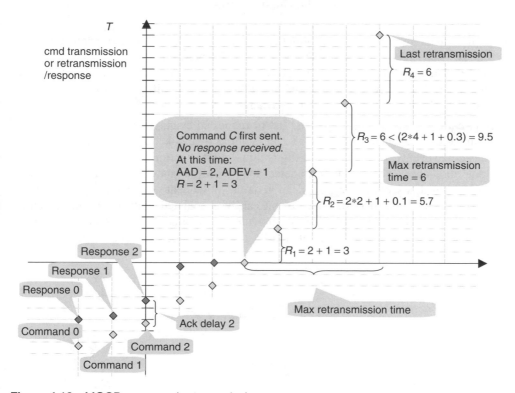

Figure 4.12 MGCP command retransmission.

4.2.2.4   MGCP commands from the call agent to the gateway

4.2.2.4.1   Endpoint configuration command (EPCF)

This command is sent by the call agent to the gateway to configure the type of bearer encoding ('B:') to expect and send on the line side (i.e., not on the VoIP side) of one endpoint or a range of endpoints. The two values defined so far are the G.711 A-law ('e:A'; e.g., Europe) or μ-law ('e:mu'; e.g., in the US). The gateway simply responds with a return code (Figure 4.13).

4.2.2.4.2   Notification request command (RQNT)

This very elaborate command requests the media gateway to watch for specific telephony events. These events can be detected in-band, such as fax tones, DTMF, continuity tones, or analog line status signals like off-hook, on-hook, flash-hook. An example of a RQNT command is given in Figure 4.14.

EndpointId and RequestIdentifier are mandatory parameters, RequestIdentifier is used to correlate the request and the notifications it triggers. In addition, the RQNT command will usually contain some of the following parameters:

- **N parameter** (notified entity): by default the response is sent to the originator of the request (same IP address and UDP port), and gateway-initiated commands are sent to the IP addresses of the call agent resolved from the call agent name. The NotifiedEntity parameter affects where notifications and other gateway-initiated commands are sent (e.g., DLCX and RSIP), until the gateway restarts.

- **R parameter:** the comma-separated list of requested events. Event names are defined in the MGCP event packages (e.g., 'hd' for the off-hook transition), digits are also considered events ([0–9#T] means digits 0 to 9, or # or a timeout of 4 s. The symbol L (long duration) can also be used to detect long DTMF signals. In this case the detected DTMF signal is sent in a first notification and a subsequent notification is sent after 2 s if the DTMF signal persists. Each event can be associated with one or more actions listed between brackets immediately after the event name. The actions can be N for immediate notification (the default if no action is specified), A for accumulate, D to

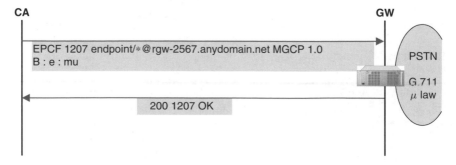

Figure 4.13   EPCF command example.

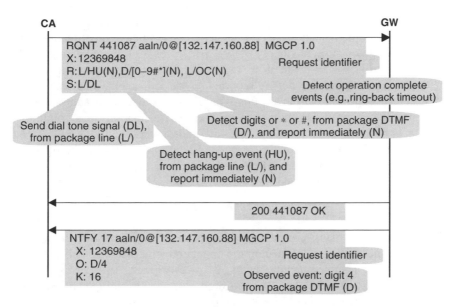

Figure 4.14 RQNT/NOTIFY command example.

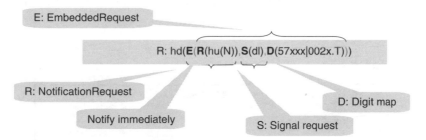

Figure 4.15 Embedded signal request and digit map example.

accumulate according to the digit map (see Figure 4.15), S to swap the active media connection to the next one if there is any, I to ignore the event, K to keep the signal active (normally, the signals present in the S line of a NotificationRequest stop at the first detected requested event, see below), and E to enable (execute) an embedded notification request. An embedded notification request (Figure 4.15) applies to the same endpoint as the NotificationRequest and consists of:

- An optional embedded RequestedEvents parameter: R(embedded RequestedEvents line). For instance, `R([0-9#T] (D), hu (N))`.

- An optional embedded SignalRequests parameter: S(signal requests). For instance, `S(dl)`.

- An optional embedded DigitMap: D(digit map). For instance, `D([0-9]. [#T])`. If an embedded digit map is absent, the current value is used.

All MGCP implementations are required to support at least one level of embedding. Most commercial implementations support exactly one. In many situations, an event will be received by a gateway immediately before it receives a notification request. Without proper handling, this would result in a 'race situation' where, depending on the exact timing, the event may or may not be reported to the call agent. The 'quarantine list', explained below, is designed to avoid this type of race situation. In addition, sometimes a call agent will request to be notified of an event corresponding to a condition that is already true, or becomes true before the gateway sends a response to the RQNT command (**glare condition**): for instance, a call agent requests a notification for the off-hook event (L/hd), but the handset is already off-hook. In this case the gateway responds with an error that indicates the current state (e.g., 401 Phone Already Off-hook, as shown in Figure 4.16).

- **D parameter** (digit map): a digit map is a set of rules telling the gateway when to accumulate or notify digits detected on the target endpoint bearer: for instance, '00T' is a digit string with a single rule and '(0T|00T|[1-7]xxx|8xxxxxxx|#xxxxxxx|*xx|91xxxxxxxxxx|9011x.T)' is a digit string with multiple rules. The syntax of each rule is derived from the Unix 'egrep' command:

  - '[1-4]' matches any digit between 1 and 4, including 1 and 4.

  - '[1-79BT]' matches any digit between 1 and 7, or 9, or B, or a timeout.

  - 'x' matches any single digit (equivalent to [0-9]).

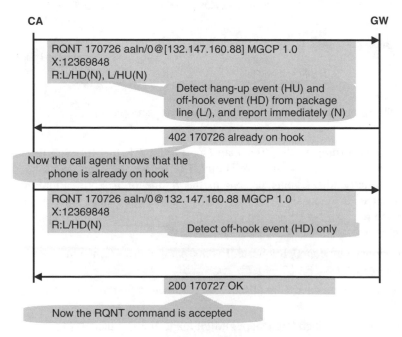

Figure 4.16 Handling of a glare condition with MGCP.

- 'x.' (the dot operator) matches any positive number of occurrences of the previously specified symbol. Therefore, 'x.' or '[0-9].' means 'any positive number of digits'.

If the timer (T) symbol is the last event required to match a rule, the timeout event is triggered if the last detected tone occurred more than 4 s ago (**critical timer**). If there are more digits to match after the timeout event, or when at least one more digit is required to match any of the digit map rules, the timeout event is triggered after only 16 s (**partial timer**). A digit string is accumulated until it matches one of the rules, or if it can no longer match any of the rules (overqualification). The accumulation mechanism is illustrated in Figure 4.17.

- **S parameter:** a request to apply a signal to the endpoint (e.g., ringing or a ring-back tone). By default **timeout signals** stop as soon as one of the events in the list of requested events is detected (unless the event explicitly states through the K action that it should be kept active).

- **Q parameter** (quarantine handling): this describes how signals received just before the notification request (quarantine list, see Figure 4.18) should be handled. The default is to process them, but the call agent can specify that they should be ignored. The parameter also specifies whether only one notification should be sent or whether multiple notifications should be sent. As soon as a gateway has sent a NOTIFY, it transitions to the 'notification pending' state and begins to accumulate events in the quarantine list, a sort of buffer of events not yet processed. The gateway continues to accumulate events in the quarantine list until the response to the NOTIFY command is received (success or failure). If the call agent specified in the Q parameter that it wanted only *one* NOTIFY in response to a RQNT, then the gateway continues accumulating events until the next RQNT. On the other hand, if the gateway can send multiple successive NOTIFY commands, then it processes the list of quarantined events (Figure 4.19). The

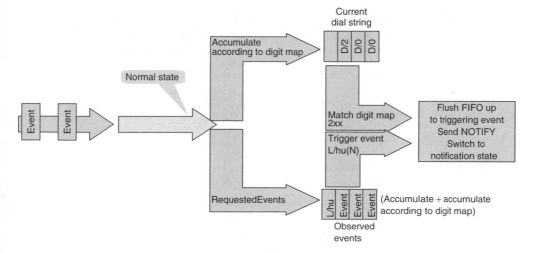

Figure 4.17 MGCP event accumulation. (normal state)

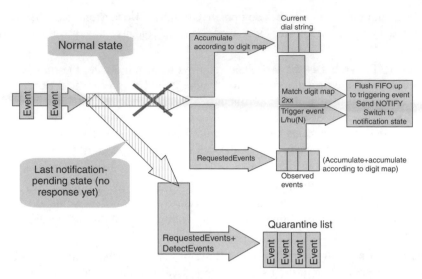

Figure 4.18 Accumulation of events in the quarantine list. If CA expects no more than one notification, continue accumulating until the next RQNT. If multiple NOTIFY commands are allowed, process quarantined events and, if a triggering condition is met, send a NOTIFY and *either* leave unprocessed events in the quarantine list *or* empty the quarantine list. Then send NOTIFY with multiple Events and dial strings.

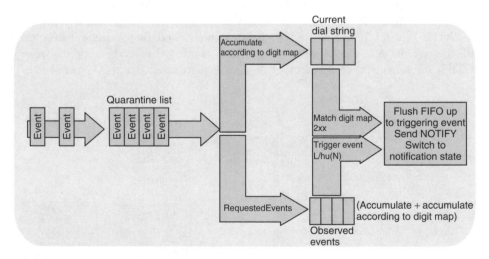

Figure 4.19 Processing the quarantine list.

gateway can then process events from the quarantine list normally as if they had just been received, using the same list of requested events, and the same digit map (the corresponding FIFOs are empty). If a triggering condition is met, the gateway goes to notification state again. Optionally, the gateway can attempt to empty the quarantine list and transmit a single NOTIFY command with multiple events, up to the last triggering

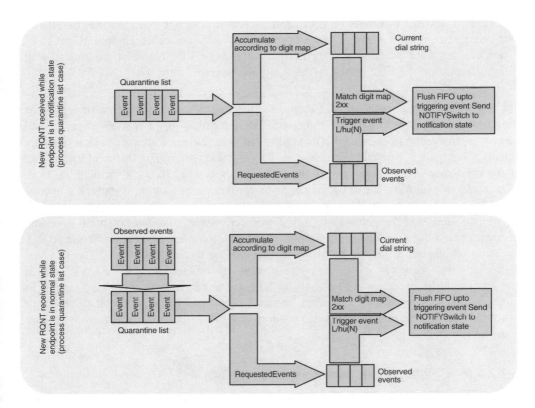

Figure 4.20 RQNT processing according to current endpoint state.

event. The gateway goes back to normal state if the complete quarantine buffer is processed without encountering a triggering event. When a new RQNT is received following a notification, if the Q parameter states that the quarantine list should be ignored, then the quarantine list, dial string and observed event FIFOs are reset. Otherwise, the quarantine list must be processed (Figure 4.20):

- If the endpoint is in notification pending state, the dial string and observed event FIFOs are reset and the quarantine list is processed.

- If the endpoint is in normal state, the dial string is reset, the observed events list (remaining events accumulated that have not yet triggered a notification) is transferred to the quarantined FIFO, and the Quarantined FIFO is processed.

If a triggering condition is reached with the new RQNT, the gateway must ensure that any previous NOTIFY has been received by the call agent before sending a new NOTIFY. The easiest way to do so is simply to wait in notification state until an ACKNOWLEDGE of the previous NOTIFY has been received. Another way is to immediately resend the pending NOTIFY with the new NOTIFY piggybacked.

- **T parameter** (detect events): this adds some events that should be detected in the quarantine list extra to the events specified in the list of requested events.

- An **encapsulated EndpointConfiguration** command, which is executed after the RQNT if it succeeds. When this command is present, the parameters of the EndpointConfiguration command are included with the normal parameters of the NotificationRequest, with the exception of the EndpointId, which is not replicated.

The RQNT command enables MGCP to offer the fine granularity of control typical of stimulus protocols (any event can be requested by the call agent), while allowing the call agent to optimize the exchanges if it does not need to be aware of all events, or if it knows already what to do if a certain event occurs. This is made possible by the combined action of the digit map (accumulation of digits), and the embedded RequestNotification. Both offer a sort of mini look-ahead program: 'Look for this event pattern, if pattern x occurs, then look for this other set of events, and apply this new signal.'

The call agent and the gateway form a distributed system that can potentially become desynchronized due to failures or, simply, race conditions. The execution of the RQNT command eliminates all desynchronizations that are due to race conditions. The fact that digit maps and requested events lists are completely replaced at each RQNT ensures that any desynchronization will last only as long as the currently accumulated events. In addition, the error messages sent when the call agent requires notification of an event which is already active (e.g., off-hook) ensures that synchronization is achieved quickly after the reboot of a gateway.

4.2.2.4.3   Create connection command (CRCX)

This command is sent by the call agent to the media gateway to create a connection on an endpoint. Several types of connections can be created.

4.2.2.4.3.1   Connections to an external media source or sink described by SDP

This is the most common case. The command (illustrated in Figure 4.21) contains the following parameters:

- A **CallId**: the call identifier is composed of up to 32 hexadecimal characters. It is unique to the call agent/gateway and identical on all connections that pertain to the same call. For the gateway, it is an opaque parameter that serves no operational purpose, but can be included in statistics and to facilitate troubleshooting.

- The target **EndpointId**. If the 'any of' wildcard is used by the call agent, the selected endpoint name will be included by the gateway in the SpecificEndpoint parameter of the response.

- Optionally, a new **Notified Entity** for the endpoint (where notifications and other gateway-initiated commands should be sent from now on).

- Optionally, **LocalConnectionOptions** which specify: the desired codecs (e.g., 'a: PCMU;PCMA;G726-32') in preference order; the MIME formats that are allowed (e.g., 'a:image/t38'); the desired packetization period in milliseconds (e.g., 'p:20–40'); the maximum bandwidth in kbps including IP/UDP/RTP overhead ('b:100–200'); the type

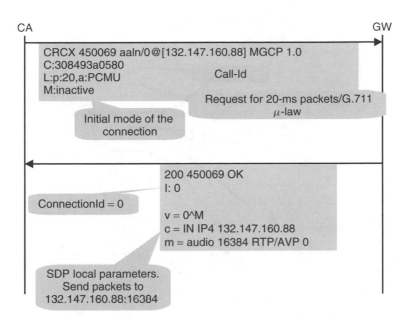

Figure 4.21   CRCX command example.

of service ('t:a2', corresponding to the DiffServ[1] code point to use—the default is 0); the activation of the echo canceler (e.g., 'e:off'—the default is active); the activation silence suppression (e.g., 's:on'—the default is active); gain control (e.g., 'gc:auto'—the default is no gain control); the security key for RTP encryption (e.g., 'k:clear:mysecret', by default there is no encryption of RTP); the network type (e.g., 'nt:IN', most gateways support a single network type); resource reservation (e.g., 'r:g' for guaranteed service, 'r:cl' for controlled load[2]).

- The **mode** of the connection. The defined modes are 'sendonly', 'recvonly' (receive only), 'sendrecv' (send and receive), 'confrnce' (conference), 'inactive', 'loopback', 'conttest' (continuity test), 'netwloop' (network loop-back), and 'netwtest' (network continuity test). Figure 4.22 illustrates the relations between the media streams of each type of connection and the corresponding endpoint. The $\sum$ symbol means that signals are mixed before transmission. Signals received from 'conference' connections are sent to all other connections that are also in 'conference' mode, and are mixed before transmission. 'sendonly', 'sendreceive', and 'conference' connections require a RemoteConnectionDescriptor (or another endpoint identifier), otherwise the media cannot be sent. A loop-back connection returns any media that are received from the endpoint back to the same endpoint (standard ITU continuity test). A continuity test connection returns a 2,010-Hz signal if a 1,780-Hz signal is received from the endpoint (used in the US). A network loop-back connection returns media received from

[1] See companion book, *Beyond VoIP Protocols*
[2] See companion book, *Beyond VoIP Protocols*, for a detailed description of the RSVP resource reservation protocol.

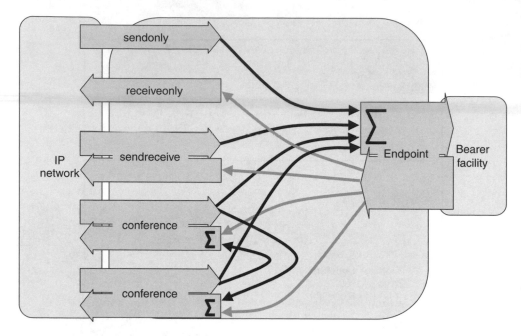

Figure 4.22   MGCP connection types.

the network back to the network. The network continuity test mode is obsolete. These connection modes are illustrated in Figure 4.23.

- Optionally, a **remote connection descriptor** that specifies using SDP where the media stream should be sent and the options for the codec to use.
- Optionally, an **encapsulated NotificationRequest command**. The parameters of the encapsulated command are simply added, apart from the EndpointID which is not duplicated.
- Optionally, an **encapsulated EndpointConfiguration command**. The parameters of the encapsulated command are simply added, apart from the endpointID which is not duplicated.

By sending an encapsulated NotificationRequest command the call agent has the ability to request the gateway to execute simultaneous actions. For example:

- Ask the residential gateway to prepare a connection, in order to be sure that the user can start speaking as soon as the phone goes off-hook.
- Ask the residential gateway to start ringing.
- Ask the residential gateway to notify the call agent when the phone goes off-hook.

This can be accomplished in a single CreateConnection command, by also transmitting the RequestedEvent parameters for the off-hook event and the SignalRequest parameter

Figure 4.23 Loop-back and continuity test connection modes.

for the ringing signal. This combination dramatically reduces the number of round trips necessary to establish a connection between two endpoints and provides more scalability options for the call agent.

After processing the CRCX command, the gateway returns:

- A ConnectionID.

- A LocalConnectionDescriptor specifying the local parameters of the connection using SDP.

- Optionally, a specific EndpointID if the endpoint was not specified in the command.

4.2.2.4.3.2 Connections to another endpoint on the same gateway

Connections to another endpoint on the same gateway occur frequently. Many manufacturers do not support any optimization for this type of connection and some events do not support it properly (this is one of the most common bugs in gateway implementations across all VoIP protocols), in which case the call agent uses the normal procedure, giving a local gateway IP address and port as the RemoteConnectionDescriptor.

But, some gateways allow endpoints to communicate locally without requiring packetization and switching through the IP network. This type of connection is similar to the previous type of connection, except that the call agent specifies a SecondEndpointID instead of a RemoteConnectionDescriptor. Such a command really creates two connections (one on each endpoint), the response provides the ConnectionIDs of both connections. The second connection is by default in sendrect mode.

Figure 4.24   MDCX command example.

4.2.2.4.4   Modify connection command (MDCX)

This command (illustrated in Figure 4.24) enables a call agent to modify a connection that has already been set up by the gateway. The parameters are the same as in the CreateConnection command, with the addition of the ConnectionID that serves to identify the target connection.

The ModifyConnection command can change all parameters of a connection: activation mode, codec, packetization period, etc.

4.2.2.4.5   Delete connection command (DLCX)

This command enables the call agent to terminate a given connection. It should be noted that if there is more than one gateway involved in a call, the call agent sends the Delete-Connection command to each of the media gateways in order to fully tear down both ends of the call.

As shown in Figure 4.25, a nice functionality provided by MGCP is that the media gateway, on termination of a connection, has to send to the call agent the following information:

- Number of packets (RTP) sent.
- Number of octets sent.
- Number of packets (RTP) received.
- Number of octets received.
- Number of packets lost.

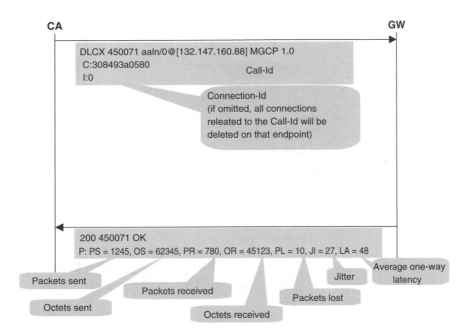

Figure 4.25 DLCX command example.

- Interarrival jitter.
- Average transmission delay.

These parameters are calculated as for RTCP (see Chapter 1).

MGCP also allows a media gateway to clear a connection on its own (e.g., in the event of connection loss or a failure). In this context the media gateway sends a DeleteConnection command to the call agent including all the connection statistics.

It is also possible for a call agent to delete all the connections of a call at the same time by omitting the ConnectionID. Note that this command does not return any individual connection statistics or call parameters.

4.2.2.4.6 Audit endpoint command (AUEP)

The call agent can use this command in order to check whether an endpoint is up and running, and to learn dynamically its capabilities (Figure 4.26). Using the 'all off' wildcard, the call agent can also learn the number of endpoints present on a given gateway.

4.2.2.4.7 Audit connection command (AUCX)

This command enables the call agent to retrieve all the parameters attached to a connection identified by a ConnectionID on an endpoint identified by its EndpointID (Figure 4.27). This can be used by a call agent to check that a connection is still active: if no information is requested, the gateway simply responds with 200 OK if the connection exists.

Figure 4.26   AUEP command example.

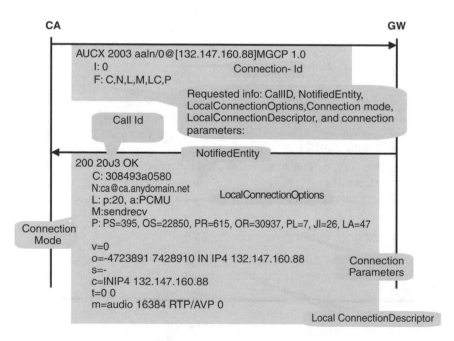

Figure 4.27   AUCX command example.

4.2.2.5 MGCP commands from the gateway to the call agent

4.2.2.5.1 Notify command (NTFY)

This command enables the media gateway to send back events that were requested by the media gateway controller. The media gateway can send one or several events in a NOTIFY command. Each notification reports events from a given endpoint (possibly a connection on an endpoint), listed in the endpoint part of the command header. The correlation between the request and the corresponding notification is provided by the RequestIdentifier (X parameter). The list of notified events is specified in the ObservedEvents (O) parameter, which is a comma-separated list of events. Events appear in the order in which they have been detected. The form of events can be:

- The event name, only if it is part of the default package (not recommended), such as hd.
- The package name and the event name: L/HD.
- The package name, event name, and ConnectionID for events detected on a connection: L/HD@134a23b.

When booting, some endpoints send without solicitation their current state (e.g., off-hook) with special request ID 0.

The NOTIFY command is acknowledged by a return code from the call agent.

4.2.2.5.2 Restart in progress command (RSIP)

This command allows a gateway to make a call agent aware of an endpoint or a group of endpoints that are going to be taken out of service. In this case the restart method can be graceful (RM:graceful), can specify a delay (RD), or can be forced (connections are lost immediately).

The message is also sent by gateways when they boot, to make the call agent aware of their presence (Figure 4.28). In this case the restart method is 'restart', and a delay can be specified until the endpoints are operational (0 is the default value if nothing is specified). Restart method 'disconnected' can also be used to alert the call agent about potential state mismatch.

For gateways that acquire an address dynamically through DHCP, the call agent has three ways to learn the IP address of the gateway:

- By looking at the source IP address of the RSIP message. This is not always reliable if the RSIP message is relayed.
- If **Dynamic DNS (DDNS)** is used in conjunction with the DHCP server, the DNS name of the gateway as advertised in the RSIP message will resolve to the current IP address of the gateway. This is a robust method and also provides the ability to recontact the gateway immediately if the call agent reboots. On reboot, the call agent, if it knows about the gateway, queries the DNS and can send an AUEP to the current IP of the gateway.

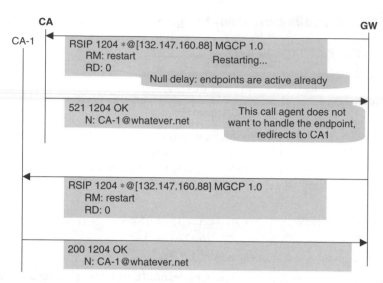

Figure 4.28 RSIP and change of call agent.

- The gateway can include its current IP address as the gateway name. This works, but makes it difficult to keep track of the gateway since the name changes with the IP address. In addition, if the call agent reboots, it will be unable to reach the gateway unless it has saved the current IP address in persistent storage.

4.2.3 Handling of fax

A new package 'fxr' for fax is being defined (draft-andreasen-mgcp-fax-xx.html). The fxr package also uses extensions of SDP for negotiation (defined in RFC 3407, 'Session Description Protocol (SDP) Simple Capability Declaration', or **simcap**). The fxr package defines new local connection options:

- 'fxr/fx:t38' for strict handling of T.38. The gateway notifies the call agent that a T.30 fax preamble is detected ('fxr/t38(start)' event) and mutes the media channel. Before starting the T.38 procedure, the gateway will check that the remote party also supports the same variant of fax transport by checking its capabilities, expressed in an extension to SDP (see below for details). The call agent is responsible for switching the connection to T.38 fax mode by sending an MDCX with 'a:image/faxt38' in the LocalConnectionOptions (L:) and providing a RemoteConnectionDescriptor with the 'm = t38' media line. Should a failure occur during the fax call, it is indicated by the 'fxr/t38(failure)' event. The end of the fax is indicated by the 'fxr/t38(stop)' event.
- 'fxr/fx:t38-loose' for loose handling of T.38. The difference from strict handling is that no confirmation of common capabilities is required from the remote end. The fax transmission attempt starts as soon as a RemoteConnectionDescriptor with a media line indicating T.38 is received.

- 'fxr/fx:off' if there is no special handling of fax.
- 'fxr/fx:gw' if the handling of fax (not necessarily T.38) is left to the gateway. This is the default mode. The gateway will send a 'fxr/gwfax(start)' event if it begins a specific fax procedure, or 'fxr/nopfax(start)' if the gateway detected a fax but decided to take no action on it. In the case of 'fxr/gwfax(start)', the call agent should remain passive until it receives a 'fxr/gwfax(stop)' event.

In the following example (Figure 4.29) the call agent configures the gateway to use strict T.38 handling on line 'aaln/0'. The gateway returns local connection parameters, as well as 'a =' elements listing its capabilities. It was necessary to extend SDP to express capabilities (RFC 3407), because the normal way of expressing support for multiple codecs in SDP also implies that media can be received immediately on these coders. Here t.38 is supported, but still cannot be received. The capability set according to RFC 3407 is identified by a serial number, incremented each time the endpoint sends a new capability set (a = sqn:<serial number>). This attribute is immediately followed by capability lines (a = cdsc:<capability number><media type> <transport> <format list>).

Note that in Figure 4.29 the call agent still does not know the capabilities of the remote endpoint (not mentioned in CRCX). If a fax was received at this point, the t38 procedure would be delayed until a proper capability descriptor for the remote endpoint is received from the call agent.

Once the call agent knows the capabilities of the remote end, it sends these capabilities in the RemoteConnectionDescriptor of a MDCX command (Figure 4.30). Now that the

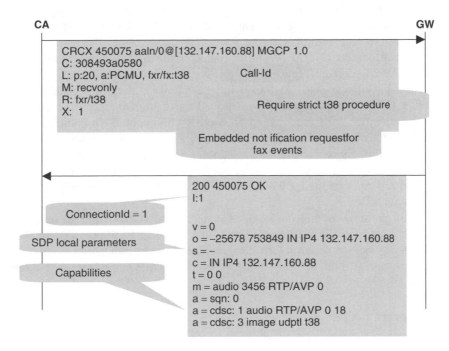

Figure 4.29 CRCX command requiring usage of the strict T.38 procedure.

CA   **GW**

Figure 4.30   Confirmation that the remote end supports T.38 fax.

local gateway knows that the remote endpoint also supports strict T.38 over UDP and, therefore, can also use that procedure. Note that for this connection the media specified in the SDP 'm =' line was G.711 and that, therefore, the media stream is activated with G.711 (we have not received a fax signal yet).

If the gateway detects a T.30 preamble characteristic of fax at any time, it reports the event to the call agent (Figure 4.31), because the call agent has requested to be notified of fxr package events. At this point the gateway mutes the audio signal and stops sending G.711 to the remote GW. The call agent immediately instructs the gateway to switch to strict T.38 mode using an MDCX command. This command does not contain a RemoteConnectionDescriptor; therefore, the previous RemoteConnectionDescriptor is still valid. Since the previous descriptor requested G.711 media (SDP 'm =' line), the GW cannot yet send T.38 data, but is prepared to receive it. Note that the reception port has not changed although the media have changed, which is the recommended behavior.

Once the call agent has obtained a RemoteConnectionDescriptor from the other gateway (Figure 4.32), it modifies the SDP media line to use T.38 (the UDP port has not changed). The local gateway begins to send T.38 datagrams to the port indicated.

The 'fxr' package also defines new statistics parameters that can be reported in response to AUCX or DLCX:

- PGS: number of fax pages sent.
- PGR: number of fax pages received.

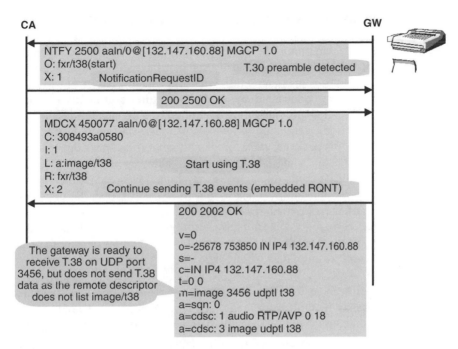

Figure 4.31 Preparing the gateway to receive T.38 media data.

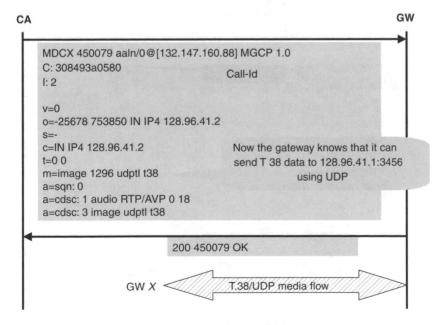

Figure 4.32 The local gateway is instructed to send T.38 data.

4.2.4   Extensions for phone user interface control

Many business phones provide multiple feature buttons (hold, retrieve, conference, mute, quick-dial, messages, etc.) and a sophisticated user interface with lamps, a large screen with symbols for activated features and call-related information, etc. Until MGCP these phones were all proprietary, controlled by the manufacturer's stimulus protocol. The standard MGCP package line provides only a limited set of capabilities to control a business phone user interface: activation of a visual message-waiting indicator, caller ID, distinctive ringing. With this package, call agent manufacturers can provide business-grade features only by heavily using audio notifications and audio menus. The MGCP handset package adds the capability to remotely activate the phone loudspeaker, thereby enabling CTI-controlled telephony applications (click to dial from the PC, operator consoles, etc.), but advanced business features remain unaddressed.

The problem of standardizing a control interface for business phones is indeed complex, because the creativity of vendors should be preserved. A good compromise can be reached by making the following assumptions (Figure 4.33):

- The phone is able to render a screen that can be described using a text syntax (e.g., XML). The screen may be built from predefined templates (cards) stored in the phone by the phone manufacturer, with replaceable parameters provided by the call agent.
- The phone has a number of named function keys, which can be associated with an endpoint-generated MGCP event sent to the call agent. No assumption is made on the function of the key. Optionally, some keys can have their function dynamically described to the user by descriptive areas on the screen or dedicated LCD labels (softkeys).

Figure 4.33   Typical IP phone LCD screen with softkeys.

- Optionally, the phone can also provide a capability to navigate through menus and select an option, or can offer numeric or alphanumeric input fields.

To date, several vendors have implemented business phone control MGCP packages based on these assumptions:

- Cisco with the BTXML2 markup language.
- Polycom with the MGCP business phone packages documented in RFC 3149.
- Swiss voice with the MGCP business phone packages documented in RFC 3149.

RFC 3149 defines:

- A feature key package (KY) which describes signals to set the key label ('KY/ls(<KeyId,Label>)') and key activation state ('KY/ks(<keyId>,<KeyState>)'). The following states have been defined: en(enabled), db(disabled), id(idle), dt(dial tone), cn(connected), dc(disconnected), rg(ringing), rb(ring-back), ho(holding), he(held). 'S: KY/ks(5,en)' in a RQNT sets key fk5 to enabled state. MGCP events are used to report the key press events (KY/fk1 to KY/fk99). These events can be requested by sending a NotificationRequest with 'R:KY/fk'.
- A business phone package (BP) in which signals are used to force speaker phone activation ('BP/hd' for off-hook, 'BP/hu' for on-hook) or play a beep ('BP/beep')
- A display XML package (XML). An XML-format signal is used to render the screen. Events are used for user input or selection. Both are prefixed 'XML/xml'. The screen control feature of the display XML package uses a special endpoint name, derived from the phone endpoint name. If the phone is called ph1@anydomain.net, the screen endpoint name will be disp/ph1@anydomain.net. This separation avoids possible problems since events deactivate signals by default, which is not the desired behavior for screen navigation. In order to request events resulting from the selection of items on screen menus, the RQNT targeted at the display endpoint must contain 'R: XML/xml'.

The XML screen description syntax of RFC 3149 defines the following widgets: input box, enumerated list box, text box, and echo box. The XML screen template can contain replaceable parameters, or tags corresponding to dynamic content (e.g., time/date or call timer). The XML format also describes the event string to send back to the call agent for each possible selection. If the main phone keypad is used to select a choice on the screen menu, the event is reported to the call agent through the XML package on the screen endpoint: the screen has precedence and only passes unused key press events to the phone endpoint subsystem (the only exception is the echo widget, which displays events but does not consume them, and can be used to echo on the screen a dialing number).

The format of the XML/xml screen signal is as follows:

```
S: XML/xml
   (<url>?<card>?$<variable1>=<value$<variableN>=<value>)
```

The <url> can be http://screenserver.anydomain.net/deck1 if the set of screen templates (called a **deck**) must be fetched on an HTTP server, or any name if it is local to the phone (provisioned). The <card> component specifies the template to select within the deck. Usually, each phone state is associated with a specific card. The variables are replaceable parameters within the card template. For instance, if deck1 is:

```
<xml>
    <card id="one">
        <p>$line1</p>
        <timer value="2"/>
        <do type="ontimer">
        <go href="#two"/>
        </do>
    </card>

    <card id="two">
        <p>$line2</p>
    </card>

    <card id="home">
        <p mode="nowrap">$dn <time align="right"></time>
        <select type="item" name="Menu" iname="StrMenu">
        <option value="1"
  onpick="post?%deck?%id?%name=%value">MENU</option>
        </select>
        </p>
    </card>

</xml>
```

If the applied signal is S: XML/xml(deck1?one?$line1=abc$line2=xyz), the phone would render:

```
<card id="one">
    <p>abc</p>
</card>
```

Then, after the timer:

```
<card id="two">
    <p>xyz</p>
</card>

<xml>
```

Figure 4.34 IP phone screen appearance showing the 'home' card.

If the applied signal is S: XML/xml(deck?home?$dn=2344), the screen would be rendered by our sample phone as shown in Figure 4.34.

If softkey 1 is pressed, the following event would be reported in the NOTIFY:

O: XML/xml(post?basic?home?Menu=1)

In addition to the functions described above, some functions must be implemented locally on the phone, such as mute, volume control, contrast control, audio path control (handset/loudspeaker/headset). RFC 3149 assumes these functions come with their own screens defined by the phone manufacturer.

Cisco BTXML2 syntax was defined after RFC 3149 and is available in conjunction with the MGCP protocol on their IP phones. It is very similar to RFC 3149 (e.g., it likewise uses a separate endpoint for screen control prefixed by 'disp/'). The main difference is that the XML description also includes feature key event mappings (Cisco phones do not have separate LCDs for each button) and provides many more widgets than those defined in RFC 3149. The display is divided in zones (similar to HTML frames), which can be described separately (Figure 4.35).

Even though the industry has not yet agreed on a common XML description format, these open control interfaces are similar enough to make it relatively easy for call agent manufacturers to support business phones. In fact, the call agent does not need to be aware of the exact XML syntax used by the phone; it interacts with the phone only by calling predefined cards with replaceable parameters, and receives named events that it needs to map to call control actions. The customization of a call agent for a specific type of phone becomes straightforward. MGCP really invented the 'open business phone'!

Figure 4.35   Screen structure of Cisco BTXML2-capable phones (the 7960 is shown).

4.3   SAMPLE MGCP CALL FLOWS

4.3.1   Call set-up

In Section 4.1, we described the two main applications of MGCP (both at the edge of the network):

- Control of analog gateways or MGCP phones in customer premises.
- Control of trunking gateways in the service provider network.

The following call flow illustrates both cases. A call is received from an SS7 network signaling transfer point (STP) by an SS7 call agent (SS7_CA); the SS7 call agent sends the call to the core VoIP network using the H.323 protocol (it could also be SIP); the core network routes the call to a residential service call agent (R_CA), and the residential service call agent rings an analog phone on a residential gateway (R_GW).

SS7 ISUP messages are always associated with a **circuit identification code (CIC)**, which relates call signaling to a specific media time slot on a given trunk (this relation is configured statically as part of the provisioning of TDM switches). The CIC enables the SS7 call agent to locate the proper media gateway and the endpoint on this gateway that terminates the specified trunk and time slot.

The first part of the call scenario is illustrated in Figure 4.36. The CRCX command instructs the endpoint to prepare for receiving G.711 μ-law media from the IP network, with a 10-ms frame size. The gateway responds by giving an IP address and port where the RTP stream should be sent. With this information, the SS7 call agent can now send

Figure 4.36 New call received from SS7 network and transmitted to the VoIP core. IAM = initial address message.

an H.323 SETUP message to the VoIP network core-routing softswitch, in this case an H.323 gatekeeper. This core-routing softswitch is responsible for finding the proper egress route, or eventually for rerouting the call to back-up routes in case of congestion or any other problem. It will also translate the calling and called party numbers if necessary, as appropriate for the destination. The SETUP message contains the RTP IP address and port in the Fast Start element, in order to expedite the media connection.

The CALL PROCEEDING message indicates that the SETUP message has been received properly and that the dialing number is complete. If the number is not complete, the core softswitch would have sent a SETUP ACKNOWLEDGE message instead, to initiate a procedure known as overlap sending, in order to accumulate more digits.

The call flow is simple because there is only one voice coder in the VoIP network and codec negotiation is not required. Variants of this call flow are needed to use the negotiation capabilities of H.323: multiple 'inactive' connections could be set up on the media gateway for each voice coder, which would provide the list of proposed logical channels for the H.323 SETUP fast Start element. If the MGCP trunk gateway cannot open the multiple inactive connections of various coders, then AUEP codec information could be used to construct an H.245 CapabilitySet, and the codec would be negotiated through a normal H.245 exchange.

In our example (Figure 4.37) the called party is managed by a residential call agent. It immediately acknowledges the SETUP message with a CALL PROCEEDING, and also establishes the H.245 dialog either with the core softswitch, or directly with the source call agent (this is not shown in the diagram). The residential gateway (RGW) is instructed to ring ('L/R0' signal) and to notify the off-hook event to the call agent. As soon as the residential gateway confirms that it is ringing, R_CA sends back the ALERTING message

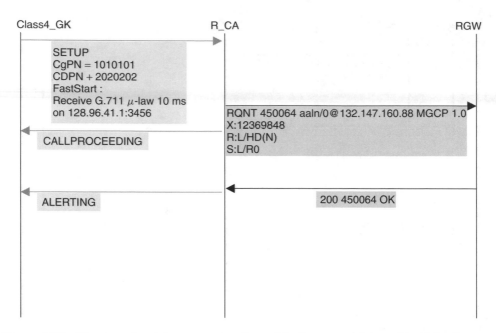

Figure 4.37 The class 4 gatekeeper routes the call to the appropriate residential call agent.

to the core softswitch. Note that within a national telephone network the actual ringing tone is never sent over the RTP media connection, it will be generated by the calling party switch when it receives the ALERTING message. However, it is still necessary to provide a receive-only media connection as soon as possible (if this can be done in the SETUP message), because in the case of a phone call to an international number the ring-back will be provided in-band through the RTP connection. In this case (not represented) the H.323 ALERTING message will contain a progress indicator (PI = 8), which instructs the originating switch not to play a local ring-back, but the remote ring-back instead.

The ALERTING message is relayed by the core softswitch to the originating SS7_CA, which sends back an address complete (ACM) ISUP message on the SS7 signaling link. In fact, the SS7_CA may also have sent the ACM immediately on receiving the CALL PROCEEDING message from the core softswitch: in SS7 ISUP the ACM means both that the number is complete and implicitly that the remote party phone is ringing. On SS7 networks the calling party may hear ring-back before the called party phone actually rings!

When calling an international network, the distant ring-back tone is sometimes provided (this also allows remote in-band error messages to be heard). In such a situation a PROGRESS or ALERTING message with a specific progress indicator (indicating that in-band audio information is being sent) would be received from the H.323 side, and the SS7 call agent should then set an equivalent indicator in the ACM or send a call progress (CPG) ISUP message. In modern telephony networks, this is normally the only situation where the ring-back tone is provided by the remote end. Figure 4.38 represents the normal case, with ringback provided by the calling party exchange. There is still no media exchanged on the VoIP network.

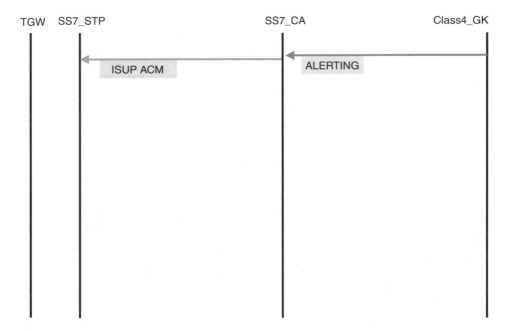

Figure 4.38 ALERTING message converted to an ISUP ACM. ACM = address complete message.

At this moment the called party answers the call. The residential gateway sends a NOTIFY back to the call agent (Figure 4.39). The call agent immediately creates a connection on the endpoint. Since the call agent already has received the voice coder settings, IP address, and port required to send media to the calling-side endpoint, this information is provided in the CRCX SDP. The residential gateway can immediately begin to send audio. In the answer to the CRCX, the residential gateway also provides an IP address and port where it will receive audio from the calling-side endpoint. This information is included in the FastStart element of the CONNECT message sent by the call agent to the class 4 gatekeeper.

The CRCX message also uses an embedded NotificationRequest instructing the residential gateway to immediately notify the call agent of any digit detected on the endpoint and, of course, if the user hangs up.

As soon as the SS7 call agent receives the CONNECT message, it relays the media information of the FastStart element to the trunk gateway, using a ModifyConnection message (Figure 4.40).

4.3.2 DTMF tones

The call flow described in Figures 4.41 and 4.42 illustrates the fact that DTMF tones are sent as signaling information, not as part of the media stream. If the user presses the '5' key of the phone, generating a DTMF tone, the gateway—as instructed—immediately

Figure 4.39   The called user picks up the handset.

Figure 4.40   The CONNECT message is converted to an ISUP ANM message. ANM = answer message.

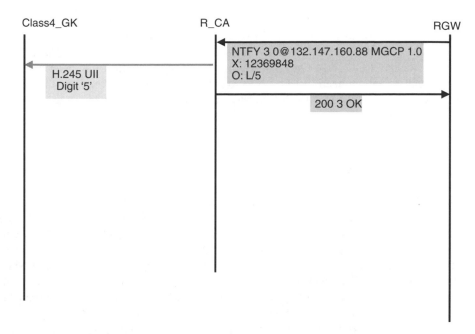

Figure 4.41   Out-of-band DTMF handling. VII = user input indication.

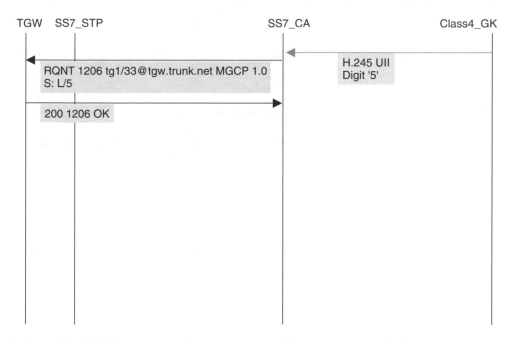

Figure 4.42   DTMF regenerated by trunk gateway. VII = user input indication.

notifies this to the residential call agent. The call agent relays this information by using an H.245 UserInputIndication message. In order to avoid having the gateway send the DTMF in-band as well, a new DD (su = false) event has been defined in the DTMF package (version 1 RFC 3660). Once the H.245 UserInputIndication has been received by the SS7 call agent, it needs to be converted back to a DTMF tone on the bearer channel. The call agent sends a RQNT to the endpoint requesting it to generate signal 'L/5' (Figure 4.42).

4.3.3 Call release

When the called user hangs up, the event is notified to the residential call agent, which sends a RELEASE COMPLETE message to the core VoIP network (note that in H.323 the media control H.245 session is released first, then the call control link). The call agent also reinitializes the gateway for the next call, by looking for the off-hook event, and then enabling an embedded NotificationRequest which applies a dial tone and waits for digits (Figure 4.43). The syntax of the requested events line means: 'accumulate all digits, *, #, and timer event according to the digit map.' The digit map is configured for a phone restricted to local service in San Jose, CA: it can only dial 6-digit numbers or an emergency number. Any other event not matching the digit map will trigger an immediate NOTIFY (e.g., '8' or '0'). A timeout of 16 s while still accumulating digits in a digit map will also trigger a NOTIFY (e.g., '123<16seconds>').

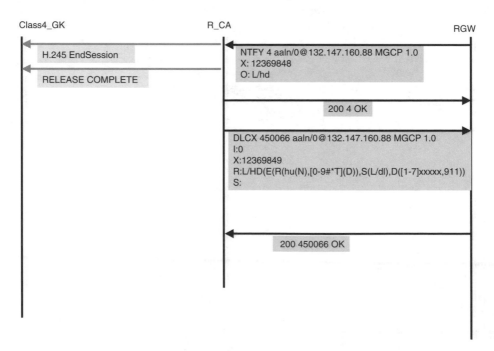

Figure 4.43   An, end-user on the MGCP residential gateway hangs up the call.

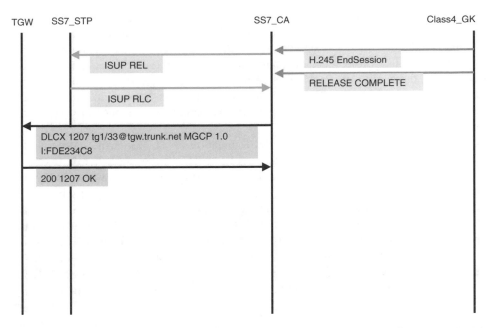

Figure 4.44 RELEASE COMPLETE (RLC) message converted to an ISUP RELEASE (REL) message.

The ISUP disconnection sequence is more complex than the H.323 or SIP disconnection sequence, and typically requires at least two messages: REL (RELEASE) indicates that the called party hung up, while RLC (RELEASE COMPLETE) indicates that the calling party also hung up (Figure 4.44). This more complex sequence is used because some networks use in-band announcements at the end of certain calls (e.g., calling cards). The RLC message instructs the device sending the in-band information to stop sending it; this message also causes the release of all circuits.

4.4 THE FUTURE OF MGCP

The PacketCable NCS specification has been standardized as IPCableCom in ITU-T SG9. This specification is also an ANSI standard (via the Society of Cable Telecommunications Engineers). The IETF MEGACO or ITU are not actively working on MGCP since the IETF MEGACO Working Group decided to work jointly with the ITU on H.248.

Despite this, MGCP is still alive and well. The reason is easy to understand: the authors of MGCP defined the scope of the protocol extremely well and, within this scope, managed to fulfill all the requirements. Manufacturers wanting to provide stimulus-controlled phones, or media gateways, cannot ask for more than MGCP provides.

In the midst of the heated debates about the best VoIP protocols with the accompanying tendency to greatly oversell or misrepresent the capabilities of each protocol, MGCP managed to stay aloof from this mixture of marketing and engineering characteristics that typified the telecom bubble era and jeopardized the quality of many specifications. Since the beginning, the development of MGCP has been driven by immediate customer requirements, while other standards were more driven by manufacturers. As a consequence, the quality of the MGCP specification is much better than that of the SIP specification or the early H.323 specifications. MGCP has been adopted by the cable industry and many manufacturers, and is currently deployed in many VoIP networks all over the world.

Although the quality of MGCP is a solid foundation for H.248, at the same time there is no great incentive for the industry to migrate to H.248, because there are very few features that the latter can provide beyond the MGCP capabilities available today. Even though RFC 3435 has introduced new extension capabilities, the few missing features can readily be added to MGCP as well. Initially, video was presented as the key feature added by H.248, with MGCP perceived as an audio-only protocol. This dates back to an early version of the RFC which stated that the protocol was not intended for video. In fact, since MGCP uses SDP, it can establish connections using any media that SDP can describe, including video. The latest versions of the MGCP RFC now acknowledge that MGCP can be used for both audio and video. There are several MGCP videophones on the market today.

MGCP is implemented on hundreds of devices from various manufacturers, from high-density voice media gateways and modem banks to two-port analog gateways, IP phones, or call agents. The flexibility of MGCP has also made possible the first comprehensive implementations of residential or Centrex features for business telephony. The good news for the future is that MGCP and H.248 are so similar that it will be very easy for all manufacturers to eventually migrate these products to H.248. In the meantime, there is no hurry to do so—if it is not broken, there's no need to fix it!

5

Advanced Topics: Call Redirection

5.1 CALL REDIRECTION IN VOIP NETWORKS

5.1.1 Call transfer, call forward, call deflection

Call transfer, as opposed to call forward, is characterized by the timing of call redirection. Call transfer redirects the call after an initial connection with the called party. Call forward redirects the call before the call is connected to the initial called party. Two flavors of call transfers exist: in consultation call transfer the redirecting party talks to the redirected-to party before transferring the call; whereas in blind call transfer the transferring party transfers the call directly without verifying whether the redirected-to party is willing to/can accept the call.

Call deflection redirects the call after the call has been presented to the called party, but before the call connects: the initial called party never enters in a conversation with the caller.

Figure 5.1 illustrates the various types of call redirection.

The call forward service is relatively simple to provide over traditional or voice over IP (VoIP) networks. Call defection and call transfer services are much more complex to provide in any network technology, but they are even more difficult to provide in VoIP networks. They raise the following issues:

- Translation of the redirected to phone number from the redirecting party format into the format appropriate for the element performing call redirection.
- Dynamic redirection of the media streams from the initial called party to the redirected-to party.

IP Telephony   O. Hersent, J.P. Petit, D. Gurle
© 2005 John Wiley & Sons, Ltd   ISBN: 0-470-02359-7

Figure 5.1 Call redirection types.

- Correlation of the initial call and the redirected call in order to generate appropriate call detail records.

This section discusses in detail the various implementation choices for call redirection services, the difficulties that appear in the context of public networks, and how these challenges can be addressed.

5.1.2   Summary of major issues

5.1.2.1 Numbering formats

Any phone call usually involves three distinct formats for calling and called party aliases:

- The **originating format**: this is the format of phone numbers that are familiar to the caller. For instance, the number of a phone in San Jose, CA will be known as 5217000 by a resident of San Jose, while the same phone will be known as 14085217000 by someone living in New York.
- The **pivot format**: this is the format used by the phone network billing system, and manipulated by the phone network administrators (e.g., when defining routes). A US

network may decide to use the full number including the area code, for instance, and may also include the country code.

- The **terminating format**: this is the format of phone numbers that are familiar to the called party.

Let's take an example to clarify: John, in San Jose, CA (+1 4085217000) calls Mark in Paris (+33158713333):

- The originating formats are: John 5217000, Mark 01133158713333.
- The pivot formats are (assuming the international format is used within the network): John 14085217000, Mark 33158713333.
- The terminating formats are : John 0014085217000, Mark 0158713333.

The issue as far as call redirection is concerned is the following: if John instructs the network to reroute the call to Mark, the network should properly understand the redirected-to phone number.

But, doesn't this seem trivial?

Look at the following example, using H.323, H.450, or SIP REFER. With these methods, a message (let's generically call it REDIRECT) is sent to the calling party, instructing him to make a new call to the supplied number.

Let's say Mark wants to redirect John to another phone number in Paris (+33158713300). For Mark, living in Paris, this phone number originating format is 0158713300, so Mark sends back to John's terminal a message REDIRECT 0158713300. John's terminal assumes this is an originating format for San Jose, CA and makes a new call to 0158713300, when it should have dialing 01133158713300. The call transfer fails ... or, worse, succeeds, but to the wrong destination.

5.1.2.2 Billing

Let's take the example of a call from A to B, with a duration of T1 seconds, redirected after the T1 seconds to C, for a duration of T2 seconds. In the case of call forward or call deflection obviously T1 = 0. How should we bill for this call?

One possibility is to bill the call as a call from A to B for T1 seconds, and a call from A to C for T2 seconds. This is the simplest choice for voice over IP where redirected calls frequently appear as two calls: A to B and A to C. Unfortunately, things are not so simple.

The reason is that such a way of billing gives B the opportunity of making a lot of money at other people's expense: he just needs to create a company that manages premium numbers C, and redirect all calls to B to this premium number C. Premium numbers are charged at a special rate, much more expensive than a traditional phone call, and the carrier shares revenues with the owner of the premium phone number. Now, anyone calling B, a 'normal' phone number, is redirected to C and actually pays for a much more expensive A to C communication! If you are not convinced of the magnitude of the potential fraud, just consider that many countries in the world get over half their state revenue from revenue sharing on international phone calls by replacing C with an international phone number. . .

In the previous scenario, all phone companies charge the redirected call as follows:

- An A to B call lasting T1 + T2 seconds.
- A B to C call lasting T2 seconds.

In other words, B pays for the redirection, and A pays as if B had never redirected the call. In the case of multiple redirections, the same procedure applies. If the call is redirected to D for T3 seconds, it is charged as three separate calls:

- An A to B call lasting T1 + T2 + T3 seconds.
- A B to C call lasting T2+T3 seconds.
- A C to D call lasting T3 seconds.

This requires the telephone switches to always be able to correlate all legs of a redirected call to compute the right sums. This is not always simple, especially with voice over IP protocols.

5.1.2.3 Call loops

Call loops are a problem in any telephone network, because they can completely jam some trunks in a matter of seconds. Usually, the prevention of call loops uses machine-generated routes with automatic loop detection and a hop counter that is decremented by each telephone switch. SS7 ISUP messages have such a hop counter. In VoIP, some vendors (e.g., Cisco Systems), have added a proprietary hop counter to the H.323 LRQ message, making it possible to avoid loops between direct-mode gatekeepers. H.323v5 adds such a counter in SETUP messages. Counters are also included in SIP INVITE messages. Finally, a switch that may redirect calls must not authorize a redirection to a number, if it knows that the call has already been redirected by that number: in H.323, the number of the last redirecting party is stored in the Redirecting Number information field of the SETUP message.

Even with these improvements, call loops remain possible:

- Calls can be looped back into the VoIP network by the TDM network through SS7 gateways, in which case the counter can be lost (depending on SS7 gateway implementation).
- Edge devices connected through user interfaces (analog, ISDN) may loop calls back to the network, in which case the hop counter is always reset. This is one of the reasons the call forward service of external calls to external extensions is usually forbidden as part of the certification program of edge devices (call forward of internal calls to external extensions does not create a call loop problem).

Manual call transfer/divert does not present the same issues because it requires human intervention for each loop.

5.1.3 Reference network configurations in the PSTN

5.1.3.1 Residential services: call forward only

5.1.3.1.1 Call forward

PSTN residential services frequently provide the call forward service. The service is implemented in the last-hop class 5 switch. Most PSTN networks cannot optimize call forwards and, therefore, the last-hop switch reinitiates a new call for the redirected call and trombones the media stream.

With this configuration the issue of numbering formats is irrelevant since the redirected-to number is interpreted by the last-hop switch, using the same numbering conventions as the redirecting party.

Note that it is important to have the call forward service provided by the switch, not by the edge device, because of the call loop issue described above. If the call forward service is performed by the switch, the call hop information is not affected. It would be reset if the call forward was performed by the edge device, leading to potential call loops.[1]

5.1.3.1.2 Call transfer/divert

Residential services typically never provide the call transfer or call divert feature without strong restrictions. This is because residential users associate billing with 'my phone is off-hook'. But, as we have described above, in order to prevent fraud, when a call is transferred from a phone this phone is still charged for the transferred portion of the call, even if it is on-hook. And now the call is under control of the calling party, and can potentially last for hours or days (note that call forward is different, because it is assumed that you forward your line to yourself while away from your home, and therefore you remain in control of call duration). It is possible to imagine restricted versions of the call transfer service for residential users (e.g., to the mobile phone, fax number, or voicemail number of the served user only).

Call transfer/divert is the key feature that differentiates Centrex from residential services. Centrex users are not charged for internal calls and, therefore, are allowed to transfer calls to other internal extensions. This remains compatible with the perception, 'I don't pay if my phone is on-hook,' since the redirected portion of the call charged to the on-hook phone is a free call. Many countries required PBXs to block the transfer of calls to external numbers to get their type approval, but this is no longer strictly enforced.

Note that you can emulate a call transfer by creating a three- way A–B–C conference if the communication remains active when conference initiator B hangs up. For this reason

---

[1] Of course, the call forward service would also stop working when the phone is unplugged! Despite these problems, some badly engineered VoIP networks still use end point-based redirection today.

residential networks always require the conference to be dropped if the initiator of the conference drops the call.

5.1.3.2 Isolated PBX: forward/transfer by the PBX

5.1.3.2.1 Call forward

Because of the loop problem described above, many countries restrict the ability for PBXs to forward external calls back toward the PSTN network. Call forward to internal extensions is usually performed by the PBX itself.

5.1.3.2.2 Call transfer

When a call transfer is performed by an isolated enterprise PBX, the call transfer is always performed locally by the PBX (Transfer by join).

The billing principles described above are still valid. For instance, if a call from random user A on the phone network is received by B in the enterprise (the communication lasts T1 seconds), and then redirected to random user C on the external phone network (this redirected communication lasts an additional T2 seconds), the phone network will see a call from A to B lasting T1 + T2 seconds (since the PBX does not notify the network in any way that a call redirection has occurred), and the network sees a call from B to C initiated by the PBX lasting T2 seconds. Therefore, the billing records generated by the public network are correct:

- Call A to B for T1 + T2 seconds.
- Call B to C for T2 seconds.

Note that if the PBX hides the real extension B, the network will use the PBX main number as the calling party number for the billing record of the redirected portion of the call. This does not create any potential for fraud, as the bill still goes to the same pocket.

Figure 5.2 Transfer performed by the transferring PBX. (Transfer by join)

5.1.3.3 Networked PBXs: network-optimized call transfer

Many heterogeneous PBX networks use that call transfer by the PBX method that has been described in Section 5.1.3.2.2 (Figures 5.2). However, this method is not optimal as call media are 'tromboned' through the redirecting PBX, and this uses more bandwidth on the corporate inter-site communications lines. If all PBXs are from the same brand, or if they support the QSIG extensions of ISDN (about one-third of PBXs are QSIG-capable), there is a possibility to optimize the call flow. The redirecting PBX will send a redirection message to the source PBX, and the source PBX is expected to re-establish the call directly to the redirected-to party (Figure 5.3). Now, the usage of transmission resources is optimized, but we must solve:

- The numbering format issue: all PBXs must operate under the same numbering format or be able to convert the redirection messages to the appropriate formats.
- The billing issue: if all internal calls are on a private network (leased lines), the service provider has no billing to do if redirected calls are restricted to internal extensions (call detail record's or CDRs, for internal calls are generated directly by each PBX). If calls are redirected to the public network, the PBX will stop using QSIG and trombone the redirected call to the public network, returning us to the case of the previous paragraph (5.1.3.2, Transfer by join). If leased lines are not used, the public network may be unable to correlate the initial call and the redirected call, and the CDRs may be generated as A–B (T1 seconds), B–C (T2 seconds), instead of A–B (T1+T2 seconds), B–C (T2 seconds). This is not always a problem if the company is billed as a single entity and does not care whether A or B is billed for the redirected call. Note also that, if private accounting systems are used at each location, their records will be inaccurate, as the B to C leg of the call should be charged to site B. But, this leg of the call is now re-originated from site A and, therefore, site B has no information on it.

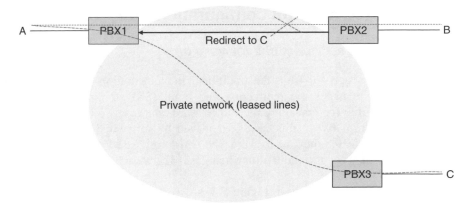

Figure 5.3 Optimized transfer.

5.1.3.4 Voice resources connected to a QSIG-capable PBX: locally optimized call transfer at the edge

In some cases (e.g., in call centers) a private installation routes calls first to a voice resource, such as an interactive voice response (IVR) server, and then the call must be redirected to another extension by the IVR. It would be a waste of resources to force the IVR to make a call to the extension and then bridge the incoming call with the new call to the extension. This would use two ports on the PBX.

Instead, most corporate IVRs are capable of sending the proper QSIG commands to the PBX, in order to ask the PBX to perform call transfer to the new extension. On analog connections, the IVR can alternatively use a DTMF command to perform call transfer. The call transfer is always performed locally by the PBX.

5.1.3.5 Network intelligent peripherals: locally optimized call transfer in the network

This is exactly the same situation as before, but within the phone service provider network. Such large-scale IVRs are called **service nodes**. When these **service nodes** need to redirect a call, sometimes they trombone the call, but some central offices also accept call transfer commands on the signaling link to the service node.

Note that there is an alternative architecture used by carriers, where carriers' central offices make use of intelligent network (IN) features and call transfer is controlled from an external **intelligent network application protocol** (**INAP**) link. In this case an external application, residing on a **service control point** (**SCP**), synchronizes call transfer and the IVR function. In this case the IVR function is called an **intelligent peripheral** (**IP**).

The **IN architecture** comes with a certain complexity, but this is the price to pay when using TDM technology to avoid tromboning as much as possible for high-volume applications, such as hosted contact centers.

5.1.4 Reference network configurations with VoIP

5.1.4.1 Residential services: call forward only

The implementation is derived from the traditional implementation in the PSTN. The edge softswitch that manages the redirecting end point (SSW_2 and end point B in Figure 5.4) is responsible for managing the call forward, and does so by initiating a new call to the redirected-to party, in this case C, managed by softswitch SSW_3. Since the softswitch is responsible for the call forward, the call loop issue is properly addressed. Figure 5.4 is an example of the multi-softswitch case where each end point is managed by a different softswitch.

Note that, although the call flow in Figure 5.4 is derived directly from the PSTN call flow, it is *much* more efficient: media flows are established directly between calling party A and redirected-to party C. Softswitch SSW_2 is only involved in the signaling path,

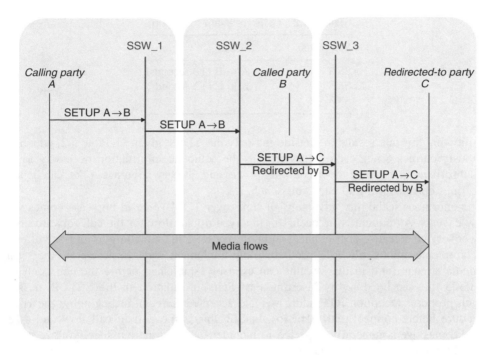

Figure 5.4 VoIP call forward scenario.

it does not handle the media stream. If SSW_2 had been a PSTN switch, the redirected call would permanently use 128 kbit/s of transmission capacity to the switch point of presence, as opposed to exactly zero in the VoIP case.

At this point it is also interesting to examine how billing should be organized. Each softswitch will probably generate its own billing records (Table 5.1).

We see that we have duplicates, and also CDRs that, depending on the softswitch support of the redirecting number information (a field in H.323 indicating the identity of the redirecting party), may be correct or not. There is a simple way to extract the correct information for these CDRs, following this simple rule:

For billing purposes, on softswitch SSW_i, keep only CDRs where the calling party belongs to the SSW_i zone.

Table 5.1 CDRs generated by a call forward in a multi softswitch environment

Softswitch	CDRs
SSW_ 1	A to B (T2 seconds)
SSW_ 2	A to B (T2 seconds)
	B to C (T2 seconds)
SSW_ 3	A to C, redirected by B (T2 seconds)

Table 5.2 CDRs relevant for billing

Softswitch	CDRs
SSW_ 1	A to B (T2 seconds)
SSW_ 2	B to C (T2 seconds)
SSW_ 3	None

Following this rule results in considering only the CDRs: given in Table 5.2, which are obviously correct. Since each softswitch has the complete information necessary to bill the subscribers in its zone, the accounting processing, in very large networks, can be split into zones corresponding to each softswitch.

The enormous scalability advantage of VoIP over TDM voice in this case comes with a few caveats. A frequent issue occurs in this type of call flow for the call forward on no answer service. Because post-connect audio delays were the most important issue of early VoIP implementations, all VoIP protocols now implement ways to accelerate the set-up of media streams, and media streams can even be established *before* the call connects. In H.323 this can be done by inserting Fast Start information in the SETUP message (which proposes reception RTP addresses and accepted codecs), or beginning the H.245 procedure before connect (early H.245). In SIP, this is the default call flow and a SIP INVITE message is generally expected to include SDP information specifying reception RTP addresses and accepted codecs.

If this type of optimized call flow is used in the case of call forward on no answer, then A and B will establish media streams before B picks up the phone (B can be an IP phone or a PSTN number behind a gateway). But after the no answer timeout, the call will be redirected to C, and C will also start streaming audio toward A. This produces a variety of results, ranging from A crashing to A playing audio toward B instead of C. If the softswitch does not control the beginning of media streams, call forward on no answer will not work properly. Note that IP phones are usually smart enough not to start streaming audio until someone picks up the handset; so, the issue is really only with end points behind analog gateways and for audio streams from A to B instead of C.

What is the correct approach? The softswitch should delay all pre-connect media information from phones that have the switch-based call forward on no answer service activated until the phone actually connects. In H.323 this means the softswitch will capture the Fast Start information inserted by B in the CALL PROCEEDING or ALERTING message, and forward it to A only if B sends back a CONNECT message. Otherwise, if B does not answer, the softswitch will redirect the call to end point C and do the same with end point C. In SIP this means delaying SDP information until the 200 OK message. Some phones do this properly anyway, but it is a lot safer if the softswitch plans ahead and is prepared to delay any pre-connect information it receives; in particular, if B is behind a PSTN gateway. There is still the possibility of B starting to stream audio toward A before CONNECT (bad IP phone implementation, or a gateway which does not know that it is connecting a call to an end-user and has no support for the in-band audio indicators of the PROGRESS message in ISUP/ISDN). In H.323 the softswitch can explicitly request the gateways not to start streaming audio before CONNECT by positioning the MediaWaitUntilConnect parameter of the H.323 SETUP message.

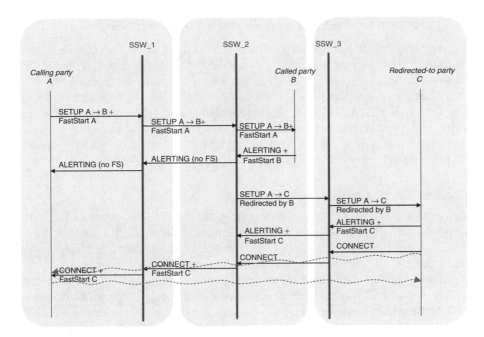

Figure 5.5 VoIP call forward on no answer scenario.

This procedure now works correctly with the call forward on no answer service, and does not create any delay in post-connect audio (Figure 5.5):

- End point B has received the RTP transport addresses that can be used to stream audio toward A in the SETUP/INVITE message. So, as soon as B picks up the handset, 'Hello' can be transmitted toward A.
- End point A will stop playing ring-back tones as soon as it receives the CONNECT/200 OK message from B. In the same message it receives the ports to start talking with B, and therefore can answer B immediately.

The softswitch should therefore use the following rules to handle pre-connect media information:

- If it does not implement the call forward on no answer service for the destination, forward the media information 'as is'.
- If it implements the call forward on no answer service for the destination, delay the media information from the destination end point until CONNECT. In H.323 also position the MediaWaitUntilConnect parameter in the SETUP message.

These rules correctly enable the forwarding of pre-connect audio when calling PSTN destinations (congestion messages, some prepaid calling card services that send back a CONNECT only after the final destination connects, etc.), but prevents any issues when calling end points that have activated the call forward on no answer service.

5.1.4.2 Isolated IP-PBX: forward/transfer by the IP-PBX

This is one of the call flows that has generated the most confusion in VoIP, because many people think of H.450 as *the* way to do call forwards and transfers in H.323, and REFER (or the deprecated BYE/ALSO method) as *the* way to do call transfers in SIP.

This is not the case at all. In fact, the situation is no different than that in the PSTN: there is QSIG (H.450 and REFER are QSIG equivalents in VoIP), targeted at private networks, and there is end point-controlled transfer. End point-controlled transfer also exists in VoIP, but it is a lot better than in traditional telephony, because there is no media tromboning.

5.1.4.2.1 Call forward

The call forward service is usually provided by the IP-PBX. Because of the call loop issue, call forward of external calls should be limited to internal PBX extensions. In any case the PBX must prevent calls from being forwarded to a destination if this destination already appears in the 'redirecting party number' field of the incoming call.

In principle, VoIP can work around this limitation by having the PBX and the softswitch collaborate for the call forward service to external extensions. If the softswitch to which the PBX is connected supports the call forward service (e.g., if it supports the Call Processing Language), in principle the forward service of external calls can also be provided by the edge softswitch, as in the residential case, without creating a call loop issue. This softswitch-based call forward will not work for calls coming from internal extensions; so, the PBX should also forward internal calls to the external extensions, which does not create call loop issues.

5.1.4.2.2 End point-controlled transfer with media optimization in H.323: NullCapabilitySet

In H.323 the call flow of Figure 5.6 can be used by any end point to control a call transfer and optimize the media path. In our example, A is the calling party, SSW_1 is the softswitch controlling end point A, B is the redirecting party (an IP-PBX in our case), and C is the redirected-to party. B is managed by SSW_2 and C is managed by SSW_3.

The call flow employs the third-party media control procedure, using a call flow known as **third-party-initiated pause and rerouting** or **NullTerminalCapabilitySet** (in short, **TCS = 0**), described in paragraph 8.4.6 of H.323v4. *All* H.323v2 (and above) end points are required to support receiving **TCS = 0** and correctly redirecting media streams as they are given a new CapabilitySet from the remote party. This is one of the most important tests to do on implementations that claim to be H.323-compliant. Not supporting this call flow is a major bug which does not allow the equipment to be connected on any H.323 network that implements transfer services (even if it is just other end points that need to do call transfers).

In the call flows in Figure 5.6 we have not represented the individual phones; the phone and the IP-PBX are represented as a single entity. Therefore, A (respectively, B and C) represents both the A (respectively, B and C) phone and the PBX-handling phone A (respectively, B and C). The method that is used by the phone to signal to the PBX that it is willing to perform a call transfer is not shown (it will be discussed in Section 5.1.5).

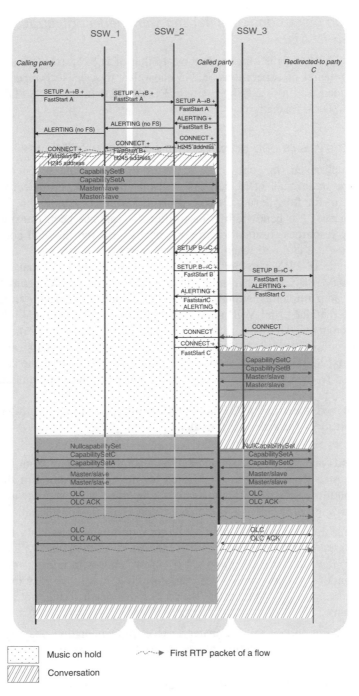

Figure 5.6 H.323 third-party-initiated pause and rerouting (controlled transfer with consultation). Most ACK PDUs are not shown and all messages are in sequence for clarity.

5.1.4.2.3   End point-controlled transfer with media optimization in SIP RE-INVITE

The call flow (Figure 5.7) is almost identical to that of H.323, except that since SIP has no formal negotiation of capabilities, H.245 capability messages disappear. The NullCapabilitySet sequence is replaced by a single RE-INVITE (a new INVITE for the same call, with new SDP information).

5.1.4.2.4   Analysis of end point-controlled transfer call flow

As already emphasized, the only real drawback to having the transfer performed by the PBX in the traditional TDM network was that the PBX permanently had to relay (trombone) the media flow during the call, using 128 kbit/s of bandwidth. With VoIP, this problem disappears! No bandwidth is used after the call has been transferred and very few processing resources (limited to copying, unchanged, the few call control messages that occur during the call and when the call terminates).

Note that some IP-PBX vendors only did a very superficial 'IP make-up' on top of a traditional TDM PBX core, by simply adding VoIP to the TDM board on their PBX chassis. These poor implementations are very hard to detect—only the transfer call flow will really differentiate 'true' VoIP implementations from quick tactical marketing adaptations of old products. These poor implementations will actually relay media streams for the entire duration of the call because the switching itself still occurs on the old TDM switching matrix! Some IP-PBXs are optimized for IP phone to IP phone calls, but not for calls transferred from one PBX to another PBX. Such obsolete implementations should be avoided.

Having removed the tromboning issue, the NullCapabilitySet/RE-INVITE call flow becomes almost ideal for a service provider:

- Billing records are correct (the second call appears to be coming from B to C, the first call is from A to B and remains established for the entire duration of the call). In the context of multiple softswitches, the same rule applies: at each softswitch, only CDRs where the calling party belongs to the softswitch zone are considered.

- Media optimization (anti-tromboning) does *not* require any optionality in the standards and is supported by all end points (as it is mandatory in the H.323v2 and SIP baseline standard). In fact, this is well supported by most vendors in current implementations. Even if a vendor does not comply with this call flow, it is fairly easy to get the vendor to support it because it is nothing less than a bug that needs fixing, not an enhancement to the standard.

- The PBX remains in control of transfer service implementation, leaving a lot of flexibility for PBX vendors to offer enhancements to the transfer service (e.g., personal music on hold).

- The PBX keeps control of the B to C call for which it is charged. If the PBX is equipped with an accounting system, the accounting system will correctly keep track of the duration of the redirected call. If the PBX crashes or becomes disconnected, all transferred calls are released. This is in fact desirable as the PBX (B) is paying for the

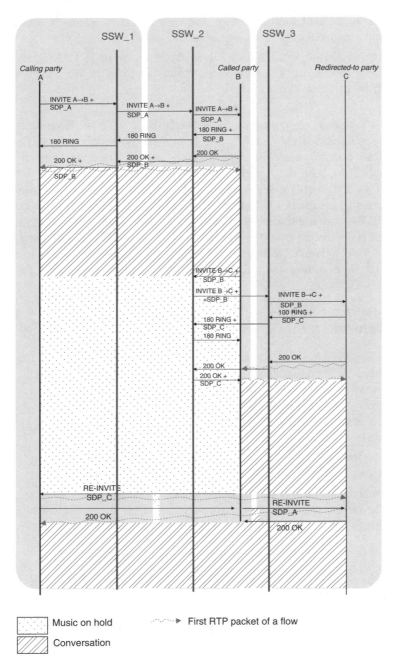

Figure 5.7 Third-Party-controlled transfer with consultation using SIP RE-INVITE. *Most ACK PDUs are not shown and all messages are in sequence for clarity.*

B to C call, so the call should be torn down if the control element is unable to keep track of and account for it.

- The numbering format issue is also solved, because the edge PBX is responsible for understanding the format of the redirected-to number from the redirecting party (which happens to be its local format).

This call flow is already supported by the major IP-PBX vendors. Because it leaves a lot of room for vendor added value, it has also been used by some vendors to enhance their PBX-based call center implementations to allow a centralized ACD to distribute calls across multiple sites, with no impact on network usage!

This call flow also works fine for VPNs and multi-site implementations, and does not require having all sites using PBXs from the same manufacturer: you can combine some sites with PBX-based implementations and some sites with Centrex implementations, where the call is controlled from the network. In fact, the ease of deployment of advanced services in a VoIP network, without tromboning, is probably the single most important reason that should convince multi-site companies to replace their TDM PBXs with IP-based PBXs or Centrex.

5.1.4.3 Networked PBXs: network-optimized call transfer

In the TDM world the key driver for implementing optimized procedures based on QSIG was to optimize the bandwidth usage on each link for closed user groups spread over multiple sites. With VoIP, this is no longer an issue, as we saw in Section 5.1.4.2.

So, there really are only two reasons for using something equivalent to QSIG transfer in VoIP (such as H.450 or SIP REFER):

- To solve the tromboning issues of poor IP-PBX implementations which still trombone media streams. The best solution is not to use them in the first place!
- To avoid relaying signaling streams through the redirecting PBX. At first glance, this seems to be a good idea, but, as we will see, sometimes it can backfire: this attempt to get the PBX to handle a couple sewer signaling messages triggers a whole lot of much more serious issues.

That being said, there is still a use for H.450/SIP REFER: this is to allow IP phones using the H.323 or SIP protocol to signal to the PBX that they are willing to perform a call transfer. This will be discussed in Section 5.1.5.

Let us describe the H.450/SIP REFER source-based call transfer call flows anyway.

5.1.4.3.1 Call flow with H.450.2

The H.450.2 implementation of call transfer, because of the QSIG heritage, is well defined and there is a reasonable level of interoperability between vendors. Call flow is explained in detail in Section 5.1.4.3.3 and summarized in Figure 5.8. H.450.2 explicitly notifies transferred-to end point C that a call transfer service is being requested. In response,

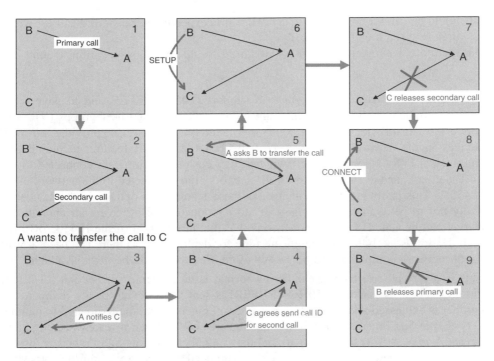

Figure 5.8 H.450.2 consultation call transfer call flow.

transferred-to end point C gives transferring end point A a reference for the call to be transferred, which is passed to transferred-to end point B. The call reference is explicitly included by transferred-to end point B in the new call generated to transferred-to end point C.

5.1.4.3.2   Call flow with SIP REFER

The SIP implementation is almost identical to the H.450.2 implementation. The REFER message tells the calling end point to make a new call to C. The exact implementation is still, at the time of writing (Q2 2003), in a state of flux, and really only works after some vendor-to-vendor tuning. Details such as which phone releases the transferred call and when, how the transferred call notifies the transferred-to party that this is a replacement for the previous call, how the transferring phone is notified of the successful transfer, etc. are still not stable. A sample call flow between Cisco SIP phones is given in the Annex.

5.1.4.3.3   Analysis of the H.323/H.450/SIP REFER call flow in public networks

As already discussed, there is really no fundamental advantage to using this procedure in the core network. It does not even save signaling messages as the H.450 activation itself requires a significant set of new messages and adds a lot of complexity to end point implementation (e.g., when implementing the second call presentation service, the end point should be careful not to accept any call if it is expecting a transferred call for which it has already given a call reference).

But there are a number of major issues raised by this call flow regarding public networks:

- It requires, obviously, all end points to support H.450.2/REFER and to have interoperable implementations. This may not be an issue in closed user groups (VPNs or networks of IP-PBXs), but this becomes a critical issue for a service provider: *all* VoIP devices in the network need to support the H.450.2/REFER feature, including SS7 call agents, IVRs, application softswitches, any customer equipment connected to the VoIP network, and even third-party VoIP networks interconnected to the network using H.450.2. This is obviously a challenge, as these features are options in the standards, likely not to be available by default in most products, and requiring an upgrade fee if they can be added. The consequence is that the introduction of the H.450.2/REFER optimization is likely to take significant time for deployment, as there is a need to wait for all network components to be ready before the service can be made available.

- This is a typical case for the numbering format issue explained in Section 5.1.2.1. If the redirection message, H.450.2 or REFER, is sent from device B, the redirected-to number is expressed relative to B. If B is in San Francisco and the redirected-to number C is also in San Francisco, it will not use the area code. When it reaches calling party device A in New York, A will try to dial the C number using its San Francisco format, and obviously the call will fail because it originated in New York and the softswitch handling New York does not understand the San Francisco format. The issue is likely to occur even in closed user groups (e.g., VPNs), when each site uses a different escape code to dial other sites using short numbers (e.g., 1–1010, for extension 1010 on site 1, where prefix '1' may have a different meaning at each site).

- If the redirected call is to another PBX C, the redirecting PBX B loses control of the redirected call, while it is still charged for the redirected portion of the call. The billing records generated by the local billing system of B will be inaccurate.

- On this call flow the redirected call naturally appears to the network elements as an A to C call. If nothing is done, the billing problem explained in Section 5.1.2.2 occurs. Instead of billing the call as an A to B call for T1 + T2 seconds and a B to C call for T2 seconds, the network will bill the call as an A to B call for T1 seconds, then a B to C call for T2 seconds, which is wrong. The network therefore needs to have some form of correlation between the A to B and A to C calls which indicates it to be a redirected call. Unfortunately, the first versions of H.450.2 overlooked this issue and the problem is now present in many implementations of H.450.2. In 2000 the H.323 implementer guides started introducing and documenting the usage of a new feature called 'call linkage' (using a new identifier called 'ThreadID'), which is used to associate one call with another and must be used in H.323v4. Unfortunately, this adds to the difficulty of introducing this feature in a multi-vendor network, as not only H.450, but the correct version of it, must be implemented everywhere. Most carriers' business plans require integration with as many PBX brands as possible, which makes this issue a critical one.

- If multiple softswitches control the network, the initial A to B call and the subsequent A to C call after redirection will be routed through a different set of softswitches.

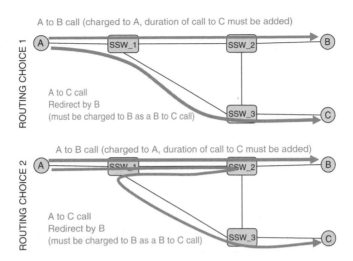

Figure 5.9 Routing dilemma in a multi-softswitch network with source-based rerouting.

There are two possible routing choices, the simplest (choice 1 in Figure 5.9) is to route the A to C redirected portion of the call as if it was a normal call. In this case the problem is that the CDRs required to charge B, instead of being generated only by SSW_2 which is managing end point B, will be potentially generated by all softswitches in the network, requiring complex CDR reconciliation across all softswitches. This prevents any possibility to scale the billing system by isolating independent regions in the network corresponding to each softswitch. A further issue is that some services (e.g., legal call interception) are likely to be located in each regional softswitch for the subscribers in its zone; so, they would require complex inter-softswitch synchronization in order to be implemented in such a network. The second choice is to always reroute the redirected portion of the call along the same path as the initial A to B call, but then to send it back to the first softswitch for optimization of the routing. This call flow allows the generation of billing records to be segmented properly and facilitates legal interception. But, it also requires three times as many ports as other implementations, and really gets ugly if a call is transferred multiple times. Another issue is that B may be allowed to call (and therefore transfer) calls to destinations unreachable from A. In order to fix this, SSW_1 would need to be aware of all the restrictions applicable to B!

- A last issue, probably the most serious issue in the context of public networks, relates to the reliability of billing. For proper billing, it is important to be able to correlate the A to B and the redirected A to C calls. Resolution of this issue using end-to-end H.450.2 depends on the redirected end point properly implementing the correlation field in the second SETUP message. This is the classical security issue known as 'third-party dependence'. In short, if you want your billing system to work properly, you need everyone in the network, all your customers, all your network peers, not just your own equipment, to include linkage information correctly. Obviously, whether the intent is malicious or just derives from unwanted situations (imagine a new firewall installed by a customer sitting between the IP-PBX and the network which does not properly reconstruct H.323 or SIP messages with all enhancements, such as call linkage information), anyone in the network can cause the network to generate wrong CDRs. All service providers have had some experience of conflicts and trust issues with customers related to billing, and will want to stay as far as possible from this situation.

Overall, having so little to gain from the H.450/REFER implementation in a public network and so much to lose, except in very specific situations, we discourage the use of H.450.2 or SIP REFER in the core network, and reserve it for very specific cases only. Obviously, H.450.2 and SIP REFER can still be used by an IP phone to signal the intention to make a call transfer to an IP-PBX; issues only occur with the use of H.450.2 or SIP REFER between customer devices and the network.

If the situation of having edge devices, using H.450/REFER redirection toward the network, cannot be avoided in an open environment, then the network softswitch should intercept the H.450/REFER message and execute the transfer locally using the NullCapabilitySet/RE-INVITE method. This works fine, but is a very intricate task for a core softswitch, actually requiring Centrex-type capabilities on the softswitch (Centrex is primarily defined by the availability of call transfer in the feature set of the switch).

5.1.4.4 Voice resources connected to a H.450/SIP REFER-capable PBX: locally optimized call transfer at the edge

This is the same as the previously described case on TDM PBXs. It works fine and does not create any of the issues discussed in Section 5.1.4.3, because this is only a local call flow.

However, since this type of call flow was primarily intended to avoid media tromboning, and there is no media tromboning in VoIP if it is properly implemented, using this optimized call flow is a lot less useful than in the TDM world, unless the voice resource or the IP-PBX are not truly native VoIP implementations (e.g., those built on TDM cores using VoIP to TDM boards to connect to IP-PBXs). In most cases, the simple call flow, where the IVR bridges the two half-calls and uses TCS $= 0$ or RE-INVITE to optimize the media path, is just as good and will cause fewer interoperability issues.

5.1.4.5 Network intelligent peripherals: locally optimized call transfer in the network

This is the same situation as above, but transposed into the public network. Again, the call flow works fine and does not create any of the numerous issues it creates when used between a user space device and a network device, but all this added complexity brings little benefit to VoIP networks.

It is expected that the service node model in VoIP networks (with the voice resource doing the call switching internally) will prevail over the intelligent network model (where the intelligent peripheral never switches the call, and only streams voice prompts). This would reverse the current situation on TDM networks, where the intelligent peripheral model dominates.

5.1.5 How to signal call transfer?

In the previous sections we have discussed in detail how to execute a call transfer. Of course, before executing a call transfer, some convention must be in place to signal to the switch executing the call transfer that the user is willing to transfer a call.

5.1.5.1 H.323 or SIP phones and residential gateways

Any device connected to a public network (IP phone, CPE gateway, etc.), if it is capable of transferring calls, must use the TCS = 0/RE-INVITE method and remain in control of the entire call during the transfer (this is the equivalent of multi-line high-end analog phones sold to small professionals). As explained in Section 5.1.3.1.1 this is because the call transfer function is never offered by the residential service provider (to avoid fraud). Therefore, transfer must be performed under the responsibility of the end point; obviously, the user interface to control this is entirely up to the end point. Note that in many countries it is even illegal to implement this service in a phone, and transfer is never offered by a public network to non-Centrex users.

When an H.323 or SIP phone is connected to an IP-PBX, the best solution is to use H.450/REFER messages to signal the transfer to the switch. Note that the switch itself can use the TCS = 0/RE-INVITE call flow toward the public network to EXECUTE the call transfer, or simply forward the H.450/REFER message to another PBX in a closed user group. The way the end point asks for a call transfer and the way the switch executes it are completely independent.

An alternative way of asking for a call transfer would be via DTMF tones (star key type of transfer code). In this case the switch would need to locally accumulate and analyse DTMF sequences. Although this is possible in theory, most SIP and H.323 IP phones have specific 'transfer' keys and use the H.450.2/REFER message to ask the switch to perform the transfer, in order to avoid having to make the appropriate transfer DTMF sequences for a given switch available in the phone. Note that choosing DTMF activation also necessitates having a specific escape sequence when transparent DTMF is required (e.g., when entering information into an interactive voice response server).

For H.323 or SIP residential gateways, the situation is a bit different and some vendors choose to let the IP-PBX perform the call transfer by analysing the DTMF digits, while other vendors analyse the DTMF digits locally and send the transfer requests to the IP-PBX using H.450/REFER.

5.1.5.2 MGCP phones and residential gateways

In a public network the transfer service is typical of Centrex offerings. In many ways, MGCP phones and CPE gateways are better suited than SIP or H.323 at offering a Centrex type of service. MGCP makes fewer assumptions on the specific logic implemented in the phone and, therefore, is more manageable in a multi-vendor environment, making it straightforward to implement new services in a homogeneous way. In addition, MGCP offers more possibilities by signaling phone-related events even before the call is active (e.g., the off-hook event), enabling such features as the announcement of the number of pending voicemail messages as soon as the handset is picked up. Finally, MGCP is now a very mature protocol, with dozens of vendors and the endorsement of large organizations, such as PacketCable® (the organization that also drives the DOCSIS standard).

The most natural way of using MGCP (with an event package line) to signal call transfers is to use DTMF activation sequences, using MGCP digit maps. This works well across all vendors of IP phones and residential gateways.

Most IP phone vendors also offer shortcut keys for the most common call actions (three-way, hold, transfer, etc.). These phone vendors allow the call agent to associate an MGCP event with each phone key. This event can be parsed by the call agent and trigger a specific action (e.g., transfer). Because of this, seen from the user's perspective, the phone behavior is identical to an H.323 or SIP phone, except that it offers more services while off-hook and can be managed much more easily by the service provider.

Once the MGCP call agent has been notified that the phone is ready to perform a call transfer (if the transferred party is not controlled by the same call agent) the call agent will use the appropriate call flow on the network side (TCS = 0 or SIP REFER), depending on the inter-softswitch protocol it uses (H.323 or SIP). The call agent may use MGCP or H.323/SIP to establish the call to the transferred-to party, depending on whether the transferred-to party is on the same MGCP call agent or not.

The call flow illustrated in Figure 5.10 is an example showing end point B managed by an MGCP call agent, using H.323 as an inter-call agent protocol. The calling party, as well as the redirected-to party, are on separate switches in the example. In Figure 5.10 the music on hold is played by the call agent (it could also be streamed by an external announcement server controlled by the call agent).

5.1.6 VoIP call redirection and call routing

5.1.6.1 Call redirection and routing in traditional voice

A TDM switch routes calls according to a route table. Most switches are able to route calls not only according to the call destination, but also according to the source of the call. In case there are multiple route tables, if a call from A to B is redirected to C, then the route table attached to call originator B will be used to route the redirected call to C.

5.1.6.2 Call redirection and routing in VoIP

What was obvious in traditional voice now becomes very tricky in VoIP. The softswitch is now very likely to have multiple route tables according to the source of the call. This is because a softswitch can potentially control end points located anywhere on the planet and

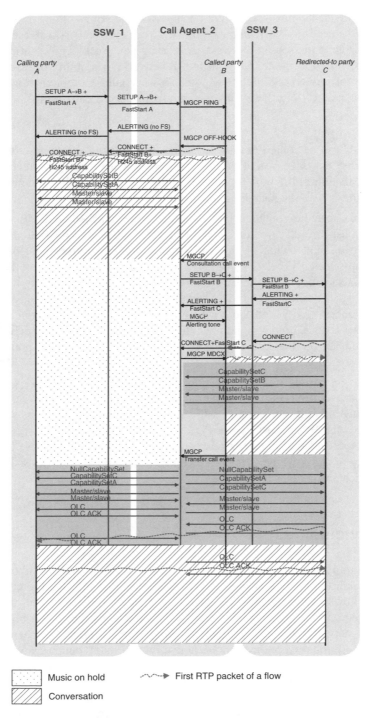

Figure 5.10 Transfer triggered by an MGCP end point, executed using the H.323 NullCapabilitySet (MGCP PDUs not detailed). Most ACK PDUs are not shown and all messages are in sequence for clarity.

in real-life deployments most softswitches control the end points of an entire country. The best route (e.g., a next-hop VoIP gateway) to New York may be very different whether you are located in San Francisco on the west coast or in New York! The criteria for selecting a route may be linked to the topology of the IP network, availability of resources in each region, quality of service considerations, etc. This becomes a problem when considering the case of call redirections. Two choices are possible for a call from A to B redirected to C:

- The call to C can be routed exactly as in the PSTN, using the route table attached to B. In this case the issue is that the media streams will in fact go directly between A and C and, therefore, the best route, if B is the source of media streams, may not be the best route if A is the source of media streams.
- The call to C can be routed using the route table attached to A. In this case the softswitch must carefully execute all the features, such as call restriction, that must be linked to B, the real owner of this call from an administrative point of view. Now, the next hop selected is likely to be the optimum choice because A is the real source of the media streams. Nevertheless, this is not always a perfect solution as B may be allowed to reach destinations that A cannot reach. For instance, most countries have premium numbers that cannot be reached from abroad. If A calls from abroad and is redirected by B to a premium number the call should work because B can call the premium number, but the softswitch is likely to have no route to the premium number attached to source A.

There is no ideal solution. In any complex deployment, involving routes spanning across countries, virtual private networks, or with strict constraints related to quality of service and the underlying IP network topology, one of the two approaches described above must be selected, and the related issues need to be addressed on a case-by-case basis. In the majority of cases we nevertheless recommend using whenever possible traditional networks, which are simpler to manage.

5.1.7   Conclusion

There is no question that the ability of VoIP to redirect calls without tromboning media is the major breakthrough of the technology. Many issues that justified the development of complex call flows in the TDM world have now been solved with much more elegance by more simple and robust VoIP call flows. This advantage alone is a big incentive to migrate TDM networks to VoIP technology: the redirect service has been presented in a Centrex or corporate telephony context above, but in many countries VoIP will be an interesting solution to the issue of local number portability (LNP), which often uses the call forward service ('onward routing' technique).

As we have seen, various call flows and technologies can potentially be used both to request a call transfer and to execute a call transfer. The complex, source-based call transfer protocols, such as SIP REFER or H.450.2, are really useful only in the corporate space to allow IP phones to request a transfer to the switch, but in the network they create many more issues than they solve when also used to execute the transfer. With VoIP, the simplest transfer call flow (using TCS = 0, or SIP RE-INVITE) is also the best, because tromboning has been removed.

Most of the call flows discussed above also demonstrate the similarity of H.323 and SIP, both protocols are identical in design and performance when deployed as a core network protocol. H.323 and SIP also present the same issues when it comes to the deployment of Centrex-type services or the control of IP phones. As already demonstrated in the TDM world where all PBXs use a stimulus protocol to control IP phones, a protocol that does not need to make assumptions about the built-in logic of a phone or a CPE is much easier to deploy and operate. In VoIP the same advantages are obtained by using MGCP (a stimulus protocol) as the protocol for controlling Centrex end points at the edge of the network. Putting all these remarks together, we can summarize as follows:

- In core service provider networks, using SIP or H.323, support of the mandatory options of the standards that cover the for dynamic redirection of media streams (TCS = 0, RE-INVITE) is sufficient to support all the redirection services already deployed in the TDM world, but much more efficiently.

- Regarding phones managed directly by a public infrastructure, H.323 or SIP are fine at the edge for IP phones or gateways providing residential telephony. But if Centrex features, among which is mid-call transfer, are also to be provided, the MGCP package line is a better choice at the edge to control IP phones and residential gateways. Of course, both solutions can be implemented at the same time on a network, and in all cases H.323/SIP remains the core protocol.

- If corporate PBXs are to be connected to the core public network, using the same call flows as the current TDM network (local transfer by the PBX) is the only practical solution because of the third-party dependence issue. H.450 or REFER can only be used in closed user groups, and even then do not bring any significant advantage when IP-PBXs are implemented properly and are capable of media anti-tromboning.

6

Advanced Topics: NAT Traversal

6.1 INTRODUCTION TO NETWORK ADDRESS TRANSLATION

6.1.1 One-to-one NAT

Network Address Translation (NAT) was initially used to protect corporate networks from people attempting to access internal networks from the Internet. Many corporations decided to use private addresses internally (e.g., addresses in the 10.X.X.X range): such addresses cannot be routed on the public Internet and therefore it is impossible to send a packet to a private address through the Internet. Of course certain computers still need to be able to receive packets from the Internet (e.g., email servers). Such computers are given a public (also called routable) address on the Internet, and the site router or firewall has to translate this public address to the private address of the server on the fly.

For instance, in Figure 6.1 the computer with internal IP address 10.3.0.4 is a mail server and needs to receive traffic from the Internet. Its private IP address is mapped to the public address 162.167.3.14 on the Internet, and therefore any packet sent to this address on the Internet will reach the site access router, which will translate the destination address to 10.3.0.4.

Computers that need to send information to the Internet with a protocol that needs to receive an acknowledge that the packets have been properly received (e.g., TCP) also need to have a mapping to a public address, otherwise the acknowledge message cannot reach the sender from the Internet. This is the case for computer 10.3.0.2. In Figure 6.1 it is sending a packet to a computer at 1.2.3.4. The source address of this packet is translated from 10.3.0.2 to 162.167.3.12 by the router, so that if the receiving computer

IP Telephony   O. Hersent, J.P. Petit, D. Gurle
© 2005 John Wiley & Sons, Ltd   ISBN: 0-470-02359-7

Figure 6.1   One-to-one NAT.

needs to send an acknowledge message, it will be able to do so by sending it to IP address 162.167.3.12. The computer 10.3.0.3 has not been given any mapping to a public address, and therefore cannot be reached from the Internet. This also saves public IP addresses since only servers and computers reaching the Internet need them.

One-to-one NAT (in short, NAT; though NAT is commonly mistaken for NAPT) essentially gives to a set of computers a 'shadow' image on the Internet with a public address. NAT works with any protocol. In Figure 6.1 we inserted port numbers x, y, z, t in the IP packets, but NAT works just as well for protocols that do not have port numbers.

6.1.2   NAPT

With one-to-one NAT, each computer accessing the Internet needs to have one public address. With the advent of the WWW, this quickly became inconvenient as virtually anyone can browse the Internet, and this would require either an HTTP proxy that relays all queries to the Internet on behalf of the users, or a public IP address mapping for everyone.

The Network Address and Port Translation (NAPT) technique makes it possible for multiple users to access the Internet without an application-level proxy, for all applications that use transport protocols that have a port information to characterize a connection in addition to the IP address (e.g., UDP and TCP).

6.1.2.1   Full-cone NAPT

Full-cone NAPT is the most common implementation. In this implementation each outgoing stream of UDP or TCP data from a given IP address and port, *irrespective of its final destination*, is allocated a port on the router's public address. In Figure 6.2 the router public address in 162.167.3.1, and the router has allocated port 2002 for HTTP/TCP communication which sends packets from computer 10.3.0.2, port 3210. The router has allocated port 2004 for SMTP/TCP communication which sends packets from host 10.3.0.4, port 5678. The port (2002, 2004) is used as an index in the NAPT table by the router when it receives IP packets at IP address 162.167.3.1 and needs to forward them to the appropriate internal IP address and port. Each NAPT entry creates a **pinhole** through the router for incoming packets from the Internet and forwards these packets to the proper host.

In the case of TCP the NAPT entry is created in the router when the first TCP segment is sent from the internal host to the Internet, and deleted when the TCP communication is closed (FIN, FIN_ACK packets), or after a very long timeout (if one of the two computers crashes).

NAPT also works with UDP-based communications if the communication protocol responds to UDP packets received from a given IP address and port by sending response packets back to the *exact* same IP address and port. The NAPT entry in the router is then created when the first UDP packet is sent out by the internal host to the Internet, and remains for a relatively short period (about 30 s, typically). This period is extended each time a packet is sent or received corresponding to this NAPT entry.

As already mentioned, a mapping entry in the NAPT table is created for each source IP and port, irrespective of the destination. This means that if host 10.3.0.2 reuses port 3210 to communicate with port 7000 on host 1.2.3.4, or even to another computer altogether, the same entry will be used. Therefore, the NAPT entry maps to a 'full cone' of connections, which all originate at the same internal IP address and port. This property is

Figure 6.2 Full-cone NAPT.

very important for some NAT workaround methods, such as **Simple Traversal of UDP Through Network Address Translators (STUN)**.

6.1.2.2 Strict NAPT

Strict NAPT is used by some firewalls to prevent hosts on the internet from using the pinhole opened by NAPT entries, which could be used by a malicious user to send IP packets to internal computers. Let's assume that someone discovers a 'killer TCP segment or UDP packet' that crashes a computer or allows someone to take control of it. In the case of full-cone NAT, a malicious user knows that, since many people are browsing the Internet, many NAPT entries are active in the router. It is relatively easy to discover the public IP address of the router (e.g., looking at DNS entries): the attacker will then send the malicious packet to all ports at that address. At each port that corresponds to an active NAPT entry, the router will forward the packet to the internal host, leading to a successful attack. This works because full-cone NAPT does not check that the source IP address and port of the packet is indeed a server in active communication with an internal host.

Now, this is not as bad as it seems: there are random serial numbers in TCP that are very hard to guess, and a full-cone implementation, which would check these numbers, is not subject to such an attack. On UDP, however, unless the NAPT function is aware of the higher level protocol properties, the attack will work. There are few applications that succumb to potentially malicious instructions over UDP, but they do exist (e.g., in 2003 a virus successfully exploited a hole in the UDP-based communication ports of Microsoft's SQL server).

Strict NAPT creates one NAPT entry for each destination IP address (or even port); therefore, the pinhole is only opened for packets coming from this IP address on port. Figure 6.3 shows the entries created in the case of two communication channels opened from the same port on host 10.3.0.2.

The terminology 'partial/restricted cone' is sometimes used to refer to NAPT implementations which check only the source IP address of packets sent to the private domain, while 'symmetric cone' applies to NAPT implementations which check both the IP address and port of received packets.

6.1.3 Issues with NAT and NAPT

NAT and NAPT both break protocols that use multiple communication channels and transmit the IP addresses of these communication channels in applicative messages. For instance, SIP, H.323, and MGCP all use one communication channel for call control and several RTP communication channels for the media. The IP addresses of the RTP communication channels are transmitted on the call control communication channel.

In the example network of Figure 6.3, if computer 10.3.0.2 opens a VoIP communication with a computer on the Internet using a SIP INVITE or an H.323 SETUP with 'Fast-Connect', it will advertise one or more RTP reception ports on IP address 10.3.0.2. If the remote computer attempts to send RTP packets to this address, these packets will be dropped by the first router on the path, because they correspond to private addresses and

Figure 6.3 Strict NAPT.

cannot be routed on the Internet. The RTP packets in the other direction (from 10.3.0.2) will get through if the remote computer is on the Internet, and therefore we end up with a half-duplex conversation.

NAPT presents another problem: it cannot work with servers on a private network. A server is a computer that receives connections from clients' machines (e.g., a web server). Since NAPT creates forwarding entries only when packets are sent from the internal network to the Internet, there will be no active entry in the case of a new incoming connection, and it will fail. The router would need to know where to route these new incoming connections.

One limited workaround is called **port forwarding** in which incoming connections use the TCP or UDP protocol to the router IP address and a given port are forwarded by the router to a given internal host. Since most servers use a well-known port, this enables the use of exactly one instance of each type of server on the internal network.

Unfortunately, an IP phone is a server, since it receives phone calls. Even worse, if there are multiple IP phones in the internal LAN, we have a situation of multiple servers of the same type, and port forwarding will not work.

6.2 WORKAROUNDS FOR VOIP WHEN THE NETWORK CANNOT BE CONTROLLED

6.2.1 Ringing the proper phone

Reaching an IP phone behind a one to one NAT function is not a problem if each IP phone has a public IP address mapping and the call setup message can be forwarded to the corresponding internal IP address. If the phone registers dynamically, either the phone must be capable of advertising the external IP address or the gatekeeper/registrar/call agent must be provided with a translation table.

The problem is more difficult in the case of NAPT. If there is a single IP phone behind a NAPT function, then port forwarding can be used to route incoming call setup messages to that phone (in the case or H.323, TCP port 1720 should be mapped to the IP phone, whereas. In the case of SIP over UDP, UDP port 5060 should be mapped to the IP phone). This solves the problem of being able to 'ring' the phone even when it is behind a NAPT function. This solution immediately extends to the case of an analog gateway with multiple telephone lines, since a single IP address is used for all the lines.

The case of multiple IP phones can be supported if the IP phones have configurable non-standard call-signaling ports (this is rarely possible), in which case each port can be mapped to a given phone.

The problem can be solved with another approach (using the pinhole maintained by the NAPT function) if the phone can maintain a permanent connection to the call control server. This can be done easily with MGCP or SIP, which are both UDP-based protocols, if the messages used for their registration function (REGISTER message in SIP, RSIP in MGCP) use the same source port as the one used to receive call control messages. This first registration message will reach the call control server, and by looking at the source IP address and port the server can learn the external IP address and port used for the NAPT mapping entry for each phone. In order to 'ring' this phone, the call server simply needs to send a call setup message (SIP INVITE, MGCP CRCX) to that self-same port (the pinhole for that phone), and the router will forward it to the correct phone. This behavior is standard in MGCP, but wasn't correctly specified in SIP RFC 3261; this was fixed in August 2003 in RFC 3581 by using a new 'rport' parameter of the Via header to force responses to the exact apparent receive port (see Section 3.2 for more details). The only issue is to keep the pinhole open, which requires having some traffic on the signaling connection every couple of minutes. Some phones can be configured to do so (e.g., REGISTER refresh), otherwise the call server can periodically send a message to the phone that should cause the phone to respond (MGCP AUEP, SIP OPTIONS, or even malformed messages that should trigger an error response).

6.2.2 Using port forwarding to solve the wrong media address problem

These methods alone do not solve the problem of internal network IP phones advertising private IP addresses for the media stream. There are also several possible solutions to this problem:

- Some IP phones have a configurable field for the "external IP address" of the NAT function (the address that will be used by the NAT function when replacing the source IP address). These phones will open a reception RTP port on the local host (e.g., 10.3.0.2:2345), but will advertise it in the call control messages with the public external address instead (e.g., 162.167.3.1:2345, note that the port is the same). This works only if the NAPT router is configured with a default port forwarding to the IP phone (i.e., all packets to 162.167.3.1 that do not correspond to an existing active connection are routed to the IP phone). In order to avoid conflicts with existing mappings, these IP phones can usually also be configured with a restricted set of ports to use for inbound RTP connections. This also allows this solution to extend to the case of several IP phones if the NAPT router can map a specific range of ports to each IP phone.

- Some IP phones will automatically detect that the source of the RTP packets they receive from the internal host is not the same as the IP address advertised for the RTP reception port. In such a case they will assume that NAPT is in place and will ignore the IP address provided, sending their own RTP packets to the source IP address present in the packets received from the IP phone on the internal network. This works because most phones use the same sending and reception RTP ports, but, unfortunately, this is not a standards-compliant implementation (see Section 6.2.4). Obviously, both phones should not be behind distinct NAPT functions.

All these solutions were used in the first PC-to-phone implementations, with only one or a couple IP software phones in a residential environment, and work well with 'techies' and early adopters who are not scared of reconfiguring their NAT router. Unfortunately, these methods cannot be used for the general public, or for such services as VoIP-based Centrex, which need to reach many phones behind a NAT function.

6.2.3 STUN

Many proposals have been made to facilitate or ideally automate the configuration of VoIP networks to successfully work even across NAT functions. The IETF Midcom Working Group is defining the specifications for a 'middlebox' access control device and a 'Midcom' control protocol. Most of these proposals work in an ideal world where all routers could be upgraded overnight, but are of little practical interest in current networks, given the size of the installed base of low-end cable or DSL NAT routers.

The STUN approach however stands out as it does not require any configuration or change of existing NAT/NAPT routers or existing call control servers. STUN is defined in RFC 3489 (Simple Traversal of User Datagram Protocol Through Network Address Translators), and is a simple query response protocol encapsulated over UDP (some security primitives also use TCP connections).

STUN is available in more and more phones, and there are several public STUN servers on the Internet. STUN provides a way for the phone to dynamically learn the external IP address and port that will be used for each communication through the NAT function. STUN also allows a host to discover the type of NAT implementation (full-cone or strict).

Each time the IP phone knows it is about to advertise an address and port that cannot be reached through the NAT/NAPT function, it first sends a STUN query from that exact IP address and port to the STUN server on the public Internet (1.2.3.4:3478 in Figure 6.4, 3478 is the well-known port of STUN servers). The response packet of the STUN server

Figure 6.4 Softphone using a STUN server to learn the external IP port used by a full-cone NAPT router for VoIP signaling or media flow.

simply indicates what is the apparent source address and port of the query (i.e., the external IP address and port allocated dynamically by the NAT function for this communication).

The IP phone is now ready to advertise the correct external address and port in the call control messages to the remote host (SIP INVITE or 200 OK response, MGCP 200 OK response to a CRCX or MDCX command, H.323 OLC or OLC ACK).

Note that this works only if the NAPT function is full-cone; otherwise, the external port allocated by the router for the STUN query will differ from the external port allocated for the RTP stream to the remote host, since they have different IP addresses. Most residential NAT functions are full-cone. In a bullet-proof implementation that works even with strict NAPT, a service provider can co-locate the STUN server with a call server that also serves as an RTP proxy. In this situation the RTP packets, and STUN queries go to the same destination and will be allocated the same external port by the router's strict NAT function. The only downside of this approach is that the RTP proxy will add some delay to the conversation and will impact the density of the call server (RTP processing is very CPU-intensive, a typical Linux kernel-mode implementation will route a maximum of about 500 media streams of 20-ms packets per GHz on a Pentium).

Another problem with STUN is that if there are multiple IP phones behind the NAPT function, these phones will advertise a public address for media streams even if one phone calls the other in the private network. This will force the router to relay media streams that could otherwise have been transported directly by the LAN. Usually, this will not create significant QoS problems as there are relatively few internal calls on small sites and the connections to the router are 10/100 Mbit/s Ethernet. However, this may create issues on large sites. This problem could easily be fixed by getting the IP phone to advertise two reception ports (one private, one public) and having the call server select the correct one on the fly (the call server knows which IP phones are behind the same NAT function); but, this would require small changes to the existing VoIP standards.

STUN will also introduce non-optimal media paths in some networks, as media are forced to flow through the routers closer to the STUN server, which are not necessarily in the shortest possible path for media streams. The same problem may happen for signaling (Figure 6.5), although this is much less critical as signaling does not have strong latency constraints.

6.2.4 Other proposals: COMEDIA and TURN

Connection-Oriented Media Transport in SDP (**COMEDIA**) is described in draft-ietf-mmusic-sdp-comedia-05.txt; it enables traversal of symmetric NAT by allowing VoIP gateways to dynamically update their destination RTP port according to the source IP and RTP port detected in the received RTP packets, instead of continuing to use the port advertised in the remote SDP. COMEDIA uses a new SDP attribute 'a = direction:role', where role can take the following values:

- 'Active', which indicates that the end point will initiate a connection to the port number on the 'm =' line of the session description from the other end point. The port number on its own 'm =' line is irrelevant, and the opposite end point must not attempt to initiate a connection to the port number specified there; instead, it is prepared to receive

The STUN request for media obtains the optimal PATH to the STUN server. The RTP path must follow the STUN path. STUN servers must be located optimally for the media path (STUN route from a phone closest to nearest edge router).

Call controller

———————— Signaling path

— — — — — — Media path

Figure 6.5 STUN may force utilization of non-optimal paths for media and/or signaling.

media on the ports from which it sends. These end points should also immediately send some media to the ports indicated in the 'm =' line of the remote end point.

- 'Passive', which indicates that the end point will accept a connection to the port number indicated on its 'm =' line of the session description. The end point will not send any media (including control packets such as RTCP) from their passive ports until they receive a packet on these ports and record the source address and port of the sender. The passive end point then assumes that the first packet received corresponds to its active peer. From this point onward, passive end points must send UDP or RTP media from the same port as the port indicated in their 'm =' line (receive port). They must also send RTCP media from the port on which they expect to receive it (typically, the RTP port number plus 1).

- 'Both', the default value, indicates that the end point will both accept an incoming connection on the port indicated on its own 'm =' SDP line and initiate an outgoing connection to the port number on the 'm =' line of the session description from the other end point. When receiving an SDP in active mode, the end point should behave as passive, and vice versa. If the end points are in both mode, then they should send data on the 'm =' line destination, and there may be two active connections if both succeed. End points should accept media both on the 'm =' line port as well as back to the sending port (in most cases end points will be designed so that this is the same).

With the COMEDIA proposal, the end point in passive mode can send media to an end point behind a symmetric NAT function, because a UDP pinhole will be opened by the media sent out from the end point behind the NAT function, and the end point in passive mode will send back audio data through this pinhole. There are still many issues in COMEDIA, notably when both end points are behind separate NAT functions.

Traversal Using Relay NAT (TURN) is another NAT traversal approach that uses the TCP/UDP pinhole opened through the NAT function to establish a bidirectional communication tunnel with a TURN server in the public network. The device D located in the private network which requires to establish a communication with the public Internet first communicates with the TURN server using the TURN protocol. As a first step D requests an IP address and port for his own use on the turn server. The TURN server allocates an IP address and port (IP_t:port_t). Device D can then advertise this IP address and port to external IP devices that need to send packets to it. When the TURN server receives packets on IP_t:port_t, it simply forwards these packets to device D using the TURN protocol. The TURN protocol can traverse the NAT function because it is based on a permanent TCP connection between device D and the TURN server, or it uses symmetric UDP.

TURN can be seen as a way to obtain a 'remote network interface' outside the NAT domain. Although there is continued interest in TURN, there is not sufficient consensus yet to formally publish it as an IETF RFC.

6.2.4.1 VoIP NAT traversal using an RTP relay

When the network cannot be controlled, when end points do not implement any NAT traversal algorithm, and when NAT functions may be any combination of full-cone or strict NAT without any specific support for VoIP, the only possibility that remains to enable VoIP calls is to attempt to use the pinholes opened by all NAT functions when sending traffic to the public network.

As indicated in Section 6.1.2.2, even strict NAPT functions will accept, once a UDP packet from the internal network has been sent to a server, a response packet from that server to the exact port that was allocated by the NAPT function as a source port for the UDP packet initially forwarded.

This property can be used by a network-based entity E in order to get UDP-based VoIP protocols to work across the NAT function (Figure 6.6):

- Server E will receive all VoIP signaling from the end points (EP_internal) behind the firewall, and send all responses to the apparent source port of the UDP packets it receives. All protocol-level indications to send the responses to a different IP address are ignored. In addition, in order to keep the pinhole open, server E or the end point need to exchange a packet at least every 30 s. When the protocol is MGCP, this can be achieved by the server independently of the end point, using AUEP commands. When the protocol is SIP, this can be client-based (e.g., REGISTER messages) or server-based (e.g., OPTION messages).

- For media streams, server E needs to intercept all VoIP signaling commands where the end point advertises the RTP reception ports where it expects to receive media, and forward these commands to EP_external, indicating itself as the reception device. E will also put itself forward to receive all media streams sent by EP_internal, in order to analyse their apparent source address, as translated by the NAT function. As soon as the apparent source address and port S_NAT:p_NAT of a media stream sent

Figure 6.6 NAT traversal using an RTP relay.

by EP_internal is known, server E will be able to forward media streams sent by EP_external to the self-same address and port S_NAT:p_NAT. The NAT function will forward these packets to the original source address and port S:p used by EP_internal to send its RTP packets; this works because virtually all end points send and receive RTP streams for each media on the same port.

Such a NAT traversal server can be implemented on any network link where VoIP signaling can be intercepted. There are obviously two natural locations:

- At the IP point of presence concentrating the traffic from the customer. The NAT traversal servers that are implemented there are sometimes called 'border session controllers' because they should be located at the edge of the network.
- Close to the call controller, or co-located with the call controller.

Because it needs to learn the apparent source IP address of media streams sent by the internal end point, the NAT traversal server needs to relay media streams. In applications where there is a high probability of calls coming from internal end points that are rerouted by the call controller to another internal end point behind the same NAT, the NAT traversal server should either optimize the RTP path by deactivating RTP relaying or be located very close to the customer site in order to minimize RTP tromboning in the IP network. RTP path optimization is not trivial, when all the call flows possible over a VoIP network are taken into account.

6.3 RECOMMENDED NETWORK DESIGN FOR SERVICE PROVIDERS

The previous sections have made it clear that NAT/NAPT problems should be avoided at all costs for large-scale deployments, as they can become a maintenance and troubleshooting nightmare (this amounts to knowing in detail and sometimes debugging many different sorts of NAT and firewall implementations). It is likely that enterprise routers, and obviously residential routers, will be unable to provide adequate support for VoIP protocols before 2005. Even when such methods work, it is likely that they will not be optimal in some networks, because most NAT routers will need to route both signaling and media, as illustrated in Figure 6.7.

All the traversal methods described above will work, but they are really workarounds and hardly capable of sustaining robust network deployment. However, it would obviously be very costly to allocate a routable, public IP address for every IP phone.

The strategies we describe below are what we believe to be the best engineering options for a service provider wanting to implement business-grade VoIP on a large-scale network.

The RTP path must follow the non-optimal signaling path, which is the short-path route to the call controller.

—————— Signaling path

- - - - - - Media path

Call controller

Figure 6.7 VoIP NAT routers may lead to non-optimal media paths.

6.3.1 Avoid NAT in the customer premises for VoIP

6.3.1.1 Business trunking and connection to PBXs and IP-PBXs

Most large corporate sites have an existing PBX and do not wish to switch to a pure internal VoIP network before the PBX has been fully depreciated. Nevertheless, such sites can benefit from a VoIP network to carry communications between corporate sites (voice VPN), or to the PSTN (arbitrage, least cost routing). In most cases the easiest way is to connect the PBX to a CAS or 5ESS (in the US) or PRI (most of Europe) VoIP gateway. More and more PBXs also have optional VoIP trunk boards that can be purchased and will packetize voice without a need for an external VoIP gateway.

Regardless of the solution that is adopted, the VoIP interface requires a single IP address to operate. Obviously, there would be no point in using NAT for this single address, and therefore the IP address of the gateway should be reachable directly from the core network. Let's call this type of address IP_GW.

6.3.1.2 IP phones

A VoIP-based Centrex site can have dozens of IP phones. The easiest way to avoid using the Centrex site NAT function is to allocate IP addresses from the network that have been reserved for VoIP usage. Let's call this type of IP address IP_PH. All communications between the VoIP network and IP_PH addresses should be routed normally by the Centrex site router, without any translation. The media path is optimized between phones as part of the same IP_PH address pool, as illustrated in Figure 6.8.

Obviously, the allocation of addresses selected by the service provider for use by IP phones should not interfere with the other IP addresses used by the existing corporate information system (PCs, printers, servers, etc.). There are several methods that can be used to meet this requirement:

- The most trivial way is to allocate IP_PH addresses in a private address pool (such as 10.X.X.X, 192.168.X.X, or 172.X.X.X). However, the customer may also be using private addresses, and therefore the service provider should select IP addresses in the private address pool that do not conflict with the network currently in place. Depending on the service provider, this may be easy or very difficult. Some service providers have a fully packaged SME IP connectivity offer where the corporate NAT router is always configured to allocate private IP addresses to internal PCs in the same address pool (say, 10.1.X.X). On such networks, all other blocks of IP addresses, (i.e., 10.2.X.X, 10.3.X.X, ..., 192.168.X.X, and 172.X.X.X can be used safely without creating conflicts). Unfortunately, many service providers didn't plan ahead for this type of problem and allow their customers to use any set of private IP addresses they like. It can then become cumbersome to ask each customer to select IP addresses that do not conflict with the network in place. For these service providers the following two alternative approaches will work.

- Instead of allocating the IP_PH addresses in a private pool, one sure way to avoid any conflict is to allocate these IP addresses in a pool of public IP addresses. For the example let's select 162.168.X.X. This block is large enough for about 65,000

All streams use direct IP routing. All paths are optimized

———— Signaling path

- - - - - - Media path

Call controller

Figure 6.8 Optimized media path between IP phones as part of the same address pool.

phones. Apparently, this defeats the object of not using one public address per IP phone. Fortunately, this is not the case as we will see below, as this pool of IP addresses can be reused in the network as many times as necessary, providing unlimited scalability with just one class B address block.

- Another option is to put all IP phones on a VLAN. The router is then instructed to route all incoming packets from the VoIP backbone (e.g., arriving in an IP tunnel), to that VLAN, resolving de facto any addressing ambiguity. Similarly, any packet sent from the VLAN of IP phones should be sent to the VoIP backbone without translation. All other data flows from other VLANs should be routed to the router NAT function and undergo normal processing. Unfortunately, this is only possible on relatively sophisticated routers.

A summary of the optimal-routing configuration for a site is given in Figure 6.9.

We have already mentioned that the pool of IP addresses allocated to IP phones could be reused many times. In fact, there is no magic in this solution, it has simply pushed the requirement for a NAT function away from customer premises equipment (which is of variable quality) into the backbone. Figure 6.10 shows an example network where the service provider only wants to use class B block 162.168.X.X and yet provide service to more than 65,000 phones.

The first customers are served from the first VoIP access pool. The IP routing domain between all the IP addresses of this first access pool must be closed; this can be achieved by using an MPLS virtual network, IP tunnels, source-based routing, router subinterfaces on specific layer 2 links (e.g., ATM permanent circuits). This is to ensure that all routing within this access domain does not interfere with any other routing table.

Figure 6.9   Separation of VoIP and other data flows.

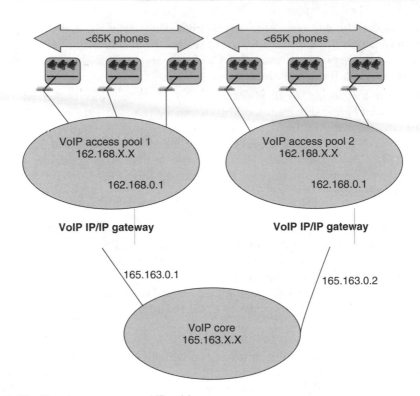

Figure 6.10   Reusing access pool IP addresses.

When the IP addresses from access pool 1 have been exhausted (this will probably be much before 65,000 IP phones have been connected because complete subnets need to be allocated to each end site set routing efficiency), a second access pool is created in its own isolated routing domain.

Since each access domain is isolated from a routing point of view, they cannot communicate. This problem is solved by the VoIP IP/IP gateway. Any phone call from a phone in access domain 1 to a phone that is not in access domain 1 will reach the VoIP IP/IP gateway, which will terminate it locally using an IP address in the access pool (e.g., 162.168.0.1), then re-originate the call using a single public IP address. In essence, the VoIP IP/IP gateway *summarizes* the complete access pool in a single public IP address. If the call needs to reach a phone in the second access zone, it will be routed by the VoIP core to the VoIP IP/IP gateway of access pool 2, which will re-originate the call within the access 2 domain using IP address 162.168.0.1.

This hierarchical access network scales indefinitely to arbitrarily large VoIP networks, by just adding more access pools. The VoIP IP/IP gateway is not a trivial function (we will discuss it in Section 6.3.2).

The method also has the advantage of cleanly separating the VoIP network and the regular data network, which helps delineate the respective responsibilities of the enterprise and the service provider in terms of security (this will be described in Section 6.3.3).

6.3.1.3 Software IP phones and PC-based videoconferencing devices

Unfortunately, the case of PC-based VoIP equipment cannot be solved by the previous method. If an IP address of type IP_PH is allocated to the PC, then any data application also running on the PC may stop working because the VoIP network isn't designed for it and will probably block non-VoIP communications. Even if it worked, this would still allow the VoIP network to reach corporate PCs, which would make the service provider responsible for possible breaches of security on corporate MIS equipment.

Therefore, PCs should be allocated IP addresses normally by the corporation, without any specific restriction for VoIP. The NAPT problem cannot be avoided. The solutions described in Section 6.2 must be used. If the corporation uses full-cone NAPT, then STUN is the best solution. It is supported by a number of VoIP software manufacturers (e.g., EyePmedia at www.eyePmedia.com). If the corporation uses strict NAT, either the company must implement a VoIP IP/IP gateway in the premises (more and more firewalls provide this feature) or the service provider must implement an RTP relay.

All solutions will end up in the same final situation; the audio or video calls originated from PCs will be mapped to VoIP calls that appear to originate from routable, public IP addresses. Similarly, it will be possible to reach all PCs by placing a VoIP call to a routable IP address (either the call control server with an RTP relay, or a public address and port on the corporate site router selected by the router NAT function—the STUN case).

Communications between PCs and IP phones can occur by routing the call to the appropriate VoIP IP/IP gateway (the gatekeeper or SIP proxy required to do this is not shown), which relays the call to the proper IP phone (as shown in Figure 6.11).

Note that the customer access router in Figure 6.11, is shown as having one link to the VoIP access pool domain and one link to the normal Internet backbone of the service provider; this was done for clarity. It can be done with a single link either using IP tunnels

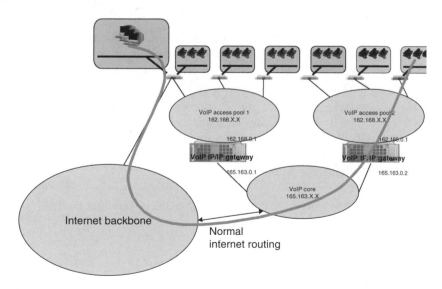

Figure 6.11 Mixed call scenario: STUN softphone to IP phone in a dedicated address pool.

or more simply by carefully configuring service provider concentration routers to route all packets with source IP addresses of type IP_PH to the VoIP access pool domain (e.g., an MPLS domain), and all other addresses according to normal internet routing tables.

As the destination of voice IP packets is to a private address (in fact, the change to existing concentration router configurations is quite minimal), a path to 162.186.x.x is added which routes by default all these packets to the closed routing domain (e.g., an MPLS domain), instead of the default Internet route. For each VoIP customer connected to this concentration router and using, for instance, 162.168.Z.x, a static route must be added to the subnet 162.168.Z.8 with the customer access router as the next hop.

6.3.1.4   *The case of data VPNs*

Large multi-site corporations using a data VPN pose a specific problem, because in most cases all corporate sites communicate on an isolated routing domain (e.g., an MPLS domain), which usually has a couple of controlled access points to the Internet, though a firewall.

From a VoIP perspective, the entire data VPN can be seen as a large site, with the access routers as the data VPN-controlled access points to the Internet.

Deploying a VoIP service in such networks (e.g., to provide a VoIP short-numbering service between sites, or least cost routing) requires some configuration:

- The service provider softswitch must be located in a specific 'resource domain', which can be reached from all VPN sites and can reach all VPN sites. Most service providers already have such a domain for their DNS and email servers. Note that the resource domain cannot be used to communicate across data VPNs, only communications to and from the shared resource domain and each data VPN is allowed. An MPLS domain can be used for this purpose.
- All VoIP gateways connected to sites' PBXs must be able to communicate with the service provider softswitch. This can be achieved simply by allocating each VoIP gateway an IP address from the resource domain. There are relatively few gateways, but this does not pose an IP address depletion problem.
- All IP phones must be provided with IP addresses in the IP_PH pool, allocated as described above. The IP_PH addresses allocated to the data VPN must be routed within this data VPN. All controlled Internet access points of this data VPN must route packets from these IP_PH addresses to the appropriate IP VoIP access pool domain (e.g., this can be a separate MPLS domain).

Figure 6.12 shows the routes that need to be enabled between the various MPLS domains typically found in a VPN environment:

- One MPLS domain per VPN customer.
- One shared resource MPLS domain.
- One MPLS domain for each VoIP access domain (shared across multiple customers).

Figure 6.12 Usage of shared resource domain and VoIP access IP address pool in a VPN service context.

VPN-to-VPN phone calls over IP can also be enabled using the VoIP IP/IP gateway: since IP communications cannot occur between each customer's data VPN, the VoIP call is first terminated to the VoIP IP/IP GW (communications to the access/shared resource domains are allowed), then the VoIP IP/IP gateway re-originates the call and routes it to the destination data VPN (communications from the access/shared resource domain to any data VPN are allowed).

If the service provider wants to provide PSTN connectivity through shared VoIP gateways, these VoIP gateways must also be part of the shared resource domain. If the calls need to be routed to a third-party VoIP network, this needs to occur through a VoIP IP/IP GW (the shared resource domain cannot communicate directly with the internet For obvious security reasons).

Providing managed VoIP services on top of data VPN services does involve strong expertise in security and IP routing in the context of isolated routing domains; but, it is one of the most successful service bundles provided by service providers to large corporate customers.

6.3.1.5 The case of residential networks

The case of residential networks is usually much simpler. Most residential networks seem to be providing at least one public address to the customer router. In most cases service providers wanting to offer residential voice will use an Integrated Access Device

(IAD) with a embedded gateway to a couple of POTS lines. In this case the VoIP gateway subsystem uses the public IP address, and there is no NAT issue. Obviously, this works just as well if the service provider only wants to allocate private IP addresses to each router; but, some VoIP IP/IP gateways may need to be deployed in the network to communicate with third-party VoIP service providers or network VoIP gateways if they use public IP addresses.

Many service providers will want to be able to offer VoIP and video services to PC users as well. In this case the STUN approach works well and has very few drawbacks, as it is highly unlikely that two PC phones behind the same residential NAT router will want to communicate. Some IAD devices also include a VoIP-friendly NAT function. In this case using STUN is obviously no longer necessary.

6.3.1.6 Smooth deployment scenario

At first it may seem scary to implement the full approach of VoIP access domains for an initial trial. This approach was introduced only to enable the reuse of one pool of IP addresses indefinitely, in order to scale the network. For small trials, however, there is no need to reuse IP addresses, and the VoIP access domain can be merged with the IP core domain. In other words, VoIP core domain components (such as a central gatekeeper or SIP proxy) can be located in the first VoIP access domain, no VoIP IP/IP gateway is initially required.

The expansion to a larger network will require the formal creation of a VoIP core domain (the core softswitch will need to be relocated) and the introduction of additional VoIP access domains. All of this can be done without having to change the IP addresses already allocated to existing IP phones, by substituting the core softswitch by an IP/IP gateway. This method enables smooth expansion of the network.

6.3.2 Media proxies

Today, many service providers still use two VoIP gateways connected back to back to provide the media proxy function. Indeed, every VoIP call will terminate at the VoIP address of the first gateway and be re-originated from the second VoIP gateway. However, this crude design is not a viable solution:

- VoIP gateways always decode the media stream to the G.711 format. If voice was originally encoded, the decoding and re-encoding of voice (known as 'tandeming') will significantly reduce the quality of voice (typically, 0.5 MOS points).
- VoIP gateways have a jitter buffer in order to prevent any gap in the TDM media stream. This jitter buffer adds a significant delay (typically, 50 ms) to the media path and adds to jitter buffer delay at the destination.
- In some call scenarios, a call may be routed from one IP domain to the other, then routed back to the original IP domain (call forward, local number portability, call transfer, etc.). The back-to-back gateway in such circumstances will continue to 'route'

the media stream, although this is clearly not the optimal path through the IP network! Again this adds unnecessary delays and jitter to the VoIP network.

- VoIP gateways only support voice (actually, H.320 video gateways do exist, but they are expensive and would degrade the video quality that can be achieved on IP networks)

- Last but not least, every call through the back-to-back gateway uses two gateway ports, which is expensive.

There are many providers of dedicated media proxies (VoIP IP/IP gateways, RTP proxies, etc.). Ideally they should support the following features:

- Support the relaying of media streams without requiring decoding and re-encoding in order to minimize delays.

- They should not have jitter buffer.

- They should add a minimal delay overhead for media processing. This does not necessarily require dedicated hardware implementations (actually, kernel implementations on operating systems, such as Linux, and high-performance network interfaces now have a performance comparable with most router).

- They should support many types of media streams (audio, T.38 fax, T.120, video).

- They should support all the call flows found in the network, not just basic calls. This includes dynamic media redirection (H.323 NullCapabilitySet, SIP RE-INVITE)

- They should automatically detect calls that are looped back to the originating IP domain and optimize the media path to stop using the proxy. This is not trivial and requires relatively sophisticated algorithms.

- They should support multiple network interfaces in order to facilitate connectivity to multiple separate networks. These interfaces can be physical or virtual (VLANs, IP tunnels).

Other nice-to-have features include some denial-of-service protection (call rate limiters, checking of media streams' token bucket profiles, mapping of DiffServ codepoints, etc.).

The media proxy function does not need to be a stand-alone product; in fact, this function works even better when combined with a call controller, because the call controller can know many more properties of the end points, such as which end points are on the same site and need to have an optimized RTP path, or which end points need to have the RTP stream fully relayed through the proxy.

Figure 6.13 shows a sample network with two enterprises A and B. Enterprise A has two sites A1 and A2. An MGCP call agent provides the IP Centrex service and includes an RTP proxy function. The call agent controls IP phones using MGCP and, when calls need to be transferred to the backbone, uses the H.323 or SIP protocols.

Since all IP phones are located on the same IP address plane (access pool 1), in theory any call from an IP phone in this domain to another IP phone in this domain can use direct RTP routing. This can be seen when a phone on site A1 calls a phone on site A2: the MGCP call controller provides each IP phone with the address of the other IP phone, so the media can be routed directly by the IP network.

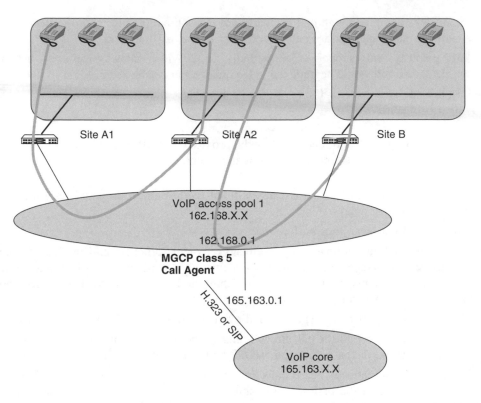

Figure 6.13 Optimization of RTP proxy usage when the softswitch has access to administrative data.

However, for security reasons, enterprise B does not want to receive media streams directly from any other enterprise, it only wants to receive media streams from the service provider (only packets from IP address 162.168.0.1 are allowed). In this case the MGCP call agent will also act as an RTP proxy and terminate, then re-originate the media stream. This is shown in the case of a call between a phone on site A2 and a phone on site B.

The example shows that the association of a call controller and a media proxy can be powerful. A separate media proxy does not have enough application-level information and would have optimized the media streams in all cases, routing RTP packets directly between enterprise B and site A1.

6.3.3 Security considerations

The allocation of controlled IP addresses to IP phones not only helps to solve the NAT issue, it also clarifies the potential security responsibility issues that can arise in a managed telephony deployment inside the enterprise. By definition, a managed VoIP service must be able to initiate IP communications to servers (IP phones) *inside* the corporate private network.

Our experience with initial deployments of IP Centrex show that some corporate MIS managers may be tempted, if they find that their corporate network has been compromised, to blame the managed telephony service provider. Therefore, it is very important for a service provider to be able to propose some mutually acceptable, simple to understand, security rules that completely isolate IP phones from the PCs on the internal network.

This is very difficult if IP phones are allocated IP addresses at random inside the corporation, but becomes much simpler if the IP addresses of all IP phones can be identified easily, as proposed.

The corporate firewall/access router should be configured to:

- Accept VoIP signaling only from the IP address of the service provider call controller.
- Accept media traffic only from IP addresses of the IP_PH pool *and* only to the IP addresses allocated to the IP phones (with an RTP filter if available in the router/firewall, otherwise a UDP filter on ports higher than 1024 should be used). This rule can be made stricter if the call controller is capable of enabling an RTP proxy for media calls coming from third-party sites, in which case the source IP address of the media streams can be restricted to just the IP address of the media proxy (or proxies).
- If the IP phones are located on a VLAN or only accessible through a specific router interface, all inbound media traffic should only be routed to this interface.

IP phones should not be allowed to communicate with any IP address that is not part of the IP_PH pool. This can be achieved by placing the phones on a separate LAN or VLAN. This prevents the unlikely but potential threat of having an IP phone compromised and serving as a relay to other machines on the LAN.

These rules are relatively simple and make it impossible for any potential attacker on the VoIP backbone to reach the computers of the enterprise by using the VoIP NAT bypass route through the corporate router. Any attack must come through the regular router NAT function (or other security policy set for calls coming from the public IP network), and therefore is not under the responsibility of the service provider. Our recommendation to service providers is to include a detailed description of the security policy and have all IP Centrex customers signing acceptance of this security policy, thereby clearing the service provider from potential future accusations.

6.4   CONCLUSION

It is frequently heard that, until IPv6 is adopted, VoIP cannot be deployed due to the lack of IP addresses. This is clearly wrong. The sophistication of the current tools allowed by IPv4 routing and the use of application-level IP/IP gateways make it possible to use only a restricted set of addresses and yet provide a service to a virtually unlimited number of users. Due to the smaller size of IPv4 packets, latency on such networks, especially on access links, will be better than on an equivalent IPv6 network and the IP overhead is also much better. Given the fact that all the tools required to provide quality of service (DiffServ, RSVP) perform just as well on IPv4 or IPv6 networks, there is really no reason today for a service provider to wait for the deployment of large-scale VoIP networks on currently deployed IP networks.

Annex

Here is the call flow between Cisco SIP phones (Cisco SIP phones reply with a "trying" message after each request for clarity these have been removed), showing which implementation choices have been made between Cisco phones:

Phone A (5559000) calls phone B (5555000)

INVITE sip:5555000@172.18.192.230 SIP/2.0
Via: SIP/2.0/UDP 172.18.192.218:5060
From: "A Phone" <sip:5559000@172.18.192.230>;tag=00070e8b5777000339a3170e-74ee1c83
To: <sip:5555000@172.18.192.230>
Call-ID: 00070e8b-577708ee-5fbe5e66-7f70fecd@172.18.192.218
Date: Thu, 13 Jun 2002 16:04:46 GMT
CSeq: 101 INVITE
User-Agent: Cisco-SIP-IP-Phone/3
Contact: sip:5559000@172.18.192.218:5060
Expires: 180
Content-Type: application/sdp
Content-Length: 271
Accept: application/sdp

v=0
o=CiscoSystemsSIP-IPPhone-UserAgent 18338 11953 IN IP4 172.18.192.218
s=SIP Call c=IN IP4 172.18.192.218
t=0 0
m=audio 29304 RTP/AVP 0 8 18 97
a=rtpmap:0 PCMU/8000
a=rtpmap:8 PCMA/8000

IP Telephony   O. Hersent, J.P. Petit, D. Gurle
© 2005 John Wiley & Sons, Ltd   ISBN: 0-470-02359-7

a=rtpmap:18 G729a/8000
a=rtpmap:97 telephone-event/8000
a=fmtp:97 0–15

SIP/2.0 180 Ringing
Via: SIP/2.0/UDP 172.18.192.218:5060
From: "A Phone"<sip:5559000@172.18.192.230>;tag=00070e8b5777000339a3170e-74ee1c83
To: <sip:5555000@172.18.192.230>;tag=003094c2e691000357031309-41c9de44
Call-ID: 00070e8b-577708ee-5fbe5e66-7f70fecd@172.18.192.218
Date: Thu, 13 Jun 2002 16:04:36 GMT
CSeq: 101 INVITE
Server: Cisco-SIP-IP-Phone/3
Contact: sip:5555000@172.18.192.221:5060
Record-Route: <sip:5555000@172.18.192.230:5060;maddr=172.18.192.230>
Content-Length: 0

SIP/2.0 200 OK
Via: SIP/2.0/UDP 172.18.192.218:5060
From: "A Phone"<sip:5559000@172.18.192.230>;tag=00070e8b5777000339a3170e-74ee1c83
To: <sip:5555000@172.18.192.230>;tag=003094c2e691000357031309-41c9de44
Call-ID: 00070e8b-577708ee-5fbe5e66-7f70fecd@172.18.192.218
Date: Thu, 13 Jun 2002 16:04:39 GMT
CSeq: 101 INVITE
Server: Cisco-SIP-IP-Phone/3
Contact: sip:5555000@172.18.192.221:5060
Record-Route: <sip:5555000@172.18.192.230:5060;maddr=172.18.192.230>
Content-Type: application/sdp
Content-Length: 220

v=0
o=CiscoSystemsSIP-IPPhone-UserAgent 11411 26110 IN IP4 172.18.192.221
s=SIP Call c=IN IP4 172.18.192.221
t=0 0
m=audio 24396 RTP/AVP 0 97
a=rtpmap:0 PCMU/8000
a=rtpmap:97 telephone-event/8000
a=fmtp:97 0–15

ACK sip:5555000@172.18.192.230:5060 SIP/2.0
Via: SIP/2.0/UDP 172.18.192.218:5060
From: "Kazoo-9 Phone"<sip:5559000@172.18.192.230>;tag=00070e8b5777000339a3170e-74ee1c83
To: <sip:5555000@172.18.192.230>;tag=003094c2e691000357031309-41c9de44
Call-ID: 00070e8b-577708ee-5fbe5e66-7f70fecd@172.18.192.218
Date: Thu, 13 Jun 2002 16:04:50 GMT
CSeq: 101 ACK
User-Agent: Cisco-SIP-IP-Phone/3
Route: <sip:5555000@172.18.192.221:5060>
Content-Length: 0

A puts B on hold:

INVITE sip:5555000@172.18.192.230:5060 SIP/2.0
Via: SIP/2.0/UDP 172.18.192.218:5060
From: "A Phone"<sip:5559000@172.18.192.230>;tag=00070e8b5777000339a3170e-74ee1c83
To: <sip:5555000@172.18.192.230>;tag=003094c2e691000357031309-41c9de44
Call-ID: 00070e8b-577708ee-5fbe5e66-7f70fecd@172.18.192.218
Date: Thu, 13 Jun 2002 16:04:53 GMT
CSeq: 102 INVITE
User-Agent: Cisco-SIP-IP-Phone/3
Contact: sip:5559000@172.18.192.218:5060
Route: <sip:5555000@172.18.192.221:5060>
Content-Type: application/sdp
Content-Length: 263

v=0
o=CiscoSystemsSIP-IPPhone-UserAgent 15014 5663 IN IP4 172.18.192.218
s=SIP Call c=IN IP4 0.0.0.0
t=0 0
m=audio 29304 RTP/AVP 0 8 18 97
a=rtpmap:0 PCMU/8000
a=rtpmap:8 PCMA/8000
a=rtpmap:18 G729a/8000
a=rtpmap:97 telephone-event/8000
a=fmtp:97 0–15

SIP/2.0 200 OK
Via: SIP/2.0/UDP 172.18.192.218:5060
From: "A Phone"<sip:5559000@172.18.192.230>;tag=00070e8b5777000339a3170e-74ee1c83
To: <sip:5555000@172.18.192.230>;tag=003094c2e691000357031309-41c9de44
Call-ID: 00070e8b-577708ee-5fbe5e66-7f70fecd@172.18.192.218
Date: Thu, 13 Jun 2002 16:04:44 GMT
CSeq: 102 INVITE
Server: Cisco-SIP-IP-Phone/3
Contact: sip:5555000@172.18.192.221:5060
Record-Route: <sip:5555000@172.18.192.230:5060;maddr=172.18.192.230>
Content-Type: application/sdp
Content-Length: 213

v=0
o=CiscoSystemsSIP-IPPhone-UserAgent 11411 26110 IN IP4 172.18.192.221
s=SIP Call c=IN IP4 0.0.0.0
t=0 0
m=audio 24396 RTP/AVP 0 97
a=rtpmap:0 PCMU/8000
a=rtpmap:97 telephone-event/8000
a=fmtp:97 0–15

ACK sip:5555000@172.18.192.230:5060 SIP/2.0
Via: SIP/2.0/UDP 172.18.192.218:5060

From: "A Phone"<sip:5559000@172.18.192.230>;tag=00070e8b5777000339a3170e-74ee1c83
To: <sip:5555000@172.18.192.230>;tag=003094c2e691000357031309-41c9de44
Call-ID: 00070e8b-577708ee-5fbe5e66-7f70fecd@172.18.192.218
Date: Thu, 13 Jun 2002 16:04:54 GMT
CSeq: 102 ACK
User-Agent: Cisco-SIP-IP-Phone/3
Route: <sip:5555000@172.18.192.221:5060>
Content-Length: 0

A calls C:

INVITE sip:5551000@172.18.192.230 SIP/2.0
Via: SIP/2.0/UDP 172.18.192.218:5060
From: "A Phone"<sip:5559000@172.18.192.230>;tag=00070e8b577700045e92261b-204bc3c6
To: <sip:5551000@172.18.192.230>
Call-ID: 00070e8b-577708ef-746a6bfe-6c96b214@172.18.192.218
Date: Thu, 13 Jun 2002 16:04:59 GMT
CSeq: 101 INVITE
User-Agent: Cisco-SIP-IP-Phone/3
Contact: sip:5559000@172.18.192.218:5060
Expires: 180
Content-Type: application/sdp
Content-Length: 271
Accept: application/sdp

v=0
o=CiscoSystemsSIP-IPPhone-UserAgent 27275 16432 IN IP4 172.18.192.218
s=SIP Call c=IN IP4 172.18.192.218
t=0 0
m=audio 29306 RTP/AVP 0 8 18 97
a=rtpmap:0 PCMU/8000
a=rtpmap:8 PCMA/8000
a=rtpmap:18 G729a/8000
a=rtpmap:97 telephone-event/8000
a=fmtp:97 0–15

SIP/2.0 180 Ringing
Via: SIP/2.0/UDP 172.18.192.218:5060
From: "A Phone"<sip:5559000@172.18.192.230>;tag=00070e8b577700045e92261b-204bc3c6
To: <sip:5551000@172.18.192.230>;tag=003094c25d94001039653a76-0c6588ad
Call-ID: 00070e8b-577708ef-746a6bfe-6c96b214@172.18.192.218
Date: Thu, 13 Jun 2002 16:04:57 GMT
CSeq: 101 INVITE
Server: Cisco-SIP-IP-Phone/3
Contact: sip:5551000@172.18.192.220:6062
Record-Route: <sip:5551000@172.18.192.230:5060;maddr=172.18.192.230>
Content-Length: 0

SIP/2.0 200 OK
Via: SIP/2.0/UDP 172.18.192.218:5060

From: "A Phone"<sip:5559000@172.18.192.230>;tag=00070e8b577700045e92261b-204bc3c6
To: <sip:5551000@172.18.192.230>;tag=003094c25d94001039653a76-0c6588ad
Call-ID: 00070e8b-577708ef-746a6bfe-6c96b214@172.18.192.218
Date: Thu, 13 Jun 2002 16:05:00 GMT
CSeq: 101 INVITE
Server: Cisco-SIP-IP-Phone/3
Contact: sip:5551000@172.18.192.220:6062
Record-Route: <sip:5551000@172.18.192.230:5060;maddr=172.18.192.230>
Content-Type: application/sdp
Content-Length: 221

v=0
o=CiscoSystemsSIP-IPPhone-UserAgent 5685 4962 IN IP4 172.18.192.220
s=SIP Call c=IN IP4 172.18.192.220
t=0 0
m=audio 16394 RTP/AVP 18 97
a=rtpmap:18 G729a/8000
a=rtpmap:97 telephone-event/8000
a=fmtp:97 0–15

ACK sip:5551000@172.18.192.230:5060 SIP/2.0
Via: SIP/2.0/UDP 172.18.192.218:5060
From: "A Phone"<sip:5559000@172.18.192.230>;tag=00070e8b577700045e92261b-204bc3c6
To: <sip:5551000@172.18.192.230>;tag=003094c25d94001039653a76-0c6588ad
Call-ID: 00070e8b-577708ef-746a6bfe-6c96b214@172.18.192.218
Date: Thu, 13 Jun 2002 16:05:02 GMT
CSeq: 101 ACK
User-Agent: Cisco-SIP-IP-Phone/3
Route: <sip:5551000@172.18.192.220:6062>
Content-Length: 0

A puts C on hold

INVITE sip:5551000@172.18.192.230:5060 SIP/2.0
Via: SIP/2.0/UDP 172.18.192.218:5060
From: "A Phone"<sip:5559000@172.18.192.230>;tag=00070e8b577700045e92261b-204bc3c6
To: <sip:5551000@172.18.192.230>;tag=003094c25d94001039653a76-0c6588ad
Call-ID: 00070e8b-577708ef-746a6bfe-6c96b214@172.18.192.218
Date: Thu, 13 Jun 2002 16:05:06 GMT
CSeq: 102 INVITE
User-Agent: Cisco-SIP-IP-Phone/3
Contact: sip:5559000@172.18.192.218:5060
Route: <sip:5551000@172.18.192.220:6062>
Content-Type: application/sdp
Content-Length: 263

v=0
o=CiscoSystemsSIP-IPPhone-UserAgent 22866 2538 IN IP4 172.18.192.218
s=SIP Call c=IN IP4 0.0.0.0

t=0 0
m=audio 29306 RTP/AVP 0 8 18 97
a=rtpmap:0 PCMU/8000
a=rtpmap:8 PCMA/8000
a=rtpmap:18 G729a/8000
a=rtpmap:97 telephone-event/8000
a=fmtp:97 0–15

SIP/2.0 200 OK
Via: SIP/2.0/UDP 172.18.192.218:5060
From: "A Phone"<sip:5559000@172.18.192.230>;tag=00070e8b577700045e92261b-204bc3c6
To: <sip:5551000@172.18.192.230>;tag=003094c25d94001039653a76-0c6588ad
Call-ID: 00070e8b-577708ef-746a6bfe-6c96b214@172.18.192.218
Date: Thu, 13 Jun 2002 16:05:04 GMT
CSeq: 102 INVITE
Server: Cisco-SIP-IP-Phone/3
Contact: sip:5551000@172.18.192.220:6062
Record-Route: <sip:5551000@172.18.192.230:5060;maddr=172.18.192.230>
Content-Type: application/sdp
Content-Length: 214

v=0
o=CiscoSystemsSIP-IPPhone-UserAgent 5685 4962 IN IP4 172.18.192.220
s=SIP Call c=IN IP4 0.0.0.0
t=0 0
m=audio 16394 RTP/AVP 18 97
a=rtpmap:18 G729a/8000
a=rtpmap:97 telephone-event/8000
a=fmtp:97 0–15

ACK sip:5551000@172.18.192.230:5060 SIP/2.0
Via: SIP/2.0/UDP 172.18.192.218:5060
From: "A Phone"<sip:5559000@172.18.192.230>;tag=00070e8b577700045e92261b-204bc3c6
To: <sip:5551000@172.18.192.230>;tag=003094c25d94001039653a76-0c6588ad
Call-ID: 00070e8b-577708ef-746a6bfe-6c96b214@172.18.192.218
Date: Thu, 13 Jun 2002 16:05:06 GMT
CSeq: 102 ACK
User-Agent: Cisco-SIP-IP-Phone/3
Route: <sip:5551000@172.18.192.220:6062>
Content-Length: 0

A transfers B to C:

REFER sip:5555000@172.18.192.230:5060 SIP/2.0
Via: SIP/2.0/UDP 172.18.192.218:5060
From: "A Phone"<sip:5559000@172.18.192.230>;tag=00070e8b5777000339a3170e-74ee1c83
To: <sip:5555000@172.18.192.230>;tag=003094c2e691000357031309-41c9de44
Call-ID: 00070e8b-577708ee-5fbe5e66-7f70fecd@172.18.192.218

Date: Thu, 13 Jun 2002 16:05:06 GMT
CSeq: 103 REFER
User-Agent: Cisco-SIP-IP-Phone/3
Contact: sip:5559000@172.18.192.218:5060
Route: <sip:5555000@172.18.192.221:5060>
Content-Length: 0
Refer-To:
sip:5551000@172.18.192.230?Replaces=00070e8b-577708ef-746a6bfe-6c96b214%40172.18.192.218
%3Bto-tag%3D003094c25d94001039653a76-0c6588ad%3Bfrom-tag%3D00070e8b577700045e92261b-
204bc3c6
Referred-By: "A Phone" <sip:5559000@172.18.192.230>

SIP/2.0 202 Accepted
Via: SIP/2.0/UDP 172.18.192.218:5060
From: "A Phone"<sip:5559000@172.18.192.230>;tag=00070e8b5777000339a3170e-74ee1c83
To: <sip:5555000@172.18.192.230>;tag=003094c2e691000357031309-41c9de44
Call-ID: 00070e8b-577708ee-5fbe5e66-7f70fecd@172.18.192.218
Date: Thu, 13 Jun 2002 16:04:56 GMT
CSeq: 103 REFER
Server: Cisco-SIP-IP-Phone/3
Contact: sip:5555000@172.18.192.221:5060
Record-Route: <sip:5555000@172.18.192.230:5060;maddr=172.18.192.230>
Content-Length: 0

Phone B calls C

NIVITE sip:5551000@172.18.192.230:5060 SIP/2.0
Via: SIP/2.0/UDP 172.18.192.221:5060
From: "5555000"<sip:5555000@172.18.192.230>;tag=003094c2e69100040d6ba94e-5f32d799
To: <sip:5551000@172.18.192.230:5060>
Call-ID: 003094c2-e69100af-66c9e3b4-04b3f0e8@172.18.192.221
Date: Thu, 13 Jun 2002 16:04:56 GMT
CSeq: 101 INVITE
User-Agent: Cisco-SIP-IP-Phone/3
Contact: sip:5555000@172.18.192.221:5060
Referred-By: "Kazoo-9 Phone" <sip:5559000@172.18.192.230>
Replaces:
00070e8b-577708ef-746a6bfe-6c96b214@172.18.192.218;to-tag=003094c25d94001039653a76-
0c6588ad;from-tag=00070e8b577700045e92261b-204bc3c6
Expires: 180
Content-Type: application/sdp
Content-Length: 270
Accept: application/sdp

v=0
o=CiscoSystemsSIP-IPPhone-UserAgent 19502 5249 IN IP4 172.18.192.221
s=SIP Call c=IN IP4 172.18.192.221
t=0 0
m=audio 24398 RTP/AVP 0 8 18 96

a=rtpmap:0 PCMU/8000
a=rtpmap:8 PCMA/8000
a=rtpmap:18 G729a/8000
a=rtpmap:96 telephone-event/8000
a=fmtp:96 0–15

SIP/2.0 200 OK
Via: SIP/2.0/UDP 172.18.192.221:5060
From: "5555000"<sip:5555000@172.18.192.230>;tag=003094c2e69100040d6ba94e-5f32d799
To: <sip:5551000@172.18.192.230:5060>;tag=003094c25d9400111fc7caba-03f8330a
Call-ID: 003094c2-e69100af-66c9e3b4-04b3f0e8@172.18.192.221
Date: Thu, 13 Jun 2002 16:05:05 GMT
CSeq: 101 INVITE
Server: Cisco-SIP-IP-Phone/3
Contact: sip:5551000@172.18.192.220:6062
Record-Route: <sip:5551000@172.18.192.230:5060;maddr=172.18.192.230>
Content-Type: application/sdp
Content-Length: 223

v=0
o=CiscoSystemsSIP-IPPhone-UserAgent 18795 10818 IN IP4 172.18.192.220
s=SIP Call c=IN IP4 172.18.192.220
t=0 0
m=audio 16396 RTP/AVP 18 96
a=rtpmap:18 G729a/8000
a=rtpmap:96 telephone-event/8000
a=fmtp:96 0–15

ACK sip:5551000@172.18.192.230:5060 SIP/2.0
Via: SIP/2.0/UDP 172.18.192.221:5060
From: "5555000"<sip:5555000@172.18.192.230>;tag=003094c2e69100040d6ba94e-5f32d799
To: <sip:5551000@172.18.192.230:5060>;tag=003094c25d9400111fc7caba-03f8330a
Call-ID: 003094c2-e69100af-66c9e3b4-04b3f0e8@172.18.192.221
Date: Thu, 13 Jun 2002 16:04:57 GMT
CSeq: 101 ACK
User-Agent: Cisco-SIP-IP-Phone/3
Route: <sip:5551000@172.18.192.220:6062>
Content-Length: 0

B notifies A that communication with C is active:

NOTIFY sip:5559000@172.18.192.218:5060 SIP/2.0
Record-Route: <sip:5555000@172.18.192.230:5060;maddr=172.18.192.230>
Via: SIP/2.0/UDP
172.18.192.230:5060;branch=855cf819-524d09ad-dbaea7b3-dc1ce9c9-1
Via: SIP/2.0/UDP 172.18.192.221:5060
From: <sip:5555000@172.18.192.230>;tag=003094c2e691000357031309-41c9de44
To: "A Phone" <sip:5559000@172.18.192.230>;tag=00070e8b5777000339a3170e-74ee1c83

Call-ID: 00070e8b-577708ee-5fbe5e66-7f70fecd@172.18.192.218
Date: Thu, 13 Jun 2002 16:04:57 GMT
CSeq: 101 NOTIFY
User-Agent: Cisco-SIP-IP-Phone/3
Event: refer
Content-Type: message/sipfrag
Content-Length: 14

SIP/2.0 200 OK

SIP/2.0 200 OK
Via: SIP/2.0/UDP
172.18.192.230:5060;branch=855cf819-524d09ad-dbaea7b3-dc1ce9c9-1,SIP/2.0/UDP
172.18.192.221:5060
From: <sip:5555000@172.18.192.230>;tag=003094c2e691000357031309-41c9de44
To: "A Phone" <sip:5559000@172.18.192.230>;tag=00070e8b5777000339a3170e-74ee1c83
Call-ID: 00070e8b-577708ee-5fbe5e66-7f70fecd@172.18.192.218
Date: Thu, 13 Jun 2002 16:05:07 GMT
CSeq: 101 NOTIFY
Content-Length: 0

C releases the call from A:

BYE sip:5559000@172.18.192.218:5060 SIP/2.0
Record-Route: <sip:5551000@172.18.192.230:5060;maddr=172.18.192.230>
Via: SIP/2.0/UDP
172.18.192.230:5060;branch=fd11b5e1-aee38059-c54b46b5-f5f62a06-1
Via: SIP/2.0/UDP 172.18.192.220:6062
From: <sip:5551000@172.18.192.230>;tag=003094c25d94001039653a76-0c6588ad
To: "A Phone" <sip:5559000@172.18.192.230>;tag=00070e8b577700045e92261b-204bc3c6
Call-ID: 00070e8b-577708ef-746a6bfe-6c96b214@172.18.192.218
Date: Thu, 13 Jun 2002 16:05:05 GMT
CSeq: 101 BYE
User-Agent: Cisco-SIP-IP-Phone/3
Content-Length: 0

SIP/2.0 200 OK
Via: SIP/2.0/UDP
172.18.192.230:5060;branch=fd11b5e1-aee38059-c54b46b5-f5f62a06-1,SIP/2.0/UDP
172.18.192.220:6062
From: <sip:5551000@172.18.192.230>;tag=003094c25d94001039653a76-0c6588ad
To: "A Phone" <sip:5559000@172.18.192.230>;tag=00070e8b577700045e92261b-204bc3c6
Call-ID: 00070e8b-577708ef-746a6bfe-6c96b214@172.18.192.218
Date: Thu, 13 Jun 2002 16:05:07 GMT
CSeq: 101 BYE
Server: Cisco-SIP-IP-Phone/3
Content-Length: 0

A releases the call to B:

BYE sip:5555000@172.18.192.230:5060 SIP/2.0
Via: SIP/2.0/UDP 172.18.192.218:5060
From: "A Phone"<sip:5559000@172.18.192.230>;tag=00070e8b5777000339a3170e-74ee1c83
To: <sip:5555000@172.18.192.230>;tag=003094c2e691000357031309-41c9de44
Call-ID: 00070e8b-577708ee-5fbe5e66-7f70fecd@172.18.192.218
Date: Thu, 13 Jun 2002 16:05:07 GMT
CSeq: 104 BYE
User-Agent: Cisco-SIP-IP-Phone/3
Content-Length: 0
Route: <sip:5555000@172.18.192.221:5060>
SIP/2.0 200 OK
Via: SIP/2.0/UDP 172.18.192.218:5060
From: "A Phone"<sip:5559000@172.18.192.230>;tag=00070e8b5777000339a3170e-74ee1c83
To: <sip:5555000@172.18.192.230>;tag=003094c2e691000357031309-41c9de44
Call-ID: 00070e8b-577708ee-5fbe5e66-7f70fecd@172.18.192.218
Date: Thu, 13 Jun 2002 16:04:58 GMT
CSeq: 104 BYE
Server: Cisco-SIP-IP-Phone/3
Content-Length: 0

Index

IP Telephony   O. Hersent, J.P. Petit, D. Gurle
© 2005 John Wiley & Sons, Ltd   ISBN: 0-470-02359-7